TELECOMMUNICATIONS NETWORK
DESIGN ALGORITHMS

McGraw-Hill in Computer Science Series

Senior Consulting Editor

C. L. Liu, University of Illinois at Urbana-Champaign

Consulting Editor

Allen B. Tucker, Bowdoin College

Fundamentals of Computing and Programming
Computer Organization and Architecture
Systems and Languages
Theoretical Foundations
Software Engineering and Database
Artificial Intelligence
Networks, Parallel and Distributed Computing
Graphics and Visualization
The MIT Electrical Engineering and Computer Science Series

Networks, Parallel and Distributed Computing

Ahuja: *Design and Analysis of Computer Communication Networks*
Filman and Friedman: *Coordinated Computing: Tools and Techniques for Distributed Software*
Hwang: *Advanced Computer Architecture: Parallelism, Scalability, Programmability*
Keiser: *Local Area Networks*
Kershenbaum: *Telecommunications Network Design Algorithms*
Lakshmivarahan and Dhall: *Analysis and Design of Parallel Computers*
Quinn: *Designing Efficient Algorithms for Parallel Computers*

TELECOMMUNICATIONS NETWORK DESIGN ALGORITHMS

Aaron Kershenbaum

IBM Thomas J. Watson Research Center

McGraw-Hill, Inc.

New York   St. Louis   San Francisco   Auckland   Bogotá
Caracas   Lisbon   London   Madrid   Mexico   Milan   Montreal
New Delhi   Paris   San Juan   Singapore   Sydney   Tokyo   Toronto

This book was set in Times Roman by Electronic Technical Publishing Services.
The editor was Eric M. Munson;
the cover was designed by Joshua Kershenbaum;
the production supervisor was Kathryn Porzio.
Project supervision was done by Electronic Technical Publishing Services.
R. R. Donnelley & Sons Company was printer and binder.

TELECOMMUNICATIONS NETWORK DESIGN ALGORITHMS

1 2 3 4 5 6 7 8 9 0 DOC DOC 9 0 9 8 7 6 5 4 3

ISBN 0-07-034228-8

Library of Congress Cataloging-in-Publication Data

Kershenbaum, Aaron.
   Telecommunications network design algorithms / Aaron Kershenbaum.
      p.   cm. — (McGraw-Hill computer science series.)
   Includes index.
   ISBN 0-07-034228-8
   1. Telecommunication systems—Design.   2. System analysis.
   3. Algorithms.   I. Title.   II. Series.
   TK5102.5.K43   1993
   621.382—dc20
                                    92-44411

Aaron Kershenbaum (Ph.D. Polytechnic University) is Manager of the Network Design and Analysis Tools Group at the IBM T. J. Watson Research Center. For twelve years prior to that, he was a Professor of Computer Science at Polytechnic University where he originated and taught many courses in telecommunications network design and supervised over 20 Ph.D. theses in that area. He was also the Director of the Network Design Laboratory at the New York State Center for Advanced Technology in Telecommunications at Polytechnic. For ten years prior to that, he was at Network Analysis Corp., where he was part of a pioneering research effort in network design. He has played an active role in the design of many major data and voice communications networks, served as an editor for several journals, published over 50 papers in the field and is a Fellow of the IEEE.

CONTENTS

PREFACE

Many people are interested in the field of network design. Networks of increasing size and complexity are being built, and the people involved in this process recognize the need to approach the problem in an organized way. As an academic discipline, network design offers the opportunity to apply optimization techniques and graph theory to problems in a classroom setting, and to build skills in using quantitative methods to solve problems.

For both the above reasons, there is an increasing number of courses, at both the graduate and undergraduate level, in the areas of voice, data, and computer networks. There are also professional and short courses given and, sometimes a sequence of courses is offered.

There are a number of ways to focus such courses, based upon the background of the students and the strengths and interests of the instructor. Most often, courses focus on specifics of the networks themselves. Different types of networks are described along with their components and major functions; and protocols and standards are discussed. In more advanced courses, techniques are presented for analyzing network performance, most often using queueing theory. Some courses go into the details of electronic communications in general, and digital communications in specific. Other courses present simple network design techniques. Most of the texts currently available in this area (and there a number of good ones) reflect these emphases.

However, the problem of network design is rarely presented in depth. This book aims to present the methodology of network design both for voice and data networks. It is suitable as a text for a second course (graduate level) in networks, as a second book for any course in networks, and as a reference for practitioners in the field, including use in professional courses.

It is not possible to quantitatively consider network design issues without also being able to analyze the performance of the networks. Analytic techniques are used to verify the feasibility of a design, and also to compare one design with another. Thus, techniques for network analysis are also covered in this text, and the instructor has considerable flexibility in creating a course with this text. There is sufficient material for a full year course, if the instructor chooses to use all the material. More commonly, however, there is room in the curriculum for only two courses in this area, and the first one will probably focus on devices and protocols. In this case, the instructor in the second course can choose from among the topics presented in this text, focussing primarily on design issues or alternately, sharing the time between design and analysis. With this curriculum, methods of analysis would also be emphasized and fewer design techniques would be presented. In the former case, of a full year course, only analytic results would be presented.

The objectives of this book are:

- To present network design problems in some depth within a general framework, which highlights the unity of these problems rather than the specifics of single problems. For example, the problem of finding optimal routes in computer networks (a central problem in computer network design) is almost always presented as a problem of minimizing delay. The emphasis is on the delay functions and their analysis via queueing theory. Conversely, routing in voice networks is presented within the context of minimizing blocking probability, with emphasis on the Erlang-B function which has very different properties than the queueing functions which govern delay. Yet both problems can be thought of, and dealt with, in essentially the same way. It is possible to solve both these problems as shortest path problems by suitably defining a length function for the network links. While it is important to present the appropriate length functions for each case, a deeper understanding of the problem can be gained by also presenting the actual routing algorithms within the unified context of shortest path problems.
- To present the solutions to network design problems as algorithms in sufficient depth to implement them and then to analyze both their effectiveness and computational complexity. We present not only the algorithms themselves, but also the data structures necessary for their implementation.
- To give the students access to an actual network design tool in a classroom setting. By using this as a teaching tool, students get a feel for real network design problems. Students can have access to the tool and be given assignments to add pieces to it. Such a network design tool (described in the Instructor's Guide) is available to instructors using this book.
- To present techniques for network analysis in the context of the design and planning of networks, and to emphasize the relationships between the two processes.
- To present the information at a level that is useful to people without a highly specialized background. This is not to say that this is an introductory text, since we assume that the reader is already conversant with networks. The first chapter gives an introduction, but this is to set the tone for the book. We neglect the traditional areas of devices, protocols, and electronic communications, which are well treated in other texts, in order to leave room for a thorough treatment of the design algorithms themselves. This also makes the book accessible both to computer scientists and electrical engineers, who often jointly populate such courses. We also avoid an advanced treatment of mathematical programming techniques. While

such techniques are important tools in the solution of these problems, they form a discipline of their own and are, rightfully, the subject of entire books. Thus, the book is accessible to people without extensive backgrounds in this area.

- To present solutions to realistic network design problems. Typically, solutions are presented in textbooks from a theoretical point of view. This is natural but leads, however, to "textbook" solutions useful either for very small problems, or which rely upon specific properties of individual problems. There are important problems still formally unsolved, such as the case where no known algorithm can guarantee a solution of reasonable quality in a reasonable amount of time. Most notable among these is the general problem of mesh network design. Such problems are usually omitted or given cursory mention at most. In the design of real networks, these problems cannot be avoided, but no one is willing to face them in any formal way, they are often dealt with in an ad hoc fashion. Instead, we include the best algorithms available for the solution of a full spectrum of network design problems, and thus provide both students and practitioners with the ability to face these problems using quantitative methods to the fullest extent possible.
- To present material not available in other texts. The plan for this book is ambitious and necessitates exclusion of unnecessary material. By concentrating on an indepth presentation of network design algorithms, it is hoped that we can do a good job on the topic. Also, the book can be useful as an auxiliary text in a more generally-based course or series of courses. Finally, it can be a useful reference book for practitioners and researchers in the field.

ACKNOWLEDGMENTS

The author would like to express his deepest gratitude to the many people who made this book possible. First to the pioneers at Network Analysis Corporation—Howard Frank, Ivan Frisch, Richard Van Slyke, Wushow Chou, Bill Rothfarb, Mario Gerla, Pat McGregor, David Rosenbaum, and many others—who began to make a science from the art of computer network design. Next, to Bob Boorstyn, Mischa Schwartz, and Leonard Kleinrock, for their contributions to this field both as researchers and teachers. Also, to Parviz Kermani and Bob Cahn, his colleagues at IBM, who have helped to bring network design to a new level and have also used a draft of this book in their courses. Also, to his many doctoral students—especially Keitha Murray, Andrew Shooman, Teresa Rubinson, Charles Palmer, and Shu-Lin Peng, whose research contributed to the material in this book and who also have provided valuable feedback. Finally, he would like to thank the many brave students at Polytechnic and elsewhere who have worked with drafts of this text, offered useful comments, and survived to tell the tale.

A special thanks to his wife, Peg, for her invaluable assistance and moral support during the preparation of this book. She has not only put up with him during good times and bad, but also made many invaluable contributions to the structure of the text. And a most special thanks to his son Joshua, who did all of the artwork for this text.

McGraw-Hill and the author would like to thank the following reviewers for their many helpful comments and suggestions: Robert Boorstyn, Berevan Consulting Limited; Wushow Chou, North Carolina State University; Anthony Ephremides, University of Maryland; Mario Gerla, University of California, Los Angeles; Alan Konheim, University of California, Santa Barbara; Galen Sasaki, University of Texas; and John Sylvester, University of Southern California.

Aaron Kershenbaum

CHAPTER
1

INTRODUCTION

This book deals with the design of telecommunications networks. Before discussing this, we briefly overview the major types of networks and their components to give a full perspective to the design issues and the techniques presented for their resolution.

1.1 TYPES OF NETWORKS

We describe some of the most common types of networks and the common functions they perform. In reality, networks do not often fall into a single type since it is the nature of networks to combine applications and architectures to form hybrids in various ways. Although this book discusses specific optimization problems in detail, the most important network design decisions are those relating to choosing a network architecture. A major goal of this book is to provide the reader with the ability to solve individual design problems quickly and reliably, to make possible a proper comparison between different architectures. If one required weeks to analyze a single architecture, it would not be realistic to compare many architectures with one another. However, if one could approximate the analysis of an architecture in an hour with the aid of a network design tool, the evaluation of dozens of alternative architectures and combinations of different architectures becomes quite feasible.

Many new types of networks have emerged over the past twenty years, and more will undoubtedly emerge in the future. It is not the intent of this book to catalogue the current "new" network architectures. However, it is important for the designer to understand the technology being considered in a network including its costs and limitations. The basic principles of network design and the basic algorithms used do not change significantly as the technology develops although specific tradeoffs may change if, for example, one type of network component becomes much less expensive. Thus, we try as much as possible to describe the network design problems generically, in terms of basic principles and device characteristics, rather than in terms of specific devices which become outdated very quickly.

1.1.1   Network Components

A detailed discussion of communication hardware is beyond the scope of this book, and the interested reader may wish to examine [Sch77][Tan81][Sta85] for a more complete treatment of this area. Nevertheless for completeness, we present an overview of the major components of a network.

1.1.1.1   FACILITIES.   We refer to the communication facilities which interconnect locations on the network as **communication channels**, **links**, **lines**, or **facilities**. These include telephone lines, coaxial cables, microwave links, satellite channels, and optical fibers. Each of these **physical media** has many interesting characteristics in terms of its ability to provide communication. We are primarily concerned with:

Cost. This includes the monthly lease cost, installation cost, cost per unit traffic, and maintenance costs. Costs vary depending on whether the facility is owned or leased. Usually, cost is modeled simply by a monthly cost figure, or, if cost is usage sensitive, by a cost per unit of usage (e.g., cost per hour or cost per bit).

Both the methodology used and the topologies selected in the design of a network are strongly influenced by the way cost varies, and it is important for the designer to take this into account. Thus, if cost varies roughly linearly with distance we choose a very different topology than if cost is relatively invariant with distance. Similarly, if there is a strong economy of scale with respect to capacity, that is, if cost goes up much more slowly than capacity, it has a profound effect on the topology selected.

There is also a significant difference between fixed monthly costs and usage sensitive costs. Often, a network incurs both types of cost and one can be traded off against the other. The typical situation is that fixed costs are used for large volumes of traffic (where a fixed cost can be justified by spreading it out over a high volume of traffic) and then let the remaining traffic flow over usage sensitive facilities.

Capacity. This is the amount of traffic the channel can carry. If the traffic is data, the units of capacity are usually bits per second (**bps**). Sometimes we speak of characters per second or messages per second. If the traffic is voice traffic, we may refer to capacity in terms of the number of simultaneous telephone calls that can be carried. If the voice traffic is digitized (i.e., turned into a bitstream) it is treated as data. In this case, one voice conversation is converted into V bps, where V is typically between 4,000 and 64,000 bps.

Sometimes one network rides on top of another and capacity is viewed differently depending on which network is being considered. One example of this occurs when a private voice network is constructed with facilities leased from a public carrier. In this case, the links of the private network become requirements for the public network. From the voice network point of view, capacity is measured in calls. From the public network point of view, capacity may be measured in terms of bits per second. A device (which may be part of either the public network or the private network) converts from one to the other.

Usually, we talk about the total capacity of a channel. In most real systems, however, part of this capacity is unavailable for communication. It is occupied by overhead of one sort or another. In that case, we may lower the total capacity and speak of **usable capacity**.

A channel may be full duplex, permitting simultaneous communication in both directions, half duplex, permitting communication in one direction at a time, or simplex, permitting communication in only one direction. It is essential that we distinguish among these cases in modelling a network.

The usable capacity of a channel is a function of the technology used in implementing the network. It is important to understand the technology well enough to make a realistic model of capacity, and to be able to determine the relationship between capacity and load. Having done so, however, we usually simply associate an effective capacity with each type of link and then use this number during the design process. The large number of alternatives that need to be considered during the design process generally precludes detailed capacity analysis during topology selection. Often, there are a number of iterations between network design and capacity analysis, refining the model and taking specifics of the data at hand into consideration.

Reliability. The fraction of time the channel is working. (This is also sometimes referred to as **availability**.) This takes into account the mean time between failures (**MTBF**) and the mean time to repair (**MTTR**). Thus, the reliability of a channel is

$$R = 1 - \frac{\text{MTTR}}{\text{MTBF}}$$

1.1.1.2 DEVICES. There are many types of devices used to construct a network often referred to generically as **nodes**. Sometimes nodes are distinguished by their functions. Types of nodes include:

Terminals. Simple devices, usually serving a single user; sources and destinations of low volume traffic. They usually include a keyboard and CRT and can also include disk drives and a printer. A personal computer, workstation, or a telephone may serve as a terminal.

Hosts. A large computer serving many users providing computing capability or access to a database. A source and/or destination of a major amount of traffic. A large workstation might be a host. These are also sometimes referred to as servers.

Multiplexors/concentrators. Devices which join the traffic on low speed lines into a single stream which can use a higher speed line. Some of these are transparent in that they work entirely at the electrical level using only hardware. Others use software and provide additional functions. A simple multiplexor has an output channel whose speed is roughly equal to the sum of the speeds of the input channels. More sophisticated multiplexors (e.g., statistical multiplexors) can compress the datastream so that the output channel has less capacity than the sum of the input channels. Likewise, concentrators usually buffer the inputs, and are able to have an output channel of a considerably lower speed than the sum of the input channels.

Local switches. Devices which allow attached facilities and devices to communicate directly with one another. Most such devices also perform other functions noted in the following discussion. There are many different types of switches. **Circuit switches** establish a fulltime connection between the input and output ports, dedicating capacity to an individual session. **Packet switches** break messages into small parts (called packets) and interleave packets from different messages. They include buffers to hold packets for a short time, thereby introducing delay, but also accommodating more bursty traffic (i.e., traffic where the peak rate is much higher than the average rate). Recently, various types of **fast packet switches**, including **frame relays** and **cell relays** have been introduced. These devices have some of the characteristics of both circuit and packet switches, in that they introduce some delay and some loss but to a lesser extent than do pure circuit or packet switches. The principal difference between frame relay switches and older conventional packet switches is that the former carry out fewer functions but are capable of handling much more data. Such tradeoffs among speed, cost, and functionality are an important part of the network design process, and give rise to the need for algorithms capable of considering many different design alternatives.

Tandem switches. Devices which interconnect nodes (attached devices described above) but also providing a path for traffic originating at other switches. These devices also perform many other functions, most notably routing described under routers.

Gateways. Tandem switches which interconnect networks with one another. These devices are generally able to handle multiple protocols (network control standards) and also can convert one protocol to another, allowing communication among different types of networks. Recently, **multiprotocol routers** have found wide use interconnecting wide area networks, and local area networks, implementing a variety of protocols.

Originally, many of these devices were predominantly, or entirely, communications hardware. Now, virtually all of them contain microprocessors and to some extent, are programmable. The boundaries between these devices are also often blurred; for example, a terminal may also do some concentration.

Again, such devices have many interesting characteristics, but our focus is on the following:

Cost. Similar to facilities; nodes have purchase costs, monthly costs, maintenance costs, and costs per unit capacity.

Capacity. The capacity of a node is usually more difficult to determine than that of a link. It is dependent upon the speed of the processor, the length of the programs running, the amount of memory, and the nature of the traffic passing through it. Nevertheless, we usually need to associate a capacity with each node in terms of the bits/sec of traffic, messages/sec, calls/sec., number of links, and aggregate speed of all attached links. Constraints on the type, amount, and mix of traffic which can pass through a node of a given type are formed on the basis of such capacity models.

√*Availability.* As with links, the fraction of time the device is working.

√*Compatibility.* The types of traffic and links that the node can handle. Some devices cannot handle certain types of traffic (e.g., analog voice). Some cannot handle links above a given speed. Also, some types of devices cannot interconnect with other devices or participate in some network architectures because the software they are using (the protocol they support) is not compatible.

1.1.2   Network Functions

The simplest way of providing for communication requirements is to provide for each one separately, but this is usually not cost-effective. Networks are constructed to share√ resources and to improve cost-effectiveness. The most obvious example of sharing is many low speed or part time communication requirements being combined to justify fulltime, high speed channels. Users benefit from the higher speed and the economy of scale of the high capacity facilities.

Less obvious, but also important, are capabilities justified within the network that could not be justified for individual requirements. These capabilities are often collectively referred to as "intelligence" within the network and they include: the ability to monitor itself (informing the user of congestion or failures), find paths for traffic through the network, modify these paths based on the state of the network, and maintain accounting records. Many of these capabilities are discussed below. Closely related to this is the fact that specialized staff for maintenance and to interface with outside vendors can be supported by the network, and these can become a resource for individual communications users.

There are many functions which may or may not be within the functional scope of the networks we design. We briefly consider some of these. Depending upon the applications the network is to handle, many of these functions may be essential parts of the services being offered or, alternatively, may just complicate its implementation, increasing cost and decreasing performance. In most cases the issue is how much of each function need be provided. Note that some of these capabilities are interdependent; that is, the network must have some of them in order to properly do others.

√*Switching.* The ability to interconnect the channels attached to each network node and to move traffic from each incoming channel to the appropriate outgoing channel when the requirement neither originates nor terminates at the node.

√*Routing.* The ability to select a path for each requirement. This capability varies widely. In some networks, only a single path is available for each requirement (fixed routing). The network may choose this path or the user may specify it. In other cases, several paths are predefined and the network chooses the best one (alternate routing).

This choice may be static based on a predefined probability, or dynamic based on the current state of the network.

Flow control. The ability to reject traffic, or slow the rate of entering traffic, in order to reduce network congestion.

Speed and code conversion. In digital networks, allowing devices using different communications codes, or operating at different speeds, to communicate. This generally involves buffering the data.

Security. The ability to prevent unauthorized access to the network and the data it carries. This may include passwords, data encryption, and even physical security (limiting access to equipment connected to the network).

Backup. The ability to react to component failures. This may include sending alarms and rerouting traffic to avoid the failed components.

Failure monitoring. The ability to keep track of which components are working. As with traffic monitoring, this is useful on a short term basis to route around failures and on a long term basis, it is useful in planning network modifications (including replacement of unreliable components).

Traffic monitoring. The ability to keep track of traffic levels, possibly by type of usage. This can be useful both on short term and long term bases. On a short term basis, it can be used to support dynamic routing and flow control. Over a long term, it can be used in network design to identify parts of the network where capacity may be productively increased or decreased.

Accountability. The ability to keep track of who uses the network. This is the basis for billing and chargeback. It differs from traffic monitoring in that it need not keep track of when the network is used but must keep track of different users. However, if separate traffic monitoring is not provided, it is possible to use accounting information for network planning.

Internetworking. Performing the functions needed to communicate with and across other networks. This includes providing routes for traffic crossing through, into, and out of the network, and allocating resources such as buffers and link capacity to traffic originating in other networks.

Network management. This includes a broad range of functions related to the management of the network. Some of these functions are mentioned in previous categories. Others include maintaining lists of users and addresses of devices, fault isolation, and keeping track of scheduled changes to the network.

The ability to provide these functions has a profound effect on the type of network built. In particular, the nodes of the network become more complex and more expensive as their functionality increases. Likewise, most of these functions imply overhead communication between nodes which requires additional link capacity. Thus, the decision to create a high functionality network is an important design decision which has a major impact on the architecture and hence the overall design of the network.

An important architectural decision, which is part of the design process, is deciding where each of these functions is carried out. As was mentioned before, some of them may be assigned to attached devices, and others may not be provided for at all. Even if the decision has been made to carry out a function within the network, there is still the issue of where to implement the capability. Of necessity, some functions are carried out at all nodes. One such funtion is the electrical interface which recognizes the presence or absence of a signal. Other functions, need only be carried out by the first and last nodes in the path.

An example of this is error control. If a message is received in error, which can be detected by sending additional information like a checksum, the message must be retransmitted or corrected. It is possible to check for errors every hop along the way or simply to check for them at the end of the path. The advantage of the former approach is that link capacity is preserved by avoiding the wasted transmission of an erroneous message over intermediate hops. The advantage of the latter approach is that processing capacity at the intermediate nodes is conserved. Thus, the tradeoff is between node and link capacity. Whichever is available at a lower cost will be traded for the other.

1.1.3   Centralized Data Networks

The simplest data networks are centralized, allowing terminals to access a single, central source of data. For example, an insurance company might maintain a central-ized database with policy, claim, and employee information. Terminals in each office around the country could access this data and update the database. Some of this access might be real time, with a user waiting for the response while other applications are off-line, with bulk data being printed for later use. Other examples of such networks are stock quotation systems, parts distribution networks, and automatic teller machine (ATM) networks. It should be noted that it is possible to use other types of networks for these applications. Also, a centralized data network may in fact be part of another more complex network. It is common to link centralized data networks for a common purpose; (e.g., the networks of different banks might be linked to provide customers with access to the facilities of other banks).

Figure 1.1 shows an example of a centralized data network. All communication is to and from the central site, which in this case is a host but might have been a network node. Concentrators and multiplexors might be used to obtain an economy of scale, with respect to transmission facilities. In some cases, the central site may function as a switch connecting all other sites to one another. Equipment at the central site can be duplicated to enhance the reliability of the network. Usually, the topology of such a network is a tree, as the terminals lack the capability to make routing decisions. In fact, the functionality of the terminals in such networks is usually kept to a minimum.

Almost all the intelligence in the network resides at the central site, and the capabilities of the equipment there determine the functionality of the entire network. From the point of view of network design, this represents one extreme in a tradeoff. By keeping the functionality of the many terminals simple, a great deal of cost and complexity in the network is avoided. We then benefit from an economy of scale by providing a single, major computational resource. On the other hand, a great deal

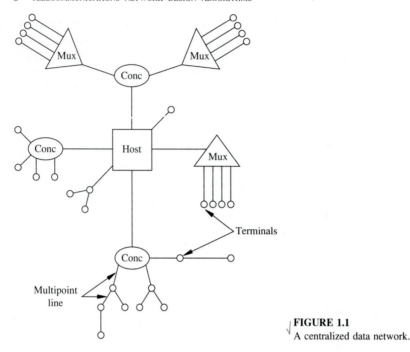

FIGURE 1.1
A centralized data network.

of additional data is usually transmitted in such a network because typically data is collected from one of many possible places, transmitted to the central site for processing and storage, and then sent back to where the data was originally collected since this is likely to be where the information is ultimately used. This makes sense for some applications, especially if there is also the need for a centralized copy of the data.

Terminals can be connected into the network on either point-to-point lines or over multipoint lines. A point-to-point line is a communication facility dedicated to a single terminal. Multipoint lines are trees. The central site controls the access to the multipoint line by polling and selecting terminals. Thus, it chooses to send messages to only one terminal at a time and to request messages from only one terminal at a time. All terminals can physically hear all messages on the line, but only the terminal for which a message is intended actually responds to the message. Thus, the load carried by a multipoint line is the total load of all the terminals sharing it.

Terminals can connect directly to the central site or may connect to a multiplexor or concentrator. Multiplexors may connect directly to the central site, via a concentrator, or even another multiplexor. Concentrators usually connect directly to the central site but can also be cascaded. Hence, there can be several levels of concentration in a network. Usually, a single network does not include too many different types of equipment because such networks are hard to manage. Thus, while we have mentioned many alternatives, usually only a few are used in any given network.

1.1.4 Distributed Data Networks

This class of network includes data networks with data from many sources and destinations. As such, they may be thought of as extensions of centralized data networks. This class also includes computer networks, where the "centers" themselves may communicate.

Figure 1.2 shows a typical distributed data network. The topology is usually a mesh, characterized by multiple paths between sources and destinations. Thus, routing is an important issue in distributed data networks which complicates the network design. The functionality of such networks is usually distributed to the devices at each node. In creating a network design it is important to understand how the functionality is distributed and what limitations exist on the utilization of network resources. For example, one may design a network with sufficient capacity to handle the offered load, but the routing algorithm used by the network may be unable to find and use that capacity. In such a case, the design cannot be called feasible. Also, the distribution of function leads to overhead traffic which requires part of the link and node capacities.

Unlike in centralized networks, it may now be possible to keep more of the data closer to its original source and move data only on request. For example, a credit card network may be designed with regional centers where credit information is kept and distributed. Since most requests for credit information are made in the region where it

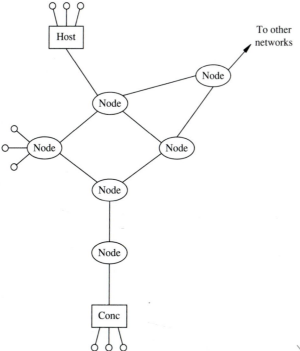

FIGURE 1.2
A distributed data network.

was originally collected. The system also has the capability of responding to requests from other regions when necessary, such as for example when a person is travelling.

In the case of a computer network, most of the users of each computer may be local. The network also has the capability of providing users of one computer with access to programs and data on other computers. It is very important in designing such networks that one understand what proportion of the usage is remote. The load on the network could easily vary by several orders of magnitude if the local users of one computer became heavy users of another one. Most real networks monitor and control such cross-usage carefully.

1.1.5   Voice Networks

These networks interconnect telephones. They can be analog, digital, or hybrids of both. Most modern voice networks are digital, and we concern ourselves primarily with these, although most of the network design techniques we discuss also apply to analog voice networks.

Most voice networks are mesh networks, supporting point-to-point communication among all pairs of nodes. Thus, they share many of the characteristics of distributed data networks from the point of view of network design. It is also possible to design a centralized voice network with all communication taking place through a single central switch. Small voice networks, especially those with limited geographic scope, are often centralized.

Most of the functions mentioned for data networks are important for voice networks as well. One of the major reasons that most modern voice networks are digital is that it is easier to carry out most of these functions in digital networks where the nodes are essentially computers which can be programmed to perform peripheral management functions.

1.1.6   Integrated Networks

There are many forms of integration. The simplest is where essentially separate networks share transmission facilities. Each network has its own nodes which carry out all necessary functions. But the capacity of high speed links is statically divided among the networks, thus realizing a greater economy of scale than the networks could individually.

More complex integration involves dynamically sharing capacity across networks; that is, if one network is momentarily quiet, another network can make use of spare capacity. This can lead to more efficient use of facilities, but requires that the nodes of the component networks communicate closely and cooperatively. In many cases, the nodes themselves are shared among component networks and across applications.

In some cases the networks are hierarchical, with different amounts of sharing at different levels in the network. Figure 1.3 is an illustration of such a network. Here, voice and data networks share transmission facilities via a multiplexor. Each network has its own switches and at the lower level appears as a separate network. There

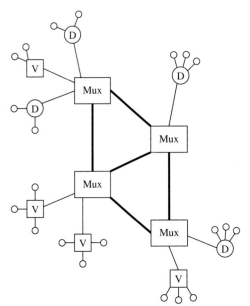

FIGURE 1.3
An integrated network.

is no dynamic sharing of transmission facilities, but it is possible to reconfigure the multiplexor and move capacity from one network to the other. In some cases this can be done easily enough to make it practical to move capacity on a time of day basis. Clearly, the design of such networks is very challenging. It involves deciding not only where to place and how to size components, but also the allocation of capacity among different applications.

In designing such networks, or considering such architectures as alternatives, it is important to understand what types of sharing are possible and how freely capacity can be shared. This relates to the capabilities of the devices used as switches and gateways. This information is input to the design process.

1.1.7 Local Area Networks (LAN)

These are networks which tie together users within a building or campus. They typically utilize high speed links that consist of coaxial cable or optical fiber. Such networks can be stand-alone or can be part of the local access portion of larger (wide-area) networks.

LANs can be interconnected in a number of ways. **Bridges** can be used to connect one LAN to another. **Routers** can also connect LANs to one another and can, additionally, perform switching functions.

The design of LANs can be considered from several different points of view. When considered alone, a LAN is a high speed network with a relatively simple topology (a star or ring) over a relatively small area (a building or a campus). When considered as part of a larger network, a LAN can be considered as a node which originates, terminates, and if routers are present, switches traffic.

LANs can also be used as the local access portion of larger networks. Thus, LANs can connect terminals within a building and these LANs can in turn be connected by routers and bridges to form a network over a campus. Campuses can be tied together into metropolitan area networks (**MANs**) and then into wide area networks (**WANs**). The boundaries between these types of networks are blurred and their respective technologies overlap.

The same basic design principles apply to all types of networks, but in practice, there are several important differences. Over short distances (e.g., within a building) propagation delay is small, permitting coordination of the different operational devices contending for the use of a communication channel. This in turn allows the efficient sharing of relatively large amounts (megabits/sec) of communication bandwidth by simple devices using simple line control procedures (protocols). Over larger distances, such sharing is also possible, but it tends to be more expensive and less efficient.

A second major difference arises from the fact that the cost of a communication channel tends to increase with distance. This reinforces the preceding conclusions that networks which operate over larger distances tend to operate at lower speeds. (The advent of fiber optics, however, is starting to change this.)

This book is primarily focused on the design of wide area networks where link cost, and hence link topology, plays a major role. Many of the design techniques presented, however, also have relevance in the design of LANs. In particular, the techniques for clustering sets of nodes within a network are important.

1.2 NETWORK DESIGN ISSUES

As in the famous story of the blind men examining the elephant, everyone interprets the network design problem from his or her own perspective. Some are most concerned with cost, others with performance, and still others with manageability. Each perspective is important and each is dealt with, to a degree, in many places. *Computer Communication Network Design and Analysis*, by Mischa Schwartz, is an excellent general reference on many of the aspects of network design.

A complete treatment of all aspects of network design, both technical and non-technical, is beyond the scope of this book, but in the following pages we present some of the major issues a network designer must consider. Together they provide a perspective on the overall problem of network design and allow for a deeper understanding of the topics covered in greater depth in the remainder of this book.

1.2.1 Justifying a Network

The most basic question is whether a network is justified at all. Some applications may be best satisfied by individual point-to-point connections to handle very specific communications requirements. Some applications may be so unusual as to not fit into any general network. Still other applications may be completely satisfied by existing common carrier offerings. While this latter case requires a network, it does not have to be designed; it need only be recognized that a network with all the right characteristics already exists. Similarly, it is possible that a private network already exists within the organization to satisfy the requirement at hand.

Extending this concept, it may be in the organization's best interest to build its network on top of existing capabilities. This is the rule rather than the exception. For example, most networks are designed using existing communication channels, usually leased from a common carrier. The carriers encourage this by offering bulk rates and **virtual network** services, where parts of the common carrier network (both links and nodes) function as if they are the user's own private network. There are many potential advantages to this. The common carrier is probably in a better position to achieve, and then share, a greater economy of scale and may be better able to maintain reliable service through experienced communications engineers on staff. On the other hand, each application may have special requirements, such as interfaces to specialized computer hardware and software outside the scope of what the common carrier offers. Viewed from this perspective, the feasibility question becomes not whether to have a network, but rather how much of a network to have, and how much of an existing network to use.

1.2.2 Scope

The scope of a network is viewed as bounded on one side by the offerings of the common carriers who provide the communications facilities from which the network is built; and on the other side, by the applications which it interconnects. There are many ways of successfully dividing the responsibility for network function among all parties involved, including those parts of the organization building the network, and the vendors providing the peripheral functions. It is essential, however, that the division be clear. For example, if a link fails and a message is lost, it is possible for the application which originated the message to retransmit it, or, alternatively, it is possible for the network to handle this responsibility itself. The key is that one or the other must retain this responsibility so that a copy of the message is saved until it is acknowledged by the destination. There is a serious problem if neither party does this, or if both do. The latter case can result in duplicate messages being sent, a problem as serious as lost messages.

The geographic scope of a network must also be considered. It might be assumed that a corporate network should satisfy all the needs of the corporation and hence extend to all corporate locations. This is not necessarily so and, in fact, trying to make the network ubiquitous throughout the corporation might be a serious design error. Locations with low volume requirements might best be served by common carrier facilities, perhaps dialing in to private network sites when necessary. Merging a domestic and international network might make both unmanageable, especially if they carry very different types of traffic. While total integration of all sites and all applications is one alternative to consider, it is not the only one.

1.2.3 Manageability

Although difficult to quantify, manageability is an important design consideration. A network comprised of many different types of facilities and which uses intricate and specialized control procedures may be, in theory, very cost effective. In practice, however, like a high performance automobile, it may be very difficult to own. It may require constant tuning and its performance may drop significantly whenever anything in its environment changes. For this reason, we prefer networks which are as homogeneous and as simple as possible. Often, slack is left in the network even though this increases cost, in order to make the network more robust and more easily managed. If a network can tolerate variability in traffic levels, they do not need to be monitored as closely. The relationship between slack and manageability, however, should be quantified as much as possible. Then, slack is added only where it actually enhances performance and the gain is measured against the added cost.

1.2.4   Network Architecture

In the following chapters, the focus is on specific problems in topological optimization. Before detailed decisions about node and link placement are made, however, the more general issue of the overall "shape" of the network should be discussed. Decisions at this level are partially art and partially science. The goal is to bring as much science as possible into play.

Will the network be a single, homogeneous mesh comprised of a single type of node and a single type of link, or will it be a hybrid of different types of equipment? It might be a hierarchical network with one type of link riding on another, or applications might be kept separate without sharing resources at several levels of the network until, at the highest level, they share common facilities. There may be uncertainty about the proper network architecture and several alternatives may need to be explored. Before significant effort is expended on exploring any detailed design, however, we need to get at least an approximate picture of which architectures are viable alternatives. Then, the most promising ones can be explored in more detail.

Another issue relating to the overall architecture of a network is whether to decompose the network into subnetworks for the sake of design and operation. It's possible, for example, to begin by clustering the nodes into regions, locate gateways within each region, and then design a backbone connecting the gateways. All communication between regions is via this backbone. Local access networks are designed within each region. This is often how wide area networks which include LANs are designed.

Alternatively, the network might be designed as a single mesh. If LANs which include routers are considered, it may be in the organization's best interest to trade off node and link capacity within the LAN against capacity in the wide area network. Thus, the wide area network may connect parts of the LAN (or different LANs on a campus), or the LAN may offload switches in the wide area network.

If the network is considered as a single mesh, it becomes more difficult to design and to operate but might function more efficiently because of the additional option available to share capacity among all nodes. This is an important design tradeoff. Unless substantial gains can be made, it is usually better to consider the designs at different levels, separately. One of the major goals of the design procedures presented in this book is to determine when the additional complexity of considering multiple levels of a network at once are justified and when they are not.

1.2.5 Switching Mode

One of the main reasons for building a network, as opposed to giving each application its own dedicated facilities, is to share resources, specifically transmission facilities. There are a number of different ways of doing this.

1.2.5.1 PACKET SWITCHING. Packet switching is often used in computer networks where individual users have need of the channel intermittently. While using the channel, the application requires high bandwidth, but most of the time, each user does not require the channel at all. As an example, a car rental site might need access to data stored in a computer in the course of servicing individual customers. When the rental clerk requests information, it is desirable that the screen fill up quickly, obtaining a screenful (several hundred characters) of information in a few seconds. Such a request occurs only once a minute, however, as during the remainder of the time the clerk is reading the information or speaking with the customer.

Such applications, characterized by a high peak-to-average requirement for capacity, are called **bursty** and are ideal candidates for **packet switching**. In packet switching, messages (from the user to the computer, or from the computer to the user) are broken into short (typically no greater than 128 character) blocks and interleaved with other messages. Thus, users queue for the channel and share it with one another efficiently.

1.2.5.2 CIRCUIT SWITCHING. Circuit switching, on the other hand, is effective for applications which make comparatively steady use of the channel. Analog voice is a prime example. In this case, it makes sense simply to dedicate the entire channel to the application for the duration of its use. In circuit switching, an end-to-end path is set up at the beginning of a session, dedicated to the application, and then released at the end of the session. It does not incur the overhead of interleaving different applications; indeed, the switch does not even have to become involved in the details of the information transfer at all.

To function efficiently, the time to set up and break down a circuit should be small in comparison with the time information is transferred. Also, as mentioned before, the application should make reasonably good use of the channel. Note that, unlike packet switching, the information need not be stored in intermediate nodes. Thus, there is no appreciable delay and much less need for storage within the network. On the other hand, since a fixed amount of dedicated capacity must be allocated to each session, it is sometimes necessary for a circuit switched network to block traffic when no additional capacity is available.

1.2.5.3 RANDOM ACCESS. An alternative to switching is **random access** where all attached devices and facilities share a single high speed channel. In the simplest mode of random access, devices transmit when they want to but listen to the channel to see if they have collided with other transmissions. If so, they retransmit after a random amount of time. A more sophisticated approach is to listen to or sense the channel before transmitting, and avoid most collisions. It is also possible to use a small portion of the channel for reservations, which are made at random times, and then dedicate the bulk of the channel's capacity to servicing the reservations; this limits collisions to the reservation portion of the channel. Random access is widely used in local area networks and satellite networks.

1.2.5.4 HYBRID SWITCHING. There are hybrid forms of switching that have some of the properties of packet switching, circuit switching, and random access. Some circuit switches queue requests until capacity is available or for a fixed amount of time. **Fast circuit switches** can set up an end-to-end path fast enough to dedicate capacity to individual packets. Some switches can move available capacity between circuit and packet switched applications.

Although at first glance it may appear that these different modes of switching require different design procedures, it will be seen that often the design problem can be unified by observing that, no matter what the switching mode, performance degrades as the utilization of a channel increases. Thus, in networks that permit queueing (waiting), delay increases as utilization increases. In networks that permit blocking, blocking increases as utilization increases. Moreover, performance degrades catastrophically when utilization passes a critical point. This leads to the unifying notion of a **usable capacity** for a channel. On one hand, high utilizations are preferred, as this contributes to cost effectiveness. On the other hand, performance becomes unacceptable when utilization passes a given point. It is often possible to create designs based on a uniform utilization limit and then, at the last stage in the design process, to do detailed analysis and adjust utilizations of individual facilities.

1.2.6 Node Placement and Sizing

A fundamental problem in the topological optimization of a network is the selection of the network node sites. This actually encompasses several problems which must be dealt with separately. The first is the choice of which sources and destinations of traffic will be 'on-net,' that is, part of the network at all. It may be that some of these do not have enough traffic to justify a dedicated connection to the network, and may, instead, access the network via common carrier facilities (e.g., dialup lines).

A second problem is the decision where to place **multiplexors**, **concentrators** and **switches**. These terms are often used to describe different types of devices. As mentioned before, they may even be equipment, like routers, which are usually associated with LANs. For simplicity, we refer to a multiplexor as a device which takes a number of low speed lines as input and combines them into a single, high speed line, with the capacity of the high speed line being equal to the aggregate speeds of the low speed lines (or a convenient value somewhat larger.) A concentrator is a device which essentially does the same thing but is also capable of compressing the data stream; for example, statistically, so that the high speed line can have a lower capacity. Thus, concentrators may introduce delay or block traffic. Finally, a switch has all these capabilities and also can form direct connections between its input lines, as well as form connections between input lines and output lines. This latter capability is called **routing** and may include the ability to set up end-to-end paths through multiple switches. Switches possessing this latter capability are referred to as **tandem** ✓ **switches**. Ordinarily, we are not concerned with the specific capabilities of network nodes, but when this is important, we try to make the nodes' specific capabilities clear.

In theory, it is possible to place nodes anywhere. In practice, the placement of nodes is usually limited to a finite set of candidate sites. Typically, these are a subset of the sites which are sources and destinations of traffic. Often they are simply the major sources and destinations. Major nodes which handle large amounts of traffic are usually placed at locations where there are trained personnel to do maintenance. Sometimes such nodes are placed at common carrier sites. The set of candidate sites is usually an input to the network design process while the actual sites selected are an output.

Closely related to the problem of selecting where to place nodes is the problem of deciding exactly what type of devices to place at each location. This includes such high level questions as whether a specific site should be a switching site or just a multiplexor site. It also includes more detailed questions on equipment configuration. For the most part, our model of a network and the algorithms we present are not complete enough to handle these latter questions in detail, but such problems must be considered at least approximately, to properly answer the more general questions.

1.2.7 Link Topology and Sizing

Link topology and sizing involves selecting the specific links interconnecting nodes, and is one of the major focuses of this book. At the highest level, this is where the architecture of the network is derived. Thus, a hierarchy that includes a backbone as well as local access networks may be defined. It is possible to permit the backbone to be a mesh while the local access networks are constrained to be trees. At a lower level, having at least tentatively decided on an architecture, a selection is made from among specific topologies, including and excluding specific links. The overall problem is often partitioned into subproblems for the different parts of the network.

As with node selection, the lowest level problem in link selection is the determination of the specific number and type of links. Here, unlike in node selection, it is easier to build accurate, detailed models of the links; and hence, it is possible to deal with the problem in greater detail. Thus, procedures are described for deciding on the exact number and type of links in a network.

1.2.8 Routing (Protocol Selection)

Routing involves selecting paths for each requirement. At the highest level, this involves selecting the routing procedure itself; that is, deciding on what type of routes are permissible. For example, the specific switches used in the network might be capable of only allowing a single route for each requirement. At the other extreme, some switches can dynamically choose the routes, based on the current state of the network, which includes the level of congestion in nodes and links. The class of routes permissible is part of the architecture of the network and is closely related to the protocols implemented within it. This book does not focus on details of network protocols; instead it is assumed that the network protocols are essentially inputs to the design process. The interested reader is referred to [Tan81], [Sta85] for more details in this area.

The selection of a protocol is closely related to the selection of an architecture for the software running in the nodes. This selection includes the operating system as well as the software which carries out network functions. Compatibility with application software running on hosts within the network is an important consideration here. A detailed discussion of this topic is beyond the scope of this book and the interested reader may wish to examine [Tan81]. The overhead introduced by the protocol affects the usable capacity of both the nodes and the links.

1.3   DATA IN SUPPORT OF NETWORK DESIGN

The most time consuming part of the entire network design process is often the assembly of the input data. As well, in subtle ways, the organization of this data imposes constraints on the design process and limits our choices from among design alternatives. A surprisingly large amount of network design is still done by hand using rules of thumb, even in large, technologically advanced organizations. One of the major reasons for this is the difficulty of assembling the data necessary to do quantitative automated network design. In collecting this data, a balance must be struck between the amount of information available with a realistic amount of effort and sufficient detail needed to support quantitative analysis.

To be useful, this input data needs to be current. This constraint, together with limitations on available manpower, limits the amount of data which can be collected. Since long range projections tend to be unreliable, it is often better to work with recent data even if it is approximate or incomplete rather than embark on a long term effort to create a comprehensive database of all necessary inputs. By the time the data is collected, it is often out of date. One of the major reasons organizations build their own communications networks is to be able to respond to changes in requirements more quickly, rather than relying on common carrier offerings. A lengthy data collection process in support of network design can negate this advantage.

One way of cutting down both the time and effort in collecting data is to build upon existing automated sources of information. For example, attached host computers may maintain accounting information used to bill for host services which can be used as sources of traffic requirements. The existence of such an ongoing data collection process, which can be relied upon to produce current data, is ideal for network design. It is often worthwhile to modify the design process or create a special interface to make use of such a data source. Another example is the use of existing databases containing useful information. Tariff services can be used as sources of link cost information.

It is important to distinguish between data which is relatively static and that which is more dynamic. Thus, the list of *locations* where an organization has communications requirements usually changes more slowly than the *magnitude* of the communications requirements themselves. Where data is less perishable, it is afford-able to collect it in greater detail.

The amount of detail maintained is a function of the objectives of the design. Fortunately, the level of detail required usually matches the level available. Thus, if the design is being done in support of a proposal or feasibility study, relatively little detail is required and relatively little is usually available. Conversely, the database used in support of design modifications of an existing network may be relatively extensive.

It is also important to be prepared to deal with inconsistencies in the data and not to attempt to obtain an unrealistic degree of precision. For example, actual message counts can be collected on an hourly basis each day for several weeks. A close analysis of this data, however, may reveal significant variations from one day to the next. These may be due to transient factors such as storms (which often result in a large number of telephone calls) or a sudden jump in the stock market. Another important example is an outage in the network itself, which can significantly distort traffic patterns. While it is important to note the variability and to take it into account in the design process, it is not safe to assume that a complete and detailed model of variability in the load can be built from a few weeks of observations. It is more reasonable to make a conservative estimate based on these limited observations.

In some cases, there may be outright errors in the data. The data collection process, either manual or automated, may be imperfect. It is generally more trouble than it is worth to completely eliminate all such errors, especially in light of the natural variability of the data, as noted above. It is, of course, important to apply some consistency checks to the data to assure, at least, gross integrity.

In some cases, it may be desirable to filter the data. In designing a voice network, for example, there may be a relatively small number of calls to and from a relatively large number of odd locations. Rather than burdening the entire network design process with tracking all these additional locations, it may be preferable to filter them out and to account for them in an approximate way. Similarly, data requirements for small applications without any accounting records may be handled approximately.

To plan and manage changes to the network, it is often necessary to maintain multiple views of the network, since the present view may include only currently installed equipment and current traffic. Then, six months in the future, the view may include planned installations and projected traffic. When such views are disjoint, they are fairly easily handled. A problem occurs, however, when they overlap. For example, there may be tracking of actual load on a network where a node was removed. This could give rise to traffic to a node no longer part of the network. It is essential that the data collection process be flexible enough to handle such situations without extensive human intervention.

This all means that the network design process must be able to function with incomplete and approximate inputs. At the same time, it should not degenerate into simply using "fudge factors" and crude rules of thumb. Therefore, as much accurate and detailed information as possible is sought given a realistic amount of effort. Finally, by understanding the network design process itself, and in particular, the sensitivity of specific design decisions to changes in the data, we try to minimize the effects of our approximations. For example, in trying to decide between installing a switch with normalized usable capacity 100 and a switch with capacity 1000, we do not care if the load is 386 or 422. In either case we will install the larger switch. On the other hand, if the switch were modular, we might be talking about installing an add-on module to increase its usable capacity from 400 to 500. In this case, with a more precise knowledge of the load, we might be able to defer the installation of the add-on module.

Each part of the design database is now described in more detail. In practice, each such database may include information specific to the applications and network at hand. Here the focus is on data which is of general use. It is also critical that there be no attempt to include data not actually available with an acceptable amount of effort. A network design procedure which demands data that is unavailable is useless. An organization is better off adapting the design process itself to work with the data available.

1.3.1 Location Data

The information needed to describe the locations in the network is discussed in this section. A number of issues common to the collection and maintenance of all the information in the network database are also highlighted.

The most basic information in the database concerns the set of locations involved in the network. This includes sources and destinations of requirements and candidate locations for network devices such as switches. At one extreme, consider each terminal in a data network and each telephone in a voice network as a location, and track detailed traffic information on this basis. Another extreme is to consolidate to the level of entire cities. Each of these alternatives, as well as a hybrid approach which deals with large buildings and small cities as locations might be appropriate in specific situations, depending upon the overall size of the network, available computational resources, and the level of detail required in the design. It is also possible to maintain multiple views of the network during the design process. For example, we might consolidate to the city level during the design of the backbone network and then partition into detailed views of each city during local access network design.

The location set chosen affects the rest of the database. Requirements are reported between locations and can only be reported in as much detail as the degree of fineness of the location set. Also, one of the most important design decisions is the on-net/off-net decision, where the choice is made as to which locations to serve via permanent facilities as opposed to dial-in or not to serve at all. An early decision not to include certain small locations in the database may, in effect be making that decision for us.

Similarly, all links go between locations. By including two sites inside the same building as different locations, we can keep track of links between them. This may be important in cases where there is a significant amount of traffic on such links. In addition to increasing the size of the problem, however, it also brings with it the obligation to obtain the cost for such links. Unlike links between geographically separate locations, however, there may be the problem of dealing with private facilities with untariffed costs that are difficult to obtain or even estimate.

The selection of the appropriate location set is critically important. If it is done incorrectly, the design may be adversely affected. It is also difficult to undo decisions in this area, because so much other data depends on location data. It is also hard to make such decisions early in the design process. Indeed, the selection of a location set is itself a significant design decision.

For all these reasons, and to retain flexibility, the original "raw" data is generally collected and maintained on the basis of as large a set of locations as is practical. This data is then consolidated onto a more manageable location set for decision making, specifically as input to the network design algorithms. Even this consolidation effort can be time-consuming, so it is best not to do it more often than necessary. This leads to multiple views of the network where the data is compressed, filtered, and refined as the design proceeds. The major concern is with the initial raw data view, which is voluminous and often contains errors. Once the initial consolidation is done, it is usually relatively easy to do further processing. Therefore, a key to success in network design is to organize the process so that this initial consolidation can be done efficiently without impacting the entire design process. The object is to minimize the effort in iterating this process.

The actual data maintained for each location varies based upon the specific network at hand. The focus here is on generally useful information and suggestions for other possibilities as well. Each location needs at least one identifier, usually several. Thus, for the sake of reporting, it is possible to use city names and even full street addresses. It is essential to identify each location uniquely and, in general, make the reports as easy to read as possible. Keep in mind, however, that space is limited, especially if terminal-based displays are used. Thus, there are usually brief (at most 4-character) identifiers for the purposes of location representation.

There is a temptation in this regard to go to the extreme of simply using a number to represent locations. This is useful for internal representations passed to the network design algorithms, but is a bad idea for externally stored data. It is difficult to check such data and to identify inconsistencies. Also, as locations enter and leave the network, it is easy for the numbering system to become confused. Therefore, it is best to use mnemonic identifiers externally and then convert them to numbers for internal representation.

Sometimes, because we rely upon external sources of data, it may be necessary to recognize different identifiers for the same location. In this case, it is necessary to map from one identifier set to another. Obviously, this is dangerous and should be avoided whenever possible. It is better to have the entire network database maintained on the basis of a single set of location identifiers.

Each location should have a set of coordinates associated with it. These are necessary to support cost calculations as well as for display purposes. Most tariff data is keyed to a coordinate system. In the U.S., telephone company vertical and horizontal (V&H) coordinates are used. Overseas, other coordinate sets, as well as latitude and longitude, may be used. In some cases, different coordinate sets may be needed for the same location. For example, interstate tariffs may be based on one coordinate set while intrastate tariffs are based on another. Some local tariffs are based on the location of the associated telephone company point of presence (POP). Intrastate tariffs are also based on the local access and transport area (LATA) associated with a location.

Fortunately, nearly all this information is available in databases keyed to the area code and exchange (AC/EX, NPA/NNX) of a location. Since the AC/EX is part of any telephone number at the location, this is easily obtained.

Properties can be associated with each location. Thus, for example, we may record a location's status as a switch site. The status may be that it is a mandatory site, an optional site, a prohibited site, a current site, or some combination of the above. Similarly, we may record the presence or absence of various applications (i.e., voice).

These properties may be filled in by the user, by the design procedures, or both. It is important to distinguish between data which can be changed by the system and data which cannot. Thus, in the previous example, a mandatory switch site is one whose status should not be changed by a design procedure, while the status of a current switch site may be changed by the system.

We usually think of, and physically represent, the data associated with a location as fields within a record. Often these fields are fixed format, although this is not absolutely necessary. It is relatively easy to change the physical form of the information; the important thing is to have the right information present in the first place.

In many parts of the design process, it is convenient to associate a physical device with a location. Thus, the device type may be one of the location properties and the location and the device become identical. There is exactly one such device at each location, and properties of the device and location are intermixed. This simplifies the implementation of various design procedures. It is, however, really appropriate only for internal views of the network. Externally, it is best to keep locations and devices separate. In particular, we may want to have several different types of devices, or several devices of the same type at the same location. Therefore, it is important to keep the distinction clear. In recording routes, for example, it is usually necessary to know which specific device is in the path. On the other hand, the cost of links between the same pair of physical locations is comparable to the cost of another link (of the same type) between the same pair of physical locations. Since link costs require a lot of space, an important savings in space can be obtained by recording them on the basis of locations rather than devices. Examine each case individually and make the best decision possible. It is important to consciously make these decisions in order to avoid confusion.

1.3.2   Traffic Requirements

The traffic requirements are usually the most voluminous part of the database that must be collected by the user. As mentioned before, it is dependent upon the choice of location set. There are many other dimensions to this data as well since it is also a function of time of day, date, and application. There may be multiple views based on projections of different lengths.

Generally, traffic requirements start with raw data which may be as detailed as records of individual sessions (telephone calls or terminal sessions). Such records might include the source, destination, starting time, date, length of session, number of packets, number of characters, type of session, application, route, and possibly other information as well. Alternatively, there might be only projections in terms of average number of sessions per terminal per day. This data is then processed into a more manageable form.

Conceptually, the data can be viewed in terms of a traffic matrix with a single number representing the magnitude of the requirement between each source and destination pair. In practice, more detail is often needed. Thus, some measure of the variability of the load on a time of day basis is needed to do performance analysis. Similarly, a message length distribution may also be needed. A breakdown by application may also be required because different applications may need different types of facilities, such as secure channels.

The matrix can have many dimensions—source, destination, time of day, application. Since each of these dimensions can range over many values, it is impractical to actually store the information in this way. Therefore, the information is compressed along several of the axes. Thus, using the original raw data, a message length distribution can be formed for each application. The length distribution can be used as input to queueing formulas. On the other hand, to compute the requirement between each source and destination, simply multiply the number of messages by the average message length. Alternatively, count the total number of characters and use this quantity directly as the requirement. In either event, the source-destination load is compressed to a single number. Similarly, it is possible to use the raw data to compute a distribution of load over the hours of the day, and use this to estimate the percentage of the daily load arriving during the busy hour. The network can then be designed to handle this busy hour load. However, in recording the requirements, only a single number is still maintained for each pair of locations.

Even this two dimensional matrix may be too voluminous if the number of locations is large. If so, compress the information still further. For some purposes, simply record the total requirement to and from a node since sometimes this is all that is available in any case. Or, the total requirement at each location can be recorded by application. Even simpler, the number of terminals at a location may be known and from this and an estimate of activity per terminal, an estimate of traffic can be obtained.

We may go further along the lines of generating a traffic matrix and spread the traffic over a given set of destinations, based on an estimate of load per destination. Thus, it is possible to compress detailed data into a simple model or, alternatively, build up an estimate of point-to-point traffic from a simple model.

All of these approaches are useful at times. However, information cannot be created. A requirements matrix built up from a simple model is nothing more than an estimate and is only as accurate as the underlying model. Nevertheless, it can serve a useful purpose, especially during the early phases of design.

Depending on the type of communications, requirements can be either directed or undirected. For example, analog voice ties up the channel in both directions and so is considered undirected. Most data requirements, on the other hand, are directed. Often, for each message going in one direction there is another message (with a different length) going in the other direction. Thus, while there is a relationship between traffic going in opposing directions, there is a difference as well, and these data requirements must be tracked separately. It is important to understand which requirements use capacity in only one direction and which use capacity in both.

A distinction should be made between the external representation of requirements on disk and their internal representation in memory. It is important for the external representation to be easily read and edited. It is also important that it not constrain later design decisions.

Usually a good external representation is a simple list of records, each containing source, destination, type of traffic, and magnitude. Locations are represented by their identifiers. Internally, additional structure is useful. Therefore, the data can be compressed into a smaller location set, a matrix created, and locations represented by indices. Through this, the necessary efficiencies are achieved for the algorithms without losing any data since the permanent record of the network on disk still has all the original information. All interface with the user is in the external format. It is usually a serious error to allow the user to edit any data in the internal format since indices are often functions of the specific design and it is easy to introduce inconsistencies into the database.

1.3.3   Link Costs (Tariffs)

Tariffs, the published rates for communication services filed by the common carriers are, next to traffic information, the most voluminous part of the database. This section of the database contains the costs of all possible links in the network. Conceptually, for leased lines, this is a three dimensional matrix indexed by pairs of locations and link types. Usually, this information is extracted from a tariff database that is kept on-line in support of network design tools.

Costs include both monthly charges and one time charges, such as installation costs. The latter can be converted to monthly costs by amortizing them over a reasonable period of time. Similarly, in the case where private communication facilities are constructed (e.g., a private microwave line) this cost is converted to a monthly cost by amortization over the expected lifetime of the facility, usually several years.

Some communication costs are usage sensitive, such as direct dial telephone costs or packet switched common carrier costs. There are also tariffs covering these. In this case we maintain cost per minute or cost per bit, as appropriate.

Tariff costs are location sensitive. Usually it is sufficient to know the area code and exchange of a location. All necessary tariff information can be looked up on this basis. A location's LATA (Local Access and Transport Area) is also needed for some intrastate tariffs. In some cases, the distance from the telephone company central office is also needed; this is usually difficult to obtain.

There are many different tariffs since different types of lines have different tariffs, different common carriers have different tariffs, and each intrastate carrier has its own tariffs. These tariffs differ not just in their cost but also in their structure.

Some are usage sensitive. Some are time of day sensitive. Some include bulk rates. Tariff information changes frequently, both in terms of the rates and structure of the tariffs.

There are services which, for a fee, provide current tariff information in machine readable form. Even with such services, however, it is difficult to keep up to date because often they lag the actual tariffs and, when the structure of a tariff changes, it may be necessary to write new interfaces between the database and existing network design tools. Finally, in some cases, special arrangements may be negotiated with common carriers that can affect the rates.

Because of all this, sometimes an approximate view of link costs is used, com-✓ promising absolute accuracy for a more realistic amount of effort and the ability to respond to network design problems in a more timely fashion.

1.3.4 Device Characteristics

The cost of each type of device is obtainable from the manufacturers or the organizations leasing it. As before convert onetime costs to monthly costs by amortization. Include a single fixed cost, a cost per unit capacity, or a combination of both.

Similarly, associate a capacity with each type of device. Again, if the device is modular, it can have both fixed capacity and add-on capacities. Throughput capability ✓ can be measured in bits-per-second, packets-per-second, or simultaneous connections. Sometimes the aggregate bit rate of all channels or the total number of ports on a switch is of interest. Note that in general the aggregate bit rate of the incident channels is larger than the total amount of traffic that the switch can handle. Another measure of capacity is the number of sessions (or calls) per minute a switch can set up.

These different capacity limitations are a function of the architecture of the device and are generally based on a model of its performance. In some cases, several capacity limitations are simultaneously active. In most cases, the work can be completed in terms of just one.

Note that to check feasibility of a design, relative to any specific capacity limitation, the corresponding information about the requirements must be available. For example, to check limitations on both the number of packets and the number of simultaneous connections, requirement information is needed both in packets-per-second and session length. It may be difficult to obtain such information, and often one capacity constraint dominates the other. Usually, a simple model of requirements is built and a single requirement magnitude is used.

> **Example 1.1.** Suppose a switch could handle 400 packets-per-second of transit traffic and could handle 50 simultaneous connections. Suppose further that the average session lasted 10 minutes and gave rise to 600 packets (one packet-per-second). It is seen that the limitation on simultaneous sessions clearly dominates—50 sessions versus 400 packets (which in this case happen to also be equivalent to sessions), and therefore, this constraint is used. Traffic could be measured in sessions or packets, but one need not use both measures. This is a considerable reduction in the overall effort to complete a design. On the other hand, if both constraints are found to be active, as would be the case if a

session contributed only 100 packets on average and the actual number varied greatly from session to session, both constraints might have to be checked, or the requirements adjusted to reflect the worst case.

1.3.5   Performance Objectives

The network design problem is usually thought of as one of minimizing cost while satisfying throughput requirements. In most of the algorithms presented in the following chapters, that is all that is explicitly considered. However, constraints on performance must also be satisfied. Thus, there is usually a limit on the tolerable delay in systems that queue, or loss in systems that block. There also may be constraints on reliability.

Such constraints can be stated as input or thought of as additional costs, to be traded off against other objectives. We choose to take the latter point of view. In reality, there are rarely hard constraints on delay, loss, or reliability. If a network had to double the cost to reduce its delay by 10%, few people would elect to reduce delay. Conversely, if delay could be reduced from 8 seconds to 4 seconds while increasing the cost by only 2%, few people would pass up the opportunity to do so. Ideally, the network design effort should yield as output the relationship between performance and cost, to allow the designer to make an informed decision about how much performance he or she is willing to pay for.

There are many ways of describing performance objectives. Fortunately, none require a great deal of input, although the way they are specified may have a profound effect on other inputs and on the algorithms used to solve the design problems.

Thus, a delay objective simply can be specified as a cost associated with average delay. This allows delay to be traded against cost. Even if the cost were nonlinear (that is, beyond a certain point, the incremental cost of delay increases) the tradeoff can still be easily made. If, on the other hand, the objective is specified in terms of a distribution on delay or in terms of separate objectives for different types of traffic, the representation of the traffic requirements and all the algorithms working with these requirements become much more complicated.

If the delay requirement is specified as a hard constraint rather than as an objective with an associated cost, the algorithms become more complicated. In general, whenever possible take the point of view that performance objectives are costs rather than constraints.

Another useful way of looking at a delay objective is as a simple constraint on utilization. Observe that the relationship between delay, D, and utilization, u, is usually governed by a function of the form

$$D = \frac{D_0}{1 - u}$$

where D_0 is a constant called the service time (e.g., the transmission time for a message). We discuss this relationship in detail in Chapter 2 when we discuss queueing formulas. Notice that as u approaches 1, D gets very large. At the other extreme, if u is close to 0, D is close to D_0 and only changes slowly as u increases.

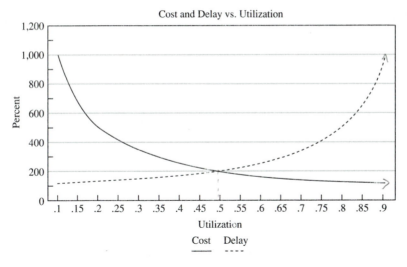

FIGURE 1.4

Similarly, the relationship between utilization and cost, C, is usually roughly governed by a function of the form

$$C = \frac{C_1}{u}$$

<small>← max cost</small>

<small>$ub : c^1$</small>

where C_1 is the cost when a utilization of 100% is permitted. This follows from the fact that the lower the utilization, the greater the number of channels needed to carry a given amount of traffic.

Combining these relationships, a direct relationship between cost and delay is obtained

$$C = \frac{C_1 D}{D - D_0}$$

<small>merge by subbing "u" wrt D.</small>

Figure 1.4 shows the relationships between cost, delay and utilization. As can be seen from examining the figure, there is a range of utilization where delay rises sharply but cost decreases only slightly. Also, there is a region where cost rises sharply and delay decreases only slightly. In practice, the objective is to design networks away from these regions to avoid trading a lot of one criterion for only a little of the other. Another corollary to this is that we may wish to reconsider any hard constraints which force us into these regions.

1.4 NETWORK DESIGN TOOLS

The major focus of this book is the development of quantitative methods for the design of networks. Thus, we are interested in studying alternatives in this area. We begin with an overview of current approaches to network design and then describe the major components of automated tools for the design of networks.

1.4.1 Approaches to Network Design

The techniques described next run the gamut from informal to formal. There is a brief overview of each type, introducing nomenclature used throughout the book. It is important to distinguish between techniques used to design a network "from scratch" as opposed to incrementally. In reality, most real network designs are incremental beginning with an existing network. Unfortunately, most of what is known and can be proven relates to techniques that design an entire network from a given set of requirements.

1.4.1.1 MANUAL DESIGN.

A surprisingly large number of networks are still designed by hand using rules of thumb, or even no "rules" at all. The most attractive aspect of this approach is its flexibility: A database doesn't have to be assembled. All sorts of unusual constraints and objectives can be taken into account. The designer can be very responsive to changes in goals and requirements because there is no "setup" time. Also, incremental as well as total designs can be done.

This approach has several major disadvantages, however. It is rarely quantitative. Often, the designer makes decisions subjectively and inconsistently, sometimes even unconsciously. Thus, it is difficult to repeat a successful design when similar circumstances arise and, likewise, difficult to learn from previous mistakes.

The approach is also usually too labor intensive to allow for the proper consideration of all alternatives. Thus, designs tend to follow the designer's preconceived notions of what they should look like rather than what is actually best in each situation. Serious design mistakes can be made this way. In particular, as the requirements evolve, there is little to motivate a corresponding change in network architecture.

Automating the design process can overcome these problems. In doing so, however, it is important to maintain the advantages of manual design as well. In particular, be sure to leave the designer "in the loop" and with the final word as to what the design should be. This is why good tools are interactive, allowing the user to retain control over the design process and still take into account information available to him but not to the tool. Indeed, this it why we call these software systems tools.

1.4.1.2 HEURISTICS.

Heuristics are design principles incorporated into algorithms and thus are automatable. Like rules of thumb, they are good ideas embodying design experience. They differ from rules of thumb, however, in that they can be made quantitative and repeatable. Thus, alternative heuristics can be compared by implementing them on the same problem, then observing which gives rise to the best solution. It is therefore possible to refine heuristics, keeping what works well and discarding what does not. (This same process goes on in manual design, but to a much lesser extent.) With automated design, many more alternatives can be tried and objectively compared. Experience can be transferred from one designer to another by allowing them to share the automated system.

Some heuristics are specific to particular types of networks. The most useful ones, and the focus in this book, are based on principles which apply across many types of networks. One of the most widely used heuristics is the **greedy algorithm**.

Confronted by a series of choices, the greedy algorithm chooses the best one it can at each stage.

Example 1.2. Suppose that a given set of locations is to be connected and that the cost of all possible connections is known. We do not care about capacity, reliability, delay, or any other performance objective. We simply want to connect the cities at a minimum cost. The greedy algorithm tells us to examine all pairs of locations not already√ connected, and make the cheapest connection possible. We continue doing this until all cities are connected. Figure 1.5 shows an intermediate step in the greedy algorithm; the link between B and C is the cheapest link connecting two locations not already connected (via a path of one or more links), and is hence the next link to be added to the network.

The greedy algorithm is a broadly applicable heuristic based on the simple observation that inexpensive networks tend to contain inexpensive links. This principle is true for all types of networks. Therefore, expect this algorithm to yield good net-√ works. Do not, however, expect it to yield the optimal solution. It is quite possible that the least expensive network does not contain some of the least expensive links. It happens that in the case of this simple unconstrained network, the greedy algorithm does in fact yield the optimal solution. This is discussed at greater length in Chapter 3. However, with added constraints, the greedy algorithm no longer yields the optimal solution.

Example 1.3. Suppose it is desired to connect a set of locations to a central site that is far away. Each location could be connected to the remote site individually, but this would be very expensive. Suppose a delay calculation is performed and it is found that it is possible for pairs of terminals to share a line, but that three or more terminals can not. The task now is to pair up the terminals using links of minimum cost. Then, connect the pairs to the central site, but ignore this latter part of the problem and focus on the first part. A typical situation is illustrated in Figure 1.6. The greedy algorithm would pair up

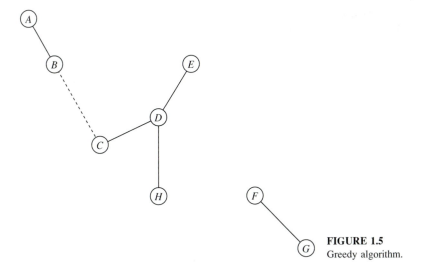

FIGURE 1.5
Greedy algorithm.

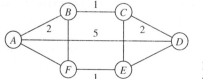

FIGURE 1.6
Greedy pairs.

B and *C*, assuming this was cheapest. It might then pair up *E* and *F*. The only choice left is to pair up *A* and *D*, which is expensive. It is seen that a better choice—*A* and *B*; *C* and *D*; and *E* and *F*—exists. The problem is that the greedy algorithm chose short links without any regard for the consequences later in the design. This is the basic problem with the greedy algorithm.

There is a class of problems for which it can be proved that the greedy algorithm yields an optimal solution. There are other classes of problems for which the greedy algorithm does an acceptable job but cannot guarantee an optimal solution. Finally, there are problems where the greedy algorithm produces unacceptable results, possibly no feasible solution at all. One goal is to find ways of classifying these problems and then determining to which class a problem of interest belongs. This is discussed at length in later chapters.

There are other general heuristics. In selecting links in a centralized network, first consider locations far from the center, based on the observation that if such locations were to be neglected, they might become "stranded," forcing the addition of expensive links to connect them to the center. This is basically what happened in the example in Figure 1.6. Another heuristic is to give preference to locations with large traffic because of the difficulty in finding feasible neighbors for such locations.

Heuristics are valuable because they allow us to obtain feasible solutions to difficult problems in a reasonable amount of time. Instead of an exhaustive exploration of the entire solution space, consideration is limited to solutions with characteristics that appear to be good. Even when it is possible to search the entire solution space and obtain a provably optimal design (i.e., exact methods exist) heuristics may still be chosen, at least in the early stages of a design, to gain the benefit of this speed up.

1.4.1.3 FORMAL OPTIMIZATION TECHNIQUES. Except for the smallest problems, it is not possible to enumerate all possible solutions and then choose the best one. A network with ten nodes, for example, has 45 potential links, each of which could be either included or excluded from the network. Even if it is assumed there is only one possible link speed, this gives rise to 2^{45} possible solutions, more than 10^{13} possibilities. The set of all possible solutions to a problem is referred to as the **solution space**.

The notion of best is expressed in formal optimization techniques by the **objective function**. The objective function associates a value with the design variables. Thus, for example, each link *i* can be included or excluded from a design. There is a cost associated with each link and to minimize cost, use the sum of the costs of the links chosen as the objective function.

While cost is the most commonly used objective function, it is not the only one. We might want to maximize reliability. In this case, associate a reliability with each possible node and link in the network and then compute the reliability of candidate networks comprised of specific nodes and links. Unlike the case of cost, however, this may be a complex calculation in its own right. The best value of the objective function is referred to as the **optimum**. As can be seen from the two previous examples, the optimum can be either a minimum or a maximum.

If the optimum is to be found for problems of realistic size, we must rely on properties of the problem which allow us to avoid looking at most of the possible solutions. One such situation was mentioned before; the greedy algorithm, which examines only a very limited number of alternatives and guarantees an optimal solution for a special class of problems. There are also other situations where a general heuristic produces optimal solutions and some of these are seen in later chapters.

There are also algorithms which always produce optimal solutions. The problem with these, however, is that they only apply to a limited class of problems; for problems outside this class, they do not work at all. One such algorithm is the **simplex method** [Dan63]. This algorithm only works for the class of problems called **linear programming problems**. Linear programs are problems where both the constraints and the objectives are weighted sums of the variables; that is, they are of the form

$$z = \sum_{i=1}^{N} a_i x_i$$

where the x_i are the N variables that we are optimizing over, and the a_i are the respective coefficients.

In this case, the solution space can be searched in a very orderly way, avoiding most of the possible solutions and reaching the optimal solution in a reasonable amount of time.

Often, a problem can be phrased as a linear program with the additional constraint that the variables must be integers. This situation arises frequently in network design; that is, when the decision is whether or not to include a link in the solution. The choice is to include the link ($x_i = 1$) or not to include it ($x_i = 0$); there is no middle ground. These problems (called **integer programming problems**) are much more difficult to solve than linear programming problems. In particular, the simplex method and its variants do not work on integer programming problems.

Another important class of problems is when the solution space and the objective are **convex**. These concepts are illustrated in Figure 1.7. Suppose we have a function of two variables, x_1 and x_2, and we want to find the minimum value of this function, subject to some constraints. Values of x_1 and x_2 which satisfy these constraints are said to be **feasible solutions**, or simply **solutions**. Suppose that the points inside the polygon are all the feasible solutions to the problem. For example, x_1 and x_2 might be the amount of traffic from two applications associated with a node in the network and the constraints represent capacity limitations on the node. This solution space is said to be convex (see Fig. 1.7a) because for any two points, a and b, inside this region, all the points on the line connecting a and b are also inside the region. Figure 1.7b

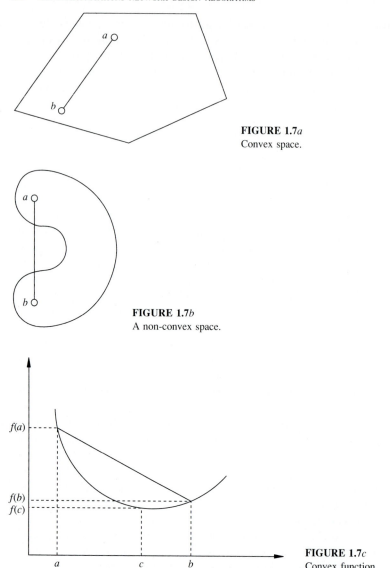

FIGURE 1.7*a*
Convex space.

FIGURE 1.7*b*
A non-convex space.

FIGURE 1.7*c*
Convex function.

is an example of a solution space that is not convex. In a convex space, movement can be made from one feasible solution to another in small steps, without leaving the feasible region. This allows the use of simple, incremental methods to search for the optimal solution.

A convex function is a function with the property that for any point c, between a and b, $f(c)$ lies below the line connecting $f(a)$ and $f(b)$. Convex functions must be defined on convex sets for this property to hold, since otherwise, $f(c)$ may not even exist. Figure 1.7c shows a function that is convex. A function that is not convex (but is defined on a convex region) is shown in Fig. 1.7b as the function's value lies

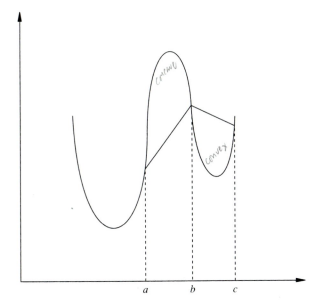

FIGURE 1.7*d*
A non-convex function.

above the line between *a* and *b*. Convex functions can be functions of many variables. In this case, the notion of a line generalizes to higher dimensional planes.

Convex functions have the property that their minimum can be found via a local search. Specifically, the minimum can be found by starting at any point and moving in a direction where the function decreases. Such methods are called **descent methods**. This is much simpler than having to search through all possible values in the solution space. Descent methods can also be used on nonconvex functions but, in general, they converge to a **local minimum** (the bottom of a valley) instead of a **global minimum** (the bottom of the deepest valley.)

If $f(c)$ lies above the line connecting $f(a)$ and $f(b)$ for all c between a and b, it is said that f is **concave**. The function shown in Fig. 1.7*d* is concave on the interval a to b and convex on the interval b to c. It is neither concave nor convex on its entire interval of definition. The maximum of concave functions is easy to find in a manner similar to that for finding the minimum of convex functions.

Note that linear functions are both convex and concave and thus their minima and maxima can be found relatively easily. This is part of the reason why linear programming problems are solvable in a reasonable amount of time. Integer programming problems, on the other hand, have solution spaces that are not convex but are isolated points at integer coordinates. None of the points on the line connecting two adjacent integers are in the solution space.

Sometimes a problem, P, is difficult to solve but is closely related to another problem, P', which is much easier to solve. P' may be P with some constraints removed. Another possibility is that P' might have the same constraints as P but its objective function might be more well-behaved, perhaps convex. P' can then be solved in the hopes of obtaining a solution, or a close approximation to a solution, to P.

Example 1.4. In designing a centralized network, multipoint lines (trees) can be formed with no more than 10 terminals on each line. The problem, P, of finding a set of trees of minimum total length all rooted at the central site and each having at most 10 nodes on them is very hard to solve exactly. The problem, P', of finding a set of trees of minimum total length without a restriction on the number of nodes in a tree can be solved using the greedy algorithm. Thus, a solution for P can be attempted by first solving P' and then adjusting the solution to satisfy the constraint on the number of nodes in a tree.

Suppose P' is formed by substituting a well-behaved objective function, $g(x)$, for the original objective function, $f(x)$, from P. If $g(x)$ has the property

$$g(x) \le f(x) \qquad \forall x$$

then it is said that $g(x)$ is a **lower bound** for $f(x)$.

Example 1.5. Consider the functions shown in Fig. 1.7e. Consider the non-convex function, $f(x)$, and the convex function, $g(x)$. Over the entire interval shown, $g(x)$ is less than $f(x)$. The minimum of $g(x)$ can be found using a descent method. This would lead to the value $x = x_2$, the minimum of $g(x)$. On the other hand, if an attempt to find the minimum of $f(x)$ via descent were tried we could end up at the bottom of the valley at x_1 or, if we were lucky, we might find the global minimum at x_3. $f(x_2)$ is a good approximation to the global minimum for $f(x)$. Furthermore, starting a local search (descent) for the minimum of $f(x)$ at $x = x_2$, would in fact, find $f(x_3)$. Thus, using $g(x)$ as a substitute for $f(x)$, at least temporarily, is a good approach in this case. The key to making this approach work is to be able to find a function, $g(x)$, which approximates $f(x)$ closely enough (at least in the vicinity of the optimum). This approach is good because when the minimum of $g(x)$ is found, $f(x)$ can be evaluated to see how close they are.

If P' is formed from P by dropping constraints and/or by substituting a lower bound for the objective function of P, then it is said that P' is a **relaxation** of P. Such P' has the useful property that the optimal solution to P' is no greater than the optimal solution to P. This can be seen by considering the value of x which is the

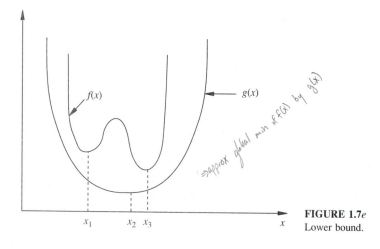

FIGURE 1.7e
Lower bound.

optimal solution to P. x is also a solution to P' since x satisfies all the constraints of P' (which are a subset of the constraints of P). Furthermore, the value of $g(x)$ is no greater than $f(x)$, since $g(x)$ is a lower bound for $f(x)$.

Therefore, it is possible to solve P with a heuristic, solve P' exactly (obtaining the optimal solution) and then compare the values of the two solutions obtained. If they are sufficiency close, say within $A\%$ of one another, no further search is necessary since at worst, the lower bound is within $A\%$ of the optimum. Note that even if the two solutions are far apart, it does not necessarily mean that the heuristic solution is far from the optimum; the lower bound may be loose. This approach can be used as the basis of a more general technique, known as **branch and bound**, described in [Geo72] and [Law76].

There are also algorithms, called **combinatorial algorithms**, which deal directly with discrete choices, typically variables which only take the values 0 and 1. Later chapters explore such methods extensively. These methods rely heavily on the structure of the problem and it is important to be able to recognize such structure when it is present. In some cases, solutions are obtained by approximating a design problem as one with sufficient structure to apply these methods.

The following chapters explore a variety of methods, both formal and informal, for the solution of network design problems. A good tool has several algorithms of varying speed and effectiveness for the solution of design problems, each useful for problems of differing magnitude, and at different stages in the design process. It is appropriate to use a time consuming, exact procedure, at the end of the design process where all other factors have been determined and the resulting design is actually implemented. Also, exact techniques are appropriate when input data is known with high precision, and when the problem is small enough to be solved in a reasonable amount of time. With regard to problem size, it is interesting to note that as a problem gets larger, exact methods often become prohibitively slow but the solution space becomes richer, making heuristics less likely to make serious mistakes. Thus again, a tool containing both can function well for all sizes of problems.

1.4.2   Structure of a Network Design Tool

A complete network design tool contains modules that allow the user to use each of the above techniques as appropriate. Also, it relieves the user of the burden of collecting and processing the inputs required by the tool to the extent that this is possible. It presents its output interactively in tabular as well as graphical form. Finally, it should allow the user to interact with it in all phases of the design process, both before and after the algorithms are run. Thus, the user gets both the first word, specifying allowable candidate nodes and links, and also the last word, editing the tool's design decisions. The tool should also include analysis routines to check the validity of the user's design decisions.

Figure 1.8 is a high level structural overview of a network design tool, showing all of the major modules. The modules are organized into four areas—the front end, database routines, algorithms, and utilities. The function of each module is briefly described.

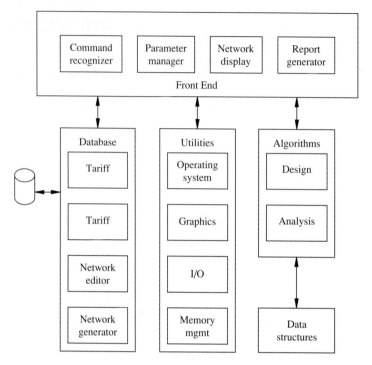

FIGURE 1.8
Structural overview.

There are many ways to implement these functions and the specific implementation chosen for a particular tool is dependent upon many factors. It is not possible to say that one approach is uniformly better than all others, but here are some general guidelines.

The first issue is the sophistication of the user. If a tool is being designed for novice users, or for a wide range of users that includes novices, it should not make heavy demands upon the user to make sophisticated decisions or to supply it with inputs which are difficult to obtain. Many of the design algorithms used in practice are parameterized and the parameters have to be specified. Sometimes, reasonable defaults can be specified and wired into the tool.

If this is not possible, the algorithm will not function reliably without them, and something else has to be done. Perhaps documentation or even training courses can be developed to help users learn how to properly set the parameters, or at least to recognize when they are not properly set. If the algorithm is fast enough, it may be possible to run it for a range of parameter values. If none of this is possible, it is best not to include the algorithm at all, and if no other reasonable algorithm is available to perform the function, then it is best not to include the function.

This may sound extreme at first, but it is not. A tool is useless if the user loses confidence in it, and the user will, rightfully, lose confidence if the tool does not reliably produce reasonable answers in all situations. It is essential to limit the tool's scope to problems that the user and the algorithms together can solve.

Here is the structure of an "ideal" tool. The reality of the situation may prevent creating such a tool, since doing so requires significant resources of many kinds. The uses of the tool help to govern the tradeoff between resources and functionality. An obvious example is the tool accompanying this book. It lacks many of the capabilities described in the following sections, but is adequate for the purposes of illustrating the design principles discussed, and requires only modest computational resources, making it accessible to most readers.

1.4.2.1 FRONT END. The Front End is responsible for both the input and output interfaces to the user. This is the part that makes the tool interactive. It presents the user with menus for selecting functions, it recognizes commands, it displays the network and associated information, it generates reports detailing designs and the results of analysis, it allows the user to manipulate parameters guiding the other modules, and finally, it allows the user to edit the network, adding and deleting nodes and links and changing their status such as making a location a backbone node site.

The Front End is dependent upon the tool's environment, both hardware and software. The more advantage is taken of the specifics of this environment, the more dependent we become upon it. Ideally, the work is within an environment which allows a "family" of implementations. For example, using C under UNIX permits an implementation on a wide variety of physical devices. It is also possible to limit device dependencies to low level utilities, thereby facilitating migration to a different environment.

Even with this, however, the issue of overall functionality must be confronted. If the tool relies upon graphics, the hardware must support graphics. It is not sufficient to simply turn off the graphics when operating in an environment that does not support graphics, if the tool does not function acceptably without graphics support. Moreover, if the tool requires color or high resolution graphics, these must be present as well. There is, of course some leeway with respect to such things as resolution, but there are limits to how far the user can be asked to compromise. For example, if the tool selects concentrator locations, it is reasonable to assume at the very least that it is capable of displaying networks with dozens of nodes.

Sometimes software can compensate for hardware. With high resolution graphics, the tool may be able to comfortably display networks with 100 nodes. With poorer resolution, the user may still be able to deal with such networks if the tool also has the capability of selectively displaying parts of the network.

A good network display capability allows the user to zoom and pan, displaying geographic segments of the network. It also allows the user to select logical parts of the network for display (e.g., the backbone or one of the local clusters). It allows the user to move nodes to clarify the display. Finally, it permits the user to obtain hard copy of the displays, via a printer or plotter. This latter capability implies the presence of a hard copy device with sufficient capability to do the job. For example, it is unfortunate if the a tool is dependent upon color when the hard copy device does not support color. Here again, it may be possible to compensate;

that is, a higher resolution plotter may be able to substitute different line types for color.

Graphics can also be used in the input. Thus, the user may wish to use a cursor to select nodes and links; that is, to change their characteristics. Some users may prefer this, but the majority of users find keyboard input faster, especially if part of the input must be entered via the keyboard in either case.

As an example, suppose the user is currently looking at a display of the network and wants to modify information about a node. There are at least three ways of selecting the node. The user might use a mouse to move a cursor to the node's location and press a button on the mouse, selecting the node. Or, the user might press an arrow key to scroll through a list of node ID's and hit enter to select it. Finally, the user might type the node ID.

Each approach has merits. The first is attractive if the user wishes to move the node; the mouse is ideal for specifying the new coordinates. The second is probably fastest, if the number of nodes is not too large, and if the list is alphabetical. The third is attractive if the keyboard must be used to enter the property, that is, a new node ID. While it is not difficult to implement all three methods, keep in mind that each mode of input is a mechanical skill learned through repetition; it is best to teach the user to rely predominantly in one. Many users prefer to use a mouse for graphical inputs and to use single keystrokes whenever possible. Notice that an arrow key with a built-in repeat is almost as good as a single keystroke.

Another case which arises frequently is the selection of an item from a small set. This includes the selection of commands, parameters, and multiple choice parameter values. This can be done most efficiently by selecting the item from a menu by hitting a key (letter or function), or by scrolling to the item and hitting enter.

The front end may include a Network Editor, which allows the user to specify nodes, links, and requirements. The usefulness of such editors is limited, however, to small amounts of data. Few users have the patience, for example, to specify a large requirements matrix, one item at a time, even with the most sophisticated editor.

An interactive editor can be very useful, however, for editing a design or moving nodes. Its usefulness can be enhanced by providing a macro capability that allows the user to enter a "pattern" of data. For example, a user might want to add a list of links to the network. Each link has two endpoints, a type and a capacity. An Add-Link macro might be provided with four windows, which the user could fill in by scrolling through permissible values. With this macro, the user might set the type and capacity, and possibly one of the endpoints, and then just vary one field to generate a sequence of links. In some cases, an "*" might be used to insert or alter a whole set of values (e.g., add a link from one node to all others).

Even with such features, however, an interactive editor can not take the place of a good bulk editor. Any system which can support a network design tool can support, and almost certainly already has, a good bulk editor (e.g., a word processor) that can be used to deal with large volumes of network input. Such software is likely to contain more features than any special purpose network editor. Even macro capabilities are likely to be better; most editors have global search and replace capabilities, for example. Furthermore, the user is probably already conversant with them.

The only strong motivation for including a sophisticated editor in the tool is integration. It is disruptive to have to switch back and forth between the tool and the editor. If the operating system supports multiple tasks at the user level, it may be possible to switch easily. If not, it may be possible to organize the use of the tool to avoid frequent switching.

Of course, in the end, the choice is the user's. Even with the most comprehensive built-in editor, the user is free to use an external editor, unless the designer of the tool was foolish enough to make the input files unreadable. It is, in general, unwise to use anything other than "clear" text files as input to a network design tool. This is discussed more fully in the section on data base functions.

Ideally, report generators should allow for output to the screen, to a printer, or to a file, at the user's discretion. The implementation should be transparent to this, treating all three as output to a file. In the case of screen output, the user should be able to scroll freely both forward and backward through the "file."

1.4.2.2 DATA BASE MODULES. To work on realistic problems, the tool must be able to deal with the bulk data needed to support the design process. The key to making this work is to rely on existing databases rather than having to manually input the data. This is not only less effort; it is more reliable.

There are services which provide machine-readable databases that contain current tariff information, typically updated on a monthly basis. Even with such a service, however, the tool must contain modules to reformat the data into a uniform format usable by the tool. Different tariffs are, in general, in different formats. Worse yet, they can rely on different information being present in the network database. For example, many tariffs function on the basis of the area code and exchange of the locations being connected. Given this, the vertical and horizontal (V&H) coordinates of the locations can be accessed in the database and the cost of a link can then be computed. Some tariffs, however, require additional information, that is, the distance from the nearest central office, which is harder to obtain. We can choose to provide a default in this latter case, sacrificing accuracy for ease of use.

Some tools can also contain a tariff generator, a module which allows the user to build a functional model of a tariff in terms of distance, line type, and speed. Such generators are useful in studying the relationships between network design and tariff structure. This is of interest to tariff designers, researchers, and students of network design. They can serve in place of actual data if necessary since most real tariffs are based on such mathematical functions. The tool accompanying this book has a tariff generator, relieving the reader of the burden of acquiring real tariffs. For a tool that is used in practice, however, there is no substitute for the actual tariffs.

Similarly, the tool should have a traffic editor to manipulate requirement information. Unlike the tariff editor, this module is dependent upon the specific user. The most important feature is to be able to interface to existing traffic databases available to that specific user. It may be appropriate to write special code to interface to existing software, producing these databases on the system where the tool is being used. Alternatively, it may be possible to use an existing bulk editor to simply reformat files available on the system. Finally, it may be necessary to create special modules.

The traffic editor should allow the user to maintain a database of requirement information, adding to it, merging files, and checking for consistency, especially with the location set. It should also facilitate the transfer of data from disk into memory. The ability to scale traffic is also useful.

Some tools may also include a traffic generator which can create requirements parametrically, based on location characteristics. This is useful where detailed traffic information is not available, or for the study of the relationship between traffic characteristics and network design. The tool accompanying this book includes a traffic generator.

The network editor mentioned before can also be extended to manage detailed node and link information. This includes location data, link data, and possibly device data as well. This function can take the tool into the domain of network management, including the management of the inventory of all facilities in the network. This is a very important function and it is also quite sensible to integrate network design and network management functions, as this helps both to bring to fruition network design decisions and to make the network design process more responsive. Unfortunately, this function is beyond the scope of this book and not discussed further.

Some tools may include a network generator which allows the user to parametrically generate random networks. This is useful for testing the algorithms, learning to use the tool, and in studying the relationship between network designs and various network characteristics (e.g., size). The tool accompanying this book has a network generator.

The database section of the tool should also allow the user to maintain sets of files associated with each design problem. The location and requirement data associated with a specific problem should be kept together, either in a single file or in a set of files identifiable as part of a single problem, so that it can be cross checked and kept consistent. Some tools also allow for a set of alternate designs to be kept together.

In designing the database section of the tool, it is important to leave the user in control of when one design permanently replaces another. It is possible to maintain many alternate versions of a design and not replace anything, but this often gives rise to confusion and so is not generally advisable. One of the strengths of an interactive network design tool is that it allows the user to explore many design alternatives, most of which are not successful. Thus, it is important to be able to generate designs without permanently storing them.

On the other hand, it is important not to lose a good design, and thus best to make it easy to save a design. The simplest approach is to give the user the option of either saving the current design under the current network name (overwriting the previous design), or to optionally rename the network (creating a new design). Most word processors offer the user this choice.

An even nicer capability, somewhat more complex to implement, is to let the user run in a "workspace" where all design alternatives are saved during a session with the option of transferring good designs into permanent storage. This allows the user to temporarily hold on to viable candidates during a session, but also inhibits the accumulation of designs which do not really pan out.

The maintenance of this workspace usually relies upon features in the language or operating system used for system implementation. Such capabilities are becoming increasingly available and even personal computers often have operating systems which allow for virtual disks which can support this function. Languages such as LISP and APL have long supported the notion of a systems environment.

1.4.2.3 ALGORITHMS. This section of the tool includes algorithms for design and analysis. The majority of this text is devoted to describing these algorithms and exploring them in detail. The focus here, however, is on the overall structure of these modules.

The major difficulty is that in an effort to create efficient, effective, and general purpose modules, it is easy to create very complex programs with subtle bugs in them. Worse yet, it is impossible to check these programs since they are performing calculations that cannot be replicated except for very small networks or very special cases. If a bug causes execution to abort, we are lucky; at least we know we have a problem. If a bug causes the program to yield a reasonable looking but infeasible design, it may not become known until the network is built. Thus, testing eliminates all the observable errors, leaving a system that only makes subtle mistakes on large problems. This is the worst of all possible situations. How can this can avoided?

There is, of course, no foolproof answer. The only hope lies in doing everything to hold down the intellectual complexity of the algorithms. The first thing this means is that given the choice between a straightforward algorithm and a more "clever" one, always choose the simpler one unless there is a *substantial* difference in efficiency or effectiveness.

A side benefit of this is that not only are simpler algorithms chosen but also ones more easily reused since the simpler algorithms tend to rely less on the specifics of a particular situation and hence are more likely to be applicable in other places. Thus, the total amount of code is decreased and greater reliance is placed on code used in many places. Such code tends to be most reliable as it is thoroughly tested through a variety of applications.

This goal is realized to a great extent by using common data structures and common operations on these data structures as the basis for the implementation of the algorithms. These basic structures and their operations are well defined with carefully implemented and thoroughly tested code to support them. These operations have clear semantics (i.e., we know exactly what they mean). More complex algorithms are then implemented in terms of these basic functions. Although this does not guarantee the algorithms are error free, it does provide a solid foundation to build on and it allows the concentration of attention in the implementation at a higher level than would be possible working in terms of the elementary statements in the implementation language.

A major decision in implementing the algorithms is whether to restrict them entirely to the main memory of the computer or, alternatively, to work with secondary storage. The latter approach gives the potential to deal with larger problems and to use algorithms that make use of more space. On the whole, however, this is usually a bad tradeoff. The loss in efficiency is devastating and tends to make the tool noninteractive. Thus, throughout this text it is assumed that the algorithms and the data they use are memory resident.

The only exception to this is that many systems in use today are in fact virtual-memory systems, where pages of memory are transparently swapped in and out. This does not affect the implementation directly and does not affect running time unless the tool consistently relies on having more memory than is physically available to it.

The specific algorithms used may vary from one tool to another based on the type of network, the type of user, and even the specific problem at hand. The concentration in this text is on presenting algorithms that are widely applicable.

1.4.2.4 UTILITIES. This section of the tool includes the utilities that interface to the operating system and are machine specific. They are building blocks that allow the remainder of the tool to be implemented more easily, more generally, and more efficiently. Often, they are supplied as part of the implementation language. They include graphics primitives which draw points, lines, and circles. They also include primitives that open and close files and test for the existence of a given file. The implementation of the tool is greatly simplified if done in terms of the primitives available as part of the host language and operating system.

Sometimes, an auxiliary package, such as a database management program or graphics library, is included in the tool as a building block. Whenever possible, it is best to use existing packages rather than writing personal code. This holds down implementation complexity and usually results in higher functionality, as such packages tend to provide a full spectrum of features. Of course, it is important to select reliable and efficient packages to build with and if possible to select packages widely available across computer systems and that will hence not impede migration.

EXERCISES

1.1. Suppose we were considering building a network to interconnect 1000 terminals located in 100 different physical locations. We are considering 9.6 Kbps lines and 56 Kbps lines. Approximately how many bytes would be required to represent each of the following? (Part of your answer includes an assumption on how the network is represented)

(*a*) the V&H coordinates of the nodes

(*b*) the costs of potential links between the terminals

(*c*) the requirements between the terminals. (Consider both the cases assuming a dense requirement matrix and assuming a sparse requirement matrix.)

BIBLIOGRAPHY

[Dan63] Dantzig, G. B.: *Linear Programming and Extensions*, Princeton Univ. Press, Princeton, 1963.

[Geo72] Geoffreon, A.: *Perspectives on Optimization*, Addison-Wesley, Reading, Mass., 1972.

[Law76] Lawler, E.: *Combinatorial Optimization: Networks and Matroids*, Holt, Rinehart & Winston, New York, 1976.

[Sch77] Schwartz, M.: *Computer Communication Network Design and Analysis*, Prentice-Hall, Englewood Cliffs, N.J., 1977.

[Sta85] Stallings, W.: *Data Computer Communications*, Macmillan, New York, 1985.

[Tan81] Tanenbaum, A. S.: *Computer Networks*, Prentice-Hall, Englewood Cliffs, N.J., 1981.

CHAPTER
2

ANALYSIS OF
LOSS AND DELAY

This chapter presents the basic techniques for the analysis of loss and delay in networks. It aims at presenting enough material for the reader to be able to carry out the basic analyses necessary to evaluate alternative designs. Entire texts have been devoted to the analysis of networks, among them [Sch77] [BG87]. The reader interested in exploring these areas in greater depth may wish to examine these references.

The analytic techniques presented in this chapter rely heavily on a knowledge of some basic aspects of probability theory and combinatorics. This book presents an overview of these areas, but again, the reader who wishes to explore these areas in greater depth may elect to examine [Fel57] [Pap65] [Liu90] to gain a deeper understanding.

2.1 ANALYSIS OF DELAY IN NETWORKS

The first step in analyzing delay in a network is obtaining a clear idea of what is to be analyzed. There are many possibilities, among them the average delay, the worst case delay, the distribution of delay (i.e., what fraction of messages have delays in a given range), and the delay between specific pairs of nodes. One may be concerned with the delay for a certain application or delay in the presence of failures. While the specific answers can, of course, differ depending upon exactly what is analyzed, the basic techniques used remain the same.

Figure 2.1 illustrates the life cycle of a typical inquiry/response message in a centralized network with polled input lines. Note that many of the factors in the life

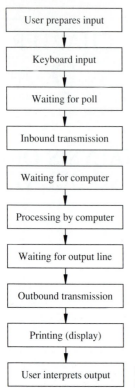

FIGURE 2.1
Message life cycle.

cycle, indeed the most time consuming ones, have little to do with the design of the network; such as the time the user spends preparing the query and interpreting the output. This message life cycle might, for example, represent a transaction at an airline ticket counter. The overall transaction might take two minutes, including the interaction between the customer and ticket agent. Of this time, less than three seconds might be spent in the network, and of that three seconds, less than two might be the time transmitting the data to and from the computer. Nevertheless, this is the time we focus on since it is the time we can control. In designing a network it is important to keep the entire message life cycle in mind and to maintain the proper perspective, remembering that it may not make much sense to completely distort the design to reduce the time for a two minute transaction by half a second.

Another important reason to study the entire message life cycle is that it gives some perspective on the message arrival process, which is one of the most important inputs to the delay analysis process. For example, a terminal cannot generate more than thirty messages per hour if the life cycle of the message is two minutes.

Examination of the message life cycle, reveals two different types of delay—service times and waiting times. Service times include transmission of messages, processing by the computer, printing, and all the activities by the user. Waiting times, also known as queueing delays, include the time spent waiting for the transmission lines and the computer. There is a direct relationship between service times and waiting times. In general,

$$T_w = \alpha T_s$$

where T_w is the waiting time, T_s is the service time, and α is a function of many factors which are discussed below. The basic idea is that messages suffer waiting time because they wait (queue) behind other messages seeking service from the same server. It is usually assumed that all messages seek the same type of service, and thus, if the number of messages in front of a given message is held constant, then the waiting time grows proportionally with the service time;

$$T_w = QT_s$$

where Q is the number of messages in the queue. Of course, if the service time increases, it is more likely that the server will fall behind and that there will be a large number of messages waiting in the queue when a given message arrives; this must be taken into account in the analysis.

First, note that all the times are random processes; messages have different lengths and the number of messages in the system varies from one instant to another. Thus, we discuss averages and distributions for all the quantities in such analyses. In reality, the variation in the message length and interarrival times is not completely random. There may be several specific types of messages in the system, each with a fairly predictable length and while individual arrivals occur in a somewhat random fashion, there are usually predictable peaks and lulls at certain times of the day. One of the decisions is to what extent the worst case is studied as opposed to average behavior. This is part of the definition of the problem.

The first step in analyzing performance in a network is to obtain an estimate of the load; that is, the number and length of the messages offered to the network. This is usually a very frustrating process as it varies due to many factors. There is a temptation to give up and not do any analysis at all because the necessary input data cannot be obtained. At the other extreme, many simplifying assumptions can be made leading to elegant analytical results, which may then bear no relationship to the real problem. The temptation is then to abandon analytical techniques altogether and simply to use rules of thumb. However, this is not necessary. Through the analysis of relatively simple models of the load, it is possible to obtain analytical results robust enough to give sufficient insight into the behavior of real networks, allowing meaningful design decisions to be made. At the same time, we learn about the limitations of these techniques and when to make conservative decisions to compensate for a lack of precise input data.

2.1.1 Elements of Probability

To analyze delays, it is important to understand some of the elements of probability. This is reviewed to make the discussion that follows self-contained. This discussion is informal, but sufficient for our purposes.

Probability is discussed in terms of events. Each event occurs or does not occur. There is an event space giving all possible events that can occur. For example, we can flip a coin. The events which can occur are a head or a tail. The event space is thus { HEAD , TAIL }. Each event has a probability, defined as the ratio of the number of times it occurs to the total number of trials in a large number of trials. Thus, if we flip a coin one million times and observe 513,496 heads, we estimate that the probability of a head is 0.513496, based on that particular experiment. On a different day (or month, depending on how fast we flip the coin!) we observe 496,905 heads and make a different estimate.

Alternatively, it could be argued, based on physical properties of coins, that the probability of a head is intrinsically 0.5. This leads to the question, "To what extent do the results of these two experiments support or contradict the statement that the intrinsic probability of a head is 0.5?" This relates to the question, "How large is large in the definition above?"

The number of heads in an experiment is a **random variable**. It varies from one experiment to the next. Suppose the experiment is performed N times. Let H_i be the number of heads observed in the ith experiment, for $i = 1, 2, \ldots, N$. Then the average, mean, or expected value of the number of heads is defined as

$$E(H) = \frac{1}{N} \sum_{i=1}^{N} H_i \tag{2.1}$$

If T is the total number of flips in an experiment, then $P(H)$, the probability of a head, is defined to be

$$P(H) = \lim_{N \to \infty} \frac{E(H)}{T} \tag{2.2}$$

Alternatively, even for $N = 1$, $E(H)$ approaches $P(H)$ as T gets very large.

Probabilities obey the following rules:

1. If events E_1 and E_2 are alternative events (like a head and a tail), the probability of E_1 occurring or E_2 (denoted $P(E_1 \vee E_2)$ occurring is $P(E_1) + P(E_2)$. Alternative events are said to be disjoint or mutually exclusive.

2. If events $E_1, E_2, \ldots, E_k$ are mutually exclusive and they are the only events which can occur (like a head and a tail, assuming we ignore the possibility of the coin landing on its edge or getting lost), their probabilities must sum to 1.

3. If events E_1 and E_2 are independent, the probability of their both occurring, denoted $P(E_1 E_2)$, is simply the product of the probabilities of their occurring separately. For example, suppose a card is drawn from a full deck. The probability of its being a heart is $1/4$ and the probability of its being a king is $1/13$. The probability of its being a heart is independent of the probability of its being a king and thus the probability of its being the king of hearts is $(1/4)(1/13) = 1/52$. This rule generalizes in a natural way to more than two events; that is, the probability of any number of independent events occurring is simply the product of their occurring separately. Care must be taken, however, to make certain that the events are indeed independent. For example, if two cards are drawn from a deck (drawing the second without replacing the first), the event that the second card is a heart is not independent of the first being a heart, since there is now one fewer heart in the deck to be drawn.

4. If events E_1 and E_2 are not disjoint, then $P(E_1 E_2)$ is greater than 0 and

$$P(E_1 \vee E_2) = P(E_1) + P(E_2) - P(E_1 E_2)$$

Note that this last relationship is a generalization of rule 1, where $P(E_1 E_2)$ is 0. This relationship can be generalized to 3 events by treating the compound event $E_1 \vee E_2$ as a single event which we will refer to as C. Thus,

$$
\begin{aligned}
P(E_1 \vee E_2 \vee E_3) &= P(C \vee E_3) \\
&= P(C) + P(E_3) - P(C E_3) \\
&= P(E_1) + P(E_2) + P(E_3) - P(E_1 E_2) - P(E_1 E_3) \\
&\quad - P(E_2 E_3) + P(E_1 E_2 E_3)
\end{aligned}
$$

Note the symmetry in this expression. It can be generalized to 4 or more events in a similar manner. When this is done, it is found that the general expression contains all combinations of events taken one at a time, two at a time, three at a time, etc., with the signs alternating on events taken a given number at a time. Thus, compound events containing an odd number of events have positive signs and compound events with an even number of events have negative signs.

If there is an underlying physical model for a random process, as there is with coins, the experiments can be thought of as confirming or contradicting the physical model. In the case of arrivals to a telecommunications network, however, experiments simply allow us to estimate an otherwise unknown mean. This leads back to the question of, "How large is large?" More specifically, how much confidence can there be in the outcome of an experiment (or a given number of experiments) of a given length. This field of mathematics is known as statistics. The reader interested in exploring this further may wish to examine [All78].

Begin the analysis by assuming that the outcome of each trial is independent of every other trial. Also assume that the underlying process is not changing while we are measuring it. In the strictest sense, these assumptions are never entirely true, but we make them because it simplifies the analysis and helps us to gain insight into the behavior of the system.

Two major factors contribute to increased confidence in the results of an experiment—a large number of samples and a low variance in the underlying process. The notion of variance relates to the way the outcomes of multiple experiments differ from one another. Suppose we walk into a health club and ask a group of 10 people engaged in high impact aerobics how much they weigh. They all respond with a number between 80 and 95 pounds, with the average (sum of the weights divided by 10) being 89 pounds. On the basis of this experiment, we form the hypothesis that people who engage in high impact aerobics weigh on the average, about 89. We would have more confidence in this hypothesis if it were based on the results of asking 20 people. Likewise, we should have less confidence if the 10 people we asked responded with weights between 70 and 160 pounds (but still with a mean of 89). In the former case, we are less vulnerable to a few atypical responses; in the latter case, we are more vulnerable.

The variance of a process is a measure of how the individual outcomes differ from the mean. Formally, if we have N observations, say the H_i in the coin flipping experiment, the observed variance is defined as

$$\sqrt{}\ V(H) = \frac{1}{N} \sum_{i=1}^{N} (H_i - E(H))^2 \tag{2.3}$$

Note that by squaring each term in the sum, the terms are all positive and do not cancel out. Thus, a process with some terms far above the mean and some far below have a large variance. Squaring has the effect of changing the units of the variance from those of the observations and of the mean. We therefore define the standard deviation, σ, to be the square root of the variance. Also, note that the value of the variance and standard deviation increase with the values of the observations. This factor is eliminated by defining the coefficient of variation to be the standard deviation divided by the mean. These definitions are summarized in Table 2.1.

While the mean and variance of a random variable tell a great deal about a random variable, they do not tell us everything. The most complete information is

TABLE 2.1
Definitions of statistical terms

x	a random variable
N	the number of trials
x_i	the outcomes of the individual trials
$E(x)$	the expectation (average, mean value, μ) of x $$= \sum_i \frac{x_i}{N}$$
$V(x)$	the variance of x $$= \frac{\sum_i (x_i - E(x))^2}{N} = \frac{\sum_i (x_i)^2}{N} - E(x)^2$$
$\sigma(x)$	standard deviation of x $$= \sqrt{V(x)}$$
$\rho(x)$	coefficient of variation of x $$= \frac{\sigma(x)}{E(x)}$$

$E(x)$ is also denoted E_x, $\mu(x)$, μ_x, or μ (if x is understood).

Similarly, $\sigma(x)$ and $\rho(x)$ may be denoted σ_x and ρ_x or even σ and ρ

given by the distribution of the random variable. The distribution is the probability associated with each possible outcome. For example, the distribution for the outcomes of rolling a single die (half a pair of dice) is

$$P_1 = P_2 = P_3 = P_4 = P_5 = P_6 = \frac{1}{6}$$

Note that the P_i must all be non-negative and must sum to 1.

Given the distribution, the mean can be computed as

$$\mu = \sum_{i=1}^{6} i P_i = 3.5$$

and the variance,

$$V = \sum_{i=1}^{6} 6(i - 3.5)^2 P_i$$

$$= \frac{1}{6}(6.25 + 2.25 + 0.25 + 0.25 + 2.25 + 6.25) = 2.92$$

and the standard deviation,

$$\sigma = \sqrt{V} = 1.71$$

The mean, μ is also known as the first moment of the random variable. μ_k, the kth moment of a random variable is defined as

$$\mu_k = \sum_{i=1}^{N} P_i(v_i)^k \tag{2.4}$$

where the random variable can take any of the N values $(v_1, v_2, \ldots, v_N)$, taking the value v_i with probability P_i. The kth central moment, c_k, is defined as

$$c_k = \sum_{i=1}^{N} P_i(v_i - \mu)^k \tag{2.5}$$

Thus, the variance is the second central moment. Note that the expression for the variance can also be rearranged for more convenient calculation

$$V = \sum_{i=1}^{N} P_i(v_i - \mu)^2 = \sum_{i=1}^{N} P_i(v_i)^2 - \mu^2 \tag{2.6}$$

Therefore, the variance is also the second moment minus the square of the first moment.

Processes in the real world can be thought of as being governed by a large and complex set of physical laws that control their outcomes precisely. Alternatively, the world can be modeled using a much simpler set of laws utilizing some degree of randomness. The latter course is usually chosen. The last piece of the puzzle in quantitatively answering questions about our confidence in the outcome of experiments involving random variables, is the precise nature of the distribution of the random variable being studied. If we are given P_i for all possible values of v_i, we can then answer questions like, "What is the probability of observing a mean greater than or equal to 4 in an experiment where the outcomes of rolling a die 3 times are recorded?"

Based on the answer to this question, it can then be decided whether or not the underlying model which claims that all outcomes are equally likely is accurate. Similarly, if we design a telephone system for an office and have been given the distribution of the number of people simultaneously making calls, we can answer the question, "How many lines should be put into the system to ensure that a line is available for an outgoing call at least 99% of the time?"

Unfortunately, the precise mean and variance is rarely known, much less the entire distribution, for processes of interest. Therefore, we make do with estimates of the mean and variance based on observations and with an assumption about the distribution which holds true in a remarkably large number of cases:

Assumption: A random process can be adequately modeled by assuming that it is normally distributed.

Before giving a formal definition of a normal distribution, a few more basic concepts must be defined. Thus far, v has been a discrete variable; that is, it took only values from a finite set. For example, the die takes values which are integers between 1 and 6. It is also possible for v to be a continuous variable, taking any value in a range, finite or infinite. Thus, the possible weights of people in the aerobics class could be any number (including fractions) in the range between 0 and 100, or even 0 and ∞. The weights can be recorded to the nearest integer or more precisely (say, to 6 decimal places) or, in theory, exactly. The probability of any specific value being observed goes to zero (i.e., can be made arbitrarily small) as we become more and more precise. Nevertheless, some values are more likely than others. This can be given by a probability density function, $p(x)$, which gives the likelihood of x taking any specific value. The density function can be expressed as a curve, as in Fig. 2.2. The height of this curve expresses the likelihood of the occurrence of each value. Analogous to the requirements on distributions, there are requirements that all values of $p(x)$ be non-negative and that the area under the curve be one.

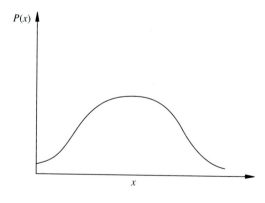

FIGURE 2.2
Probability density function.

In fact, there is a close relationship between density functions and distributions. Given any two values of x, say x_1 and x_2, the area under the density curve between x_1 and x_2 is the probability of x taking a value between x_1 and x_2. Thus, we can speak of the weights of people in the health club as being continuous variables with a given density, or as discrete variables with a given distribution. Analogous to the density curve in Fig. 2.2, is the bar graph in Fig. 2.3. As can be seen, the latter is simply a discrete version of the former.

The area under a curve can be defined by the integral:

$$P(x_1 \leq x \leq x_2) = \int_{x_1}^{x_2} p(x) \tag{2.7}$$

The integral is simply a continuous version of a sum. It can be thought of (and even evaluated) as the limiting case of a bar graph with the width of the individual bars going to zero.

2.1.2 The Normal Distribution

A random variable, x, is said to be normally distributed if its density function has the form

$$p(x) = \frac{1}{\sqrt{2\pi}\sigma} e^{\frac{(x-\mu)^2}{2\sigma^2}} \tag{2.8}$$

This somewhat formidable relation gives rise to the well-known bell shaped density function shown in Fig. 2.4. The highest part of the bell is at the mean, μ, and the

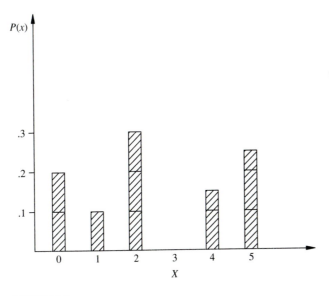

FIGURE 2.3
Probability distribution.

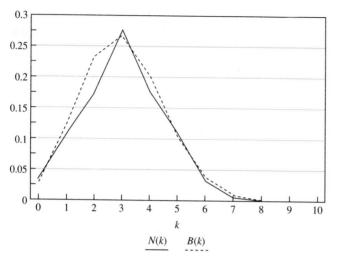

$N(k)$ —— $B(k)$ -----

FIGURE 2.4

width of the bell is controlled by σ. Thus, a random variable can be completely summarized by giving its mean and variance and by assuming that the variable is normally distributed.

Given this curve, it is now possible to answer any question about the random variable. For instance, suppose we want to know the probability of the RV, taking a value larger than $\mu + 3\sigma$. This is just the area under the curve from $\mu + 3\sigma$ to ∞. The integral of $p(x)$ does not have a closed form (i.e., it cannot be evaluated directly like x^2 or $\sin(x)$), but it has been tabulated. Table 2.2 is a brief tabulation, suitable for the discussion which follows.

It can also be evaluated using a computer. Given library functions (i.e., functions which are part of the library that comes with nearly all programming languages), **power(b,p)**, **acos(x)**, and **exp(x)** for power (b raised to the power p), arccosine (angle in radians whose cosine is x), and exponential (e raised to the power x), respectively, a one line function can be written to evaluate **normal(x , mu , sigma)**:

```
double normal( x , mu , sigma )
   return( ( 1 / ( sigma * power( 2*acos(-1.0) , 0.5 ) ) ) *
           exp( power( (x-mu) , 2.0 ) /
             ( 2 * power( sigma , 2.0 ) ) ) )
```

With this, it is then possible to compute the area under any part of the normal curve, say from **x1** to **x2**, by evaluating the normal function at intervals of **s** (e.g., 0.001) and adding the areas under rectangles of width **s**.

```
double normal_area( x1 , x2 , mu , sigma )
   sum <- 0.0
   for( x = x1 to x2-s step s )
      sum <- sum + s * normal( x+(s/2) , mu , sigma )
   return( sum )
```

TABLE 2.2
Selected values from the normal curve

	$\mu = 0$ $\sigma = 1$	
x	$N(x)$	$N(0, x)$
0.00	0.3989	0.0000
0.25	0.3867	0.0987
0.50	0.3522	0.1915
0.75	0.3013	0.2734
1.00	0.2421	0.3413
1.25	0.1828	0.3944
1.50	0.1296	0.4332
1.75	0.0864	0.4599
2.00	0.0540	0.4773
2.25	0.0318	0.4878
2.50	0.0176	0.4938
2.75	0.0091	0.4970
3.00	0.0044	0.4987
3.25	0.0020	0.4994
3.50	0.0009	0.4998
3.75	0.0004	0.4999
4.00	0.0001	0.5000

Note that the function is evaluated at the center of the interval. If s is small, and the function is well-behaved (does not change value too rapidly) this gives a good approximation to the area under the curve. As can be seen from examining the bell curve, the normal function is well-behaved and this method, which is called Simpson's Rule, works well enough for our purposes. The values in Table 2.2 are obtained in this way. It is reassuring to note that the area under the curve between x equal to 0 and x equal to 4 is 0.5 (to 4 decimal places). The curve is symmetric about the mean value (as can also be seen from direct examination of the function itself; that is, $f(\mu + x) = f(\mu - x)$) and the function itself for x equal to 4 is already very small, indicating that the area under the curve beyond x equal to 4 is very small.

The information in Table 2.2 can be used to estimate the area under the curve between any values of x_1 and x_2 by adding or subtracting areas between 0 and x for appropriate values of x.

Example 2.1. Suppose that the I.Q.'s of students in telecommunications courses are normally distributed with a mean of 140 and standard deviation of 10. What is the probability of the mean I.Q. in a given class being between 120 and 150?

The answer consists of the area under the curve from two standard deviations below the mean to one standard deviation above the mean. Examining Table 2.2 and noting that the table is calibrated in units of standard deviations, the area two standard deviations below the mean is 0.4773, the same as two standard deviations above. The area one standard deviation above is 0.3413. Thus the total area is 0.8186. To find the probability of the mean I.Q. being between 150 and 160, we need the area between one and two standard deviations and the difference of these two numbers, a probability of 0.1360.

By examination of Table 2.2, observe also that the probability of the observed mean being within one, two, or three standard deviations of the actual mean is 0.683, 0.955 and 0.997, respectively. These are useful numbers to keep in mind in maintaining a perspective on any observed results as they indicate how likely it is for there to be a given amount of deviation from the mean.

2.1.3 The Binomial Distribution

The normal distribution is a good approximation to most random variables when the number of observations is large. We again ask the question, "How large is large?" Suppose we flip a coin N times. Is the distribution of the number of heads normal? Let's examine this case in more detail.

> **Example 2.2.** What is the distribution of the number of heads if we flip a coin three times, assuming the probability of a head is 0.5?
> There are eight possible outcomes (compound events) of the three flips:
>
> $$HHH, HHT, HTH, HTT, THH, THT, TTH, TTT$$
>
> Since heads and tails are mutually exclusive events, and it is assumed they are the only possible outcomes of a flip, their probabilities must sum to 1. Thus, the probability of a tail is 1/2. The probability of each of the eight compound events is thus
>
> $$(1/2)(1/2)(1/2) = 1/8$$
>
> Note that these eight compound events are also mutually exclusive, are also the only events that can occur in three flips, and that their probabilities also sum to 1.
> The probability of three heads is given by the HHH event and is 0.125. The probability of two heads is given by events HHT, HTH, and THH, three mutually exclusive events with total probability of 0.375. Similarly, the probabilities of one head and zero heads are 0.375 and 0.125, respectively.
> Note that the probability of one head is the same as the probability of two heads and that the probability of zero heads is the same as the probability of three heads. This is not a coincidence since when we get k heads we get $3 - k$ tails, and the probability of a head is the same as the probability of a tail. Hence the probability of k tails (which is the same as the probability of $3 - k$ heads) should be the same as the probability of k heads.

Suppose the coin is flipped ten times. We can carry out the same analysis, enumerating elementary events as before. There are, however 2^{10} (= 1024) elementary events since each flip can result in two outcomes. The calculation is quite tedious. It is more efficient to observe that all events that have k heads are denoted by strings with k Hs and $10 - k$ Ts and each such event has probability $1/2^{10}$. To determine the probability of k heads, we need only compute the number of ways to put k Hs into a string of length ten. The first H can be in any of the ten positions. The second can be in any of the remaining nine positions. The third in any of the remaining eight, etc. There are then

$$(10)\,(9)\,(8)\cdots(10-k+1)$$

ways of placing the k Hs. Note that the same strings are being counted many times, because all the cases which placed Hs in the same positions in different orders as different cases were counted. For example, suppose k is 2 and we placed Hs in positions 3 and 9; we counted this event twice—once placing the first H in position 3 and the second in position 9, and then again when we counted the first in position 9 and the second in position 3. In general, we counted the same event occupying k specific positions with Hs

$$k(k-1)\cdots 1$$

times. Therefore, the correct count of the number of ways of placing k Hs in a string of length 10 is

$$\frac{(10)(9)\cdots(10-k+1)}{k(k-1)\cdots 1}$$

Denoting the product of the numbers from 1 to k as $k!$ (read k factorial), this expression can be rewritten

$$\frac{10!}{k!(10-k)!}$$

Note that this is the same if we replace k by $10-k$; since $k = 10 - (10 - k)$. The distribution of k repetitions of an event with two possible outcomes is called a binomial distribution, and the numbers above are called binomial coefficients. We denote the binomial coefficient corresponding to k heads out of ten flips as $C(10, k)$. This is the number of ways of choosing k things out of ten. More generally, we refer to the number of ways of choosing k things out of n as $C(n, k)$ and compute it as

$$C(n, k) = \frac{n!}{k!(n-k)!} \tag{2.9}$$

The probability of k heads in n flips is then

$$P(n, k) = \frac{C(n, k)}{2^n} \tag{2.10}$$

since the probability of each compound event is $1/2^n$. Note that because of the symmetry of $C(n, k)$ the probability of k heads is the same as the probability of $n - k$ heads.

This result can be generalized further to the case where the probability of the two elementary events are unequal. Thus, suppose the probability of a person being on the phone is 0.1 and there are n people in an office. The probability of k specific people being on the phone and the other $n - k$ not being on the phone is

$$(0.1)^k(0.9)^{n-k}$$

and the probability of any k people being on the phone is

$$C(n, k)(0.1)^k(0.9)^{n-k}$$

Note that now the probabilities of k and $n - k$ are no longer the same. Thus, we have the most general form of the binomial distribution. Given two disjoint events with probabilities p and $(1 - p)$, the probability of the first occurring k times and the second occurring $n - k$ times in n trials is

$$P(n, k, p) = C(n, k) p^k (1 - p)^{n-k}$$

Let us now examine the statistics of a binomial random variable. What is the expected number of heads in three flips? The probabilities of 0, 1, 2, and 3 heads are 0.125, 0.375, 0.375, and 0.125, respectively.

$$
\begin{aligned}
E(H) &= \sum_{k=0}^{3} k P_k \\
&= (0)(0.125) + (1)(0.375) + (2)(0.375) + (3)(0.125) \\
&= 1.5
\end{aligned}
$$

That is, on average, half the flips are heads. This is not a surprising result since we have already observed that

$$P(k) = P(n - k)$$

and we can define events of the type "k heads or $n - k$ heads". Each such event "averages" $n/2$ heads because the two component events within it are equally likely, and these are the only events which occur, that is, every event (k heads) has an equally likely mate ($n - k$ heads) to balance it out and bring the average to

$$(k + n - k)/2 = n/2$$

Thus, in general the expected number of heads in n flips is $n/2$.

This same result can be obtained in another, more general way. Consider a single flip. The expected number of heads is simply

$$(1/2)(0) + (1/2)(1) = 1/2$$

Each of the n flips is an independent event. The mean of the sum of two (or more) independent random variables is simply the sum of the means of the individual random variables. This makes sense because one set of events (e.g., the flip of one coin) has nothing to do with the other. The expected number of heads in n flips is thus $n/2$. Actually, the mean of the sum of any number of random variables is equal to the sum of their means, even if they are not independent.

This second way of obtaining the mean has nothing to do with the probability of a head being $1/2$, although the value of the mean ($n/2$) does, of course. If the probability of a head had been p, by the same reasoning, the mean would be np. Thus, returning to the telephone example, the expected number of people on the phone can be found at any time. Since the probability of any given person being on the phone is p and it is assumed the people use the phone independently, the average number of people on the phone would be np. We denote this as

$$E(n, p) = np$$

Now consider the variance of a binomial random variable. Again, look at a single coin flip. The variance of the number of heads is

$$V(H) = P(0)(0 - E(H))^2 + P(1)(1 - E(H))^2$$

$$1/2(0 - 1/2)^2 + 1/2(1 - 1/2)^2 = 1/4$$

Again, the sum of the variances of n independent random variables is the sum of the individual variances and we find that the variance in the number of heads in n flips is $n/4$.

This result also generalizes to the case where the probability of a head is p:

$$V(H) = (1 - p)(0 - p)^2 + p(1 - p)^2$$
$$= p^2 - p^3 + p - 2p^2 + p^3$$
$$= p(1 - p)$$

This checks for the case of coin flipping where $p = 1/2$. Also, note that $p(1 - p)$ takes its maximum value when $p = 1/2$. This makes sense too. Suppose the event occurs with probability 0.01 (or, equivalently, 0.99). There is then very little variance. On the other hand, if p is close to 0.5, there is less certainty of the outcome.

The variance of the sum of n independent events is also the sum of the individual variances. Again, n things occur and one has nothing to do with the other. Thus, the variance of the number of heads in n flips is $n/4$ and the variance is

$$V(n, p) = np(1 - p)$$

when the probability of a head is p. Table 2.3 gives the distribution of the number of heads in ten flips.

Now it is possible to "design" telephone systems! Returning to the telephone example, we can now decide how many lines to order to ensure that everyone who

TABLE 2.3
Probability of H heads in 10 flips

H	$P(H)$	$P(\geq H)$
0	0.00098	1.00000
1	0.00977	0.99023
2	0.04395	0.98926
3	0.11719	0.94531
4	0.20508	0.82813
5	0.24609	0.62305
6	0.20508	0.37696
7	0.11719	0.17188
8	0.04395	0.05469
9	0.00977	0.01074
10	0.00098	0.00098

wants to talk has a reasonable chance of doing so. We wish to decide how few lines can be supplied and still provide reasonable service to the users of the system. The more lines there are, the more the system costs. The fewer there are, the more likely it is that someone will want a line and find all are in use.

Table 2.4 gives the distribution of the number of people talking (or, alternatively, the number of lines in use) for $n = 10$ and $p = 0.1$. It gives both the probabilities for exactly k and for k or more.

The only way to guarantee that no one will have to wait for a line is to give everyone a separate line. Table 2.4, however, illustrates that the probability of five or more people talking is only 0.00164. Thus, a considerable amount could be saved by supplying four lines instead of ten. In the rare (roughly 1 in 600) event that five or more people wanted to talk, some could wait. Indeed, the fourth line is used only 1% of the time and if we were willing to allow people to wait with probability 0.0128, we could supply only three lines.

There is a subtle point here. All these statistics are based on an assumption that the probability that someone wants to talk is p, a constant, not a function of the number of lines we supply. In fact, if we supply fewer lines, some people who want to talk will not be able to do so and will have to wait. They will, presumably, try again at a different time. Thus, the calls are not independent of one another. For the moment, we ignore this, assuming that the system is designed so that few people wait, and that this effect is negligible. In the following sections, however, we will explore all this more carefully.

An estimate of the required number of lines can be obtained even more quickly by approximating the binomial distribution by a normal distribution. The mean and standard deviation of our binomial distribution are

$$E = np = 1.0$$

and

$$\sigma = \sqrt{np(1 - p)} = 0.95$$

TABLE 2.4

Probability of k of 10 people being on the phone given the problem that a person is on the phone = 0.1 $C(N,K)\alpha^k(1-\alpha)^{N-K}$

x	$P(k)$	$P(\geq k)$
0	0.34868	1.00000
1	0.38742	0.65132 $\leftarrow 1 - .34868$
2	0.19371	0.26390
3	0.05739	0.07019
4	0.01116	0.01280
5	0.00149	0.00164
6	0.00014	0.00015
7	0.00001	0.00001

respectively. Thus, each additional line moves roughly one standard deviation from the mean. Recalling (or looking at Table 2.2) that three standard deviations cover about 99% of the cases, we need three lines if we want to ensure that only about 1% of the people need to wait. It is also seen from an examination of Table 2.2, that at four standard deviations, the probability of waiting has dropped to 0.0001. All this is consistent with the more exact analysis of the actual distribution.

Figure 2.4 shows the binomial distribution for $n = 10$ and $p = 0.1$ together with the normal distribution for $\mu = 1$ and $\sigma = 0.95$. As can be seen, they are sufficiently similar to one another to allow us make observations like were done earlier. In this case, however, they are not sufficiently close to make detailed design decisions. If we needed to know whether to order three lines or four lines to assure that at most 1% of the people wait, a more detailed analysis is required. We would also have to take into account the effect of waiting callers, as was mentioned before. As the number of users increases, the fit, by a normal distribution, gets much better; the effects of waiting callers remain, but can be dealt with.

2.2 THE M/M/1 QUEUE

We now turn to the analysis of waiting time, specifically, of analyzing the numbers of messages (or people) that have to wait and how long they will have to wait. The general problem is quite complex and the interested reader is directed to Cooper [Coo72] and Kleinrock [Kle75] for a more detailed treatment of this subject. Here the discussion is confined to a basic set of cases, sufficient to handle most problems that arise during the course of designing a network. The more detailed treatment given in the references above makes use of more specific data than is usually available during the design process about the distribution of message lengths and interarrival times.

Begin by assuming that message arrivals occur independently of one another. While this in practice is in fact not quite true, the assumption gives rise to a model which is instructive and useful. Also assume that the probability that a message continues (i.e., gets longer) is independent of how long it currently is. This is like saying that the length of a message is independent of itself, or that the time someone spends on the phone, continuing a conversation, is independent of the current length of the call. This process can be thought of as one where a (biased) coin which comes up heads with probability p is flipped after each character of a message (or each second of a call). If the coin comes up heads, the message continues; otherwise it ends. The larger p is, the longer the average length of the message.

Examine the distribution of message lengths more closely. Suppose we begin by flipping the coin. With probability $1 - p$, we immediately get a tail and stop without any characters in the message. Thus, the message has length 0 with probability $1 - p$. The message has length 1 if we get a head on the first flip and a tail on the second. This occurs with probability $p(1 - p)$. Continuing in this way, we find that the probability of the message having length k is $p^k(1 - p)$.

Note that this is in fact a probability distribution. If the probability of the message having length k is denoted by P_k, then it must be shown that all P_k are nonnegative (this is obvious since both p and $1 - p$ are nonnegative), and that the sum of the probabilities of all possible message lengths is 1. To prove the latter, denote the sum by S. Thus,

$$S = \sum_{k=0}^{\infty} P_k \tag{2.11}$$

$$= \sum_{k=0}^{\infty} p^k (1 - p) \tag{2.12}$$

$$= (1 - p)(1 + p + p^2 + p^3 + \cdots) \tag{2.13}$$

Suppose both sides of the above equation are multiplied by p. We then have

$$pS = (1 - p)(p + p^2 + p^3 + p^4 + \cdots) \tag{2.14}$$

subtracting Eq. 2.14 from Eq. 2.13, we have

$$S(1 - p) = (1 - p) \tag{2.15}$$

that is, all the terms, except for the 1, in the infinite sums in the brackets cancel out and for $p < 1$, we can divide by $1 - p$ and get

$$S = 1 \tag{2.16}$$

The case $p = 1$ is special. It means heads are flipped repeatedly and the message length is infinite; not a realistic case.

Therefore, it is seen that S is indeed 1 and that this distribution, which is known as a geometric distribution is indeed a probability distribution. Now consider the average length of a message. The average message length is given by

$$A = \sum_{k=0}^{\infty} k P_k \tag{2.17}$$

$$= \sum_{k=0}^{\infty} k(1 - p)p^k \tag{2.18}$$

$$= (1 - p)\left(\sum_{k=1}^{\infty} p^k + \sum_{k=2}^{\infty} p^k + \sum_{k=3}^{\infty} p^k + \cdots \right) \tag{2.19}$$

The collection of sums inside the parentheses comes from the observation that the p_k term appears k times. Continuing to rewrite this expression,

$$A = (1 - p)p \sum_{k=0}^{\infty} p^k \sum_{m=0}^{\infty} p^m \tag{2.20}$$

$$= (1 - p)p \sum_{k=0}^{\infty} p^k \sum_{m=0}^{\infty} p^m \tag{2.21}$$

$$= \frac{(1 - p)p}{(1 - p)(1 - p)} = \frac{p}{1 - p} \tag{2.22}$$

Thus, the average message length, $A(p)$, for a geometric distribution grows with parameter p as is shown in Fig. 2.5. As p approaches 1, $A(p)$ rises quickly towards ∞.

The geometric distribution has an interesting and useful property. Suppose we ask, "What is the probability that a message has a length of at least three characters, given that it has a length of at least one character?" This is a question about conditional probability. The probability of the occurrence of event A conditioned on the occurrence of event B is defined as

$$P(A|B) = \frac{P(AB)}{P(B)} \tag{2.23}$$

that is, as the probability that both A and B occur divided by the probability that B occurs. Note that if A and B are independent events then $P(AB)$ is simply $P(A)P(B)$ and

$$P(A|B) = \frac{P(A)P(B)}{P(B)} = P(A) \tag{2.24}$$

Now, the probability that a message is at least three characters long (event A) is simply the probability of getting three heads followed by any other combination of flips. Thus, this probability is simply p^3. Similarly, the probability that a message is at least one character long (event B) is simply p. The probability that a message is both at least three characters long and at least one character long (events A and B) is p^3 since if it is at least three characters long it, is also at least one character long. Note that events A and B are not independent here; quite the contrary, all occurrences of event A are included in the occurrences of event B. The conditional probability

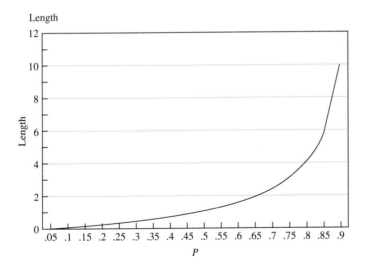

FIGURE 2.5

then is

$$P(A|B) = \frac{p^3}{p^1} = p^2$$

Therefore, the probability that a message is at least three characters long given that it is at least one character long is exactly equal to the (unconditional) probability that it is at least two characters long.

In fact, the probability of a message being at least $j + k$ characters long is p^{j+k}. The probability of a message being at least j characters long is p^j; and the probability of its being at least $j + k$ characters long, given it is at least j characters long, is p^k, the same as the unconditional probability of its being at least k characters long.

Thus it is seen that the geometric distribution exhibits the property of being memoryless; that is, the probability of a message continuing for another k characters is unaffected by its having already included j characters, for any values of j and k.

The geometric distribution is discrete; that is, the message lengths can only take integer values. Suppose messages are allowed to take any nonnegative value. A coin could then be flipped after each δ characters, letting δ become arbitrarily small. In the limit as δ approaches zero, a probability density function would be obtained of the form

$$p(x) = \frac{1}{L} e^{-x/L} \tag{2.25}$$

where e is the base of the natural logarithms (approximately equal to 2.718) and is defined by

$$e = \lim_{n \to \infty} (1 + 1/n)^n \tag{2.26}$$

This distribution is known as the **exponential distribution**. The average value of x is L. The exponential distribution is also memoryless; that is, the distribution of the length of time to the next arrival is unaltered by the length of time which has already elapsed since the previous arrival. The distribution of the length of time between arrivals, when the arrivals occur independently, is exponential. Figure 2.6 shows the density functions for exponential distributions with $L = 0.5$, 1, and 2.

The memoryless property of exponential distributions allows the analysis of the distribution of waiting times in systems where interarrivals and message lengths are exponential. Specifically, suppose we have a communication channel that operates at the rate of B bits per second. A message of length x can be transmitted in x/B seconds. We refer to the channel as a server and to the transmission time as the service time. Suppose that the message lengths are exponentially distributed with length L bits, and that the message interarrival times are also exponentially distributed with length G. If a message arrives when the channel is idle (i.e., not in the midst of transmitting another message), the message is transmitted immediately. If the channel is busy when the message arrives, the message is placed in a queue where it waits (possibly behind other messages) until the channel becomes idle (and then begins serving the next message). Now, we wish to investigate the distribution of waiting times for messages and the total time a message spends in the system, both waiting and being transmitted.

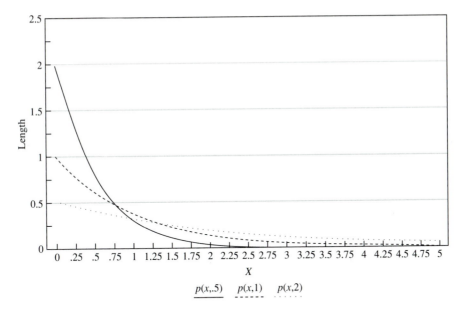

$$p(x,.5) \qquad p(x,1) \qquad p(x,2)$$

FIGURE 2.6

This system is called an M/M/1 queue. The M stands for Markovian, after the mathematician Markov. The first M indicates that the arrivals are independent of one another; that is, that the interarrival times are distributed exponentially. The second M indicates that the service times are also exponentially distributed. The 1 indicates that there is 1 server. Cases can also be considered where the arrivals or service time distributions are constants (indicated by a *D*), general (indicated by a *G*), or any other specific distribution. The case of multiple servers can also be considered. It is possible that the queue is of only finite size and after the queue fills up, then arriving messages are lost. It is possible that the population generating the messages is finite and that the arrival rate is a function of the number of messages currently in the system. Or, the service discipline may include priorities, where one class of messages is served before another. Many of these cases have been analyzed, although in some cases the analysis is quite complex. The references mentioned earlier include much of this analysis. Some of these cases are considered later. For the moment, however, we consider only the M/M/1 queue, the basic case. The system can be analyzed in terms of its state, specifically the state of the server (idle or busy), and the state of the queue (number of messages waiting). Each time a message arrives or departs (finishes service) the system changes state. A diagram can be drawn illustrating the transitions among states. Figure 2.7 is such a diagram. Each state is represented by a circle with a number inside indicating the total number of messages in the system, both waiting and in service. Since the server is never idle if there are any messages waiting, we know that if k, the number of messages in the system, is greater than 0, then one message is in service and the remaining $k - 1$ are in the queue.

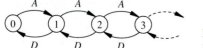

FIGURE 2.7
State transition diagram for M/M/1 queue.

The links between states in Fig. 2.7 are marked with values indicating the relative rate at which the system makes the transition between the states. The A's leading upward to states with a larger number of messages correspond to the rate of arrivals (msg/sec) to the system. Since messages arrive independently of one another, the rate of arrivals is the same for any state of the system. Similarly, since the length of service is exponentially distributed, the rate of departure from all states (except the empty state, 0, where there is none to depart) is D (msg/sec).

The value of A is related to G, the average interarrival time. Specifically,

$$A = \frac{1}{G}$$

For example, if G, the average interarrival time between messages, is ten seconds, then A, the average rate of arrivals of message is 0.1 msg/sec.

Similarly, D, the departure rate of messages is related to the service time, which is in turn related to the message length, L, and the channel speed, B. Specifically, the service time, S, (transmission time) for a message of length B bits on a channel of speed B bits/sec, is

$$S = \frac{L}{B}$$

and the average rate of departures per second, D, is

$$D = \frac{1}{S}$$

such that, if $S = 5$ seconds, the departure rate, D is 0.2 msg/sec and the average interdeparture time is 5 seconds.

The utilization, U, of the server is defined as the fraction of time that it is busy. Clearly, the higher the arrival rate, the higher the utilization; the higher the departure rate (equivalently, the faster the service), the lower the utilization rate. In fact, the utilization is seen to be simply,

$$U = \frac{A}{D}$$

In the last example, if messages arrive at an average rate, A, of one every 10 seconds and require 5 seconds of service (hence D is 5) then the server is busy 50 percent of the time; that is,

$$U = 5/10 = 0.5$$

We now turn to the question of the relative amount of time the system spends in each state (i.e., equivalently, the probability of the system being in each state). There are two possibilities. One is that the arrival rate is greater than the departure rate. In this case, the system falls behind in its service, and the number of messages in the system increases without bound. The second possibility is that the average rate of arrivals is smaller than the rate of departures. In this case, we expect that the number of messages in the system will vary, sometimes increasing and sometimes decreasing, but always eventually returning (at least for a short time) to 0.

We know that if the utilization of the server is less than 1, then it must be idle some of the time and so there must occasionally be no messages in the system to serve. Thus, over a long period of time, the number of arrivals must equal the number of departures. Also, over a long period of time, the number of transitions to the right from state k to state $k + 1$ must equal the number of transitions to the left from state $k+1$ to k. Such a system is said to be in equilibrium. For such systems, the probability of being in a given state is well defined; it is just the fraction of the total time that the system spends in that state.

The observation about the number of transitions to the left and right leads to the relation that for any two states, k and $k + 1$,

$$AP_k = DP_{k+1}$$

where P_k is the probability of being in state k. This follows from the fact that the number of transitions out of a state is the rate of transitions times the amount of time spent in the state, which is in turn directly related to the probability of being in the state. The equation is called a balance equation and is the basis for computing the probabilities of the system being in each state.

Rearranging the equation, we have

$$P_{k+1} = \frac{A}{D} P_k = U P_k \tag{2.27}$$

Thus,

$$P_1 = U P_0 \tag{2.28}$$

$$P_2 = U P_1 = U^2 P_0 \tag{2.29}$$

$$\vdots$$

$$P_k = U P_{k-1} = U^k P_0 \tag{2.30}$$

This gives P_k in terms of P_0 for all k. To find P_0, observe that all the P_k must sum to 1. Thus,

$$1 = S = \sum_{k=0}^{\infty} P_k \tag{2.31}$$

$$= P_0(1 + U + U^2 + U^3 + \cdots) \tag{2.32}$$

We recognize the sum in parentheses as $1/(1 - U)$ and have:

$$P_0 = 1 - U \tag{2.33}$$

This is not a surprising result since it says that the probability of the system being idle is 1 minus the utilization of the server.

We now know P_k exactly for all k,

$$P_k = (1 - U)U^k \tag{2.34}$$

So, P_k is geometrically distributed. The "coin" corresponds to the arrival and departure of messages. In this case, p, the probability of a "head" is U.

We now have the distribution of waiting times in the system. With probability $(1-U)$, a message arrives to an empty system and does not wait at all. With probability $U(1-U)$, a message arrives to a system with 1 message in service and waits for that one message to complete. Since the service time is exponential, however, the length of time that message has been in service does not affect the time it now takes to complete, and the average wait is the average service time, T_s. Continuing with the same reasoning, we find that with probability P_k a message waits for k other messages to complete service and waits for time equal to kT_s.

So, the waiting time is geometrically distributed and, from Eq. 2.22 above, which gives the average value of a geometric distribution, T_w, the average waiting time is given by:

$$T_w = \frac{UT_s}{(1 - U)} \tag{2.35}$$

The waiting time is just the service time multiplied by a factor of $U/(1 - U)$. For U close to 0, there is very little waiting time. As U approaches 1, the waiting time becomes infinite. When U is large, a small increase in U results in a large increase in T_w. This phenomenon is often observed on highways; once traffic becomes heavy, any increase in load results in a much longer wait.

The total time in the system, T, including both waiting and service time, can now be found:

$$T = T_s + T_w = \frac{T_s}{1 - U} \tag{2.36}$$

Thus, the average time in the system is just the service time inflated by a factor of $1/(1 - U)$. Again, as U approaches 1 this becomes very large. Figure 2.8 shows how the average time in an M/M/1 queue in the system grows with utilization, U.

Observe that the average number of messages in the system is simply the arrival rate times the average length of time in the system; that is,

$$N = AT \tag{2.37}$$

where A is the average arrival rate. This relationship is known as Little's Formula or Little's Law [Lit61]. This formula is true for a wide variety of queueing systems, not just M/M/1 queues.

Since U is A/D , and the average service time, T_s, is $1/D$,

$$N = \frac{AT_s}{(1 - U)} = \frac{U}{1 - U} \tag{2.38}$$

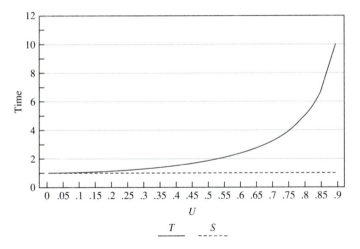

FIGURE 2.8

Also, Q, the average number of messages in the queue, is given by

$$Q = AT_w = \frac{U A T_s}{(1 - U)} = \frac{U^2}{(1 - U)} \tag{2.39}$$

Example 2.3. Messages (independently) arrive to a system at the rate of 10 per minute. Their lengths are exponentially distributed with an average of 3600 characters. They are transmitted on a 9600 bps channel. A character is 8 bits long.

(a) What is the average service time?
   The service time is the message length divided by the channel speed,

$$T_s = \frac{(3600 \text{ char})(8 \frac{\text{bits}}{\text{char}})}{9600 \frac{\text{bits}}{\text{sec}}} = 3 \text{ sec}$$

(b) What is the arrival rate (A in the discussion above)?
   The arrival rate is 10 messages/min = $\frac{1}{6} \frac{\text{msg}}{\text{sec}}$

(c) What is the service rate (D in the discussion above)?
   The service rate is the reciprocal of the service time, $\frac{1}{3} \frac{\text{msg}}{\text{sec}}$ in this case.

(d) What is the utilization of the server (U in the discussion above)?
   The utilization of the server is the arrival rate divided by the service rate, in this case,

$$U = \frac{\frac{1}{6} \frac{\text{msg}}{\text{sec}}}{\frac{1}{3} \frac{\text{msg}}{\text{sec}}} = 0.5$$

Note that the utilization is always a dimensionless quantity.

(e) What is the probability that there are 2 messages in the system?

$$P_2 = (1 - U)U^2 = 0.125$$

(f) What is, Q, the average number of messages in the queue?

$$Q = U^2/(1 - U) = 0.5$$

(g) What is N, the average number of messages in the system?

$$N = U/(1 - U) = 1$$

Note that

$$N = Q + U$$

This is true by definition since U is the average number in service, Q is the average number in the Q, and a message in the system is either in service or in the queue.

(h) What is the T_w, the average waiting time (time in the queue)?
   The waiting time is the number in the queue divided by the arrival rate. Here,

$$T_w = (0.5) \left(\frac{1}{6} \right) = 3.0 \text{ sec}$$

(i) What is the average time in the system?

$$T = T_w + T_s = 6.0 \text{ sec} \tag{2.40}$$

2.3 OTHER QUEUEING SYSTEMS

2.3.1 The M/M/2 Queue

Suppose there are two identical channels and that messages have exponential lengths and arrive independently. This is an M/M/2 queue in the terminology given in the previous section. The analysis above extends in a straight forward way to this case. Figure 2.9 illustrates the transitions between states. Notice that it is almost the same as Fig. 2.7, the only difference being that starting at state 2, the leftward transitions have rate $2D$ instead of D. This is because when there are two or more messages in the system, both servers are busy and the departure rate is twice as high.

Again, each state is labelled with the number of messages in the system. Now, however, state 1 is interpreted to mean that one of the servers is busy and the other is idle and we don't care which one. There are really two states with one message in the system—the state with only server 1 busy and the state with only server 2 busy. If, when an arrival comes to an empty system and the message is given to one of the servers in an unbiased fashion, these two states are equivalent and we need not distinguish between them. This simplifies the analysis of the system.

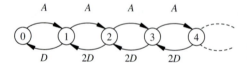

FIGURE 2.9
State transition for M/M/2 queue.

We can again set up balance equations and find the probabilities of being in each of the states. Again, use U to denote the quantity A/D. Note, however, that as will be seen, U is no longer the utilization of a server. We have

$$AP_0 = DP_1 \tag{2.41}$$

So, as before

$$P_1 = UP_0 \tag{2.42}$$

Now,

$$AP_1 = 2DP_2 \tag{2.43}$$

so,

$$P_2 = \frac{U^2}{2}P_0 \tag{2.44}$$

We also have, for $k > 2$, that

$$AP_k = 2DP_{k+1} \tag{2.45}$$

so,

$$P_k = \frac{U^k}{2^{k-1}}P_0 \tag{2.46}$$

Again, P_0 can be found by summing the probabilities of being in each state,

$$1 = S = P_0\left(1 + U + \frac{U^2}{2} + \frac{U^3}{4} + \cdots\right) \tag{2.47}$$

After some algebraic rearrangement,

$$P_0 = \frac{2 - U}{2 + U} \tag{2.48}$$

P_0 is the probability that the system is empty, that is, the probability that both servers are idle, but it is no longer simply $1 - U$. U is no longer the average utilization of a server. Since the servers are treated equally, each is equally likely to serve a message and each of the servers is given half of the messages to serve. Thus, the arrival rate to each server is $A/2$ and the average utilization of a server is

$$\frac{A}{2D} = \frac{U}{2} \tag{2.49}$$

We see that P_0 is less than $(1 - U/2)$. This is not surprising since $(1 - U/2)$ is the probability that 1 server is idle and P_0 is the probability that both servers are idle.

Given P_0, we now know P_k for all k. It is then possible to find the average length of the queue and the average waiting time. The average queue length is given by

$$Q = \sum_{k=2}^{\infty}(k - 2)P_k \tag{2.50}$$

$$= \sum_{k=2}^{\infty}(k - 2)P_0U\frac{U^{k-1}}{2^{k-1}} \tag{2.51}$$

Letting $j = k - 2$, means

$$Q = \sum_{j=0}^{\infty} j P_0 \frac{U^2}{2} \left(\frac{U}{2}\right)^j \tag{2.52}$$

which is recognized again as a geometric distribution, giving

$$Q = \frac{(2 - U)}{(2 + U)} \frac{U^2}{2} \frac{U/2}{(1 - U/2)^2} \tag{2.53}$$

$$= \frac{U^3}{(4 - U^2)} \tag{2.54}$$

The average number of messages in the system can now be found. A message is either in service or in the queue. Since each server has a utilization of $U/2$, the average number of messages in service at each server is also $U/2$ and the total average number in service is U. Therefore, the total in the system (in the queue or in service) is

$$N = U + \frac{U^3}{4 - U^2} = \frac{4U}{4 - U^2} \tag{2.55}$$

The average time in the system, both waiting and being served can now be found. From Little's Formula, the average number of messages in the system is simply the arrival rate times the average length of time in the system; that is,

$$N = AT$$

Since U is A/D, and the average service time, T_s, is $1/D$, we then get,

$$T = \frac{4T_s}{4 - U^2} = \frac{T_s}{1 - (U/2)^2} \tag{2.56}$$

Compare this result with the average time in the system for an M/M/1 queue at the same utilization. The utilization of the M/M/1 queue is U, while the utilization of the M/M/2 queue is $U/2$. Note that both U and A have doubled for the M/M/2 queue, relative to their values for the M/M/1 queue, at the same utilization. Examining the expression for T, we find that there is an added factor of $(1 + 0.5U)$ in the denominator. Thus, as U varies between 0 and 2 (the only values for which the system is stable; any larger value would result in messages arriving faster than they could be served) we pick up a factor of between 1 and 2 in total time in the system. Figure 2.10 shows the total time in the system as a function of average utilization $(U/2)$ assuming $D = 1$ (i.e., the average service time is 1.) For the sake of comparison, we also show on the same set of axes the total time in the system for an M/M/1 system with the same average utilization (not the same U).

2.3.2 The M/M/m Queue

Suppose now that there are m identical servers. The state transition diagram is shown in Fig. 2.11. Here, the departure rate increases as the number of messages in the

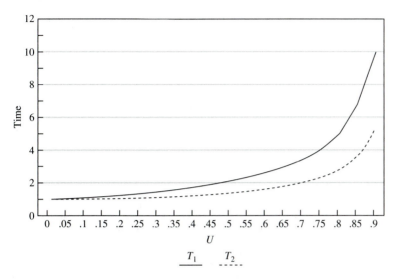

FIGURE 2.10

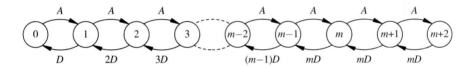

FIGURE 2.11
State transition for M/M/m queue.

system increases, since the number of busy servers increases too, until the number of messages is m. Then, all servers are busy and the departure rate remains constant at mD. The state probabilities are related by the same types of balance equations as before:

$$A P_0 = D P_1 \qquad\qquad P_1 = U P_0 \qquad\qquad (2.57)$$

$$A P_1 = 2 D P_1 \qquad\qquad P_2 = \frac{U^2}{2} P_0 \qquad\qquad (2.58)$$

$$A P_2 = 3 D P_3 \qquad\qquad P_3 = \frac{U^3}{6} P_0 \qquad\qquad (2.59)$$

$$A P_{m-1} = m D P_m \qquad\qquad P_m = \frac{U^m}{m!} P0 \qquad\qquad (2.60)$$

$$A P_{m+k} = m D P_{m+k+1} \qquad P_{m+k+1} = \frac{U^m}{m!} \left(\frac{U}{m}\right)^{k+1} P_0 \qquad (2.61)$$

P_0 is found again by summing the state probabilities:

$$1 = P_0 \left(\sum_{k=0}^{m-1} \frac{U^k}{k!} + \sum_{k=m}^{\infty} \left(\frac{U}{m} \right)^k \right) \tag{2.62}$$

So we have,

$$\frac{1}{P_0} = \left(\sum_{k=0}^{m-1} \frac{U^k}{k!} + \frac{U^m}{m!(1 - U/m)} \right) \tag{2.63}$$

The average queue length, Q, is given by

$$Q = P_0 \frac{U^m}{m!} \frac{U/m}{(1 - U/m)^2} \tag{2.64}$$

Note that the expression for Q for the M/M/2 queue is exactly of this form. N, the total number of messages in the system, is given by

$$N = Q + U \tag{2.65}$$

since, as before, the total number of messages in service is U, and the average total time in the system is, again by Little's Formula,

$$T = \frac{N}{A} = \frac{Q + U}{A} \tag{2.66}$$

Table 2.5 gives a comparison among the values of P_k for various values of k, as the utilization and the number of servers vary. Table 2.6 gives the average number of messages in the queue for various number of queues and various utilizations. Table 2.7 gives the average time in the system for various number of queues and various utilizations.

Example 2.4. 24 terminals share a 9600 bps line. Each terminal sends an average of 10 msg/min over the line. The message lengths are exponentially distributed with an average length of 2000 bits.

(a) What is the average time that a message spends in the system?
   The service time, T_s, is the transmission time for a message, which is the length of the message divided by the speed of the channel. Thus, T_s is 0.208 sec. The utilization, U, is the total arrival rate divided by the service rate. The total arrival rate is

$$A = (24 \text{ ter})(10 \text{ msg/min/ter})(1/60 \text{ min/sec})(2000 \text{ bits/msg})$$

$$= 8000 \text{ bits/sec}$$

So, $U = 0.833$ and the total time in the system is

$$T = \frac{T_s}{1 - U} = \frac{208 \text{ sec}}{0.167} = 1.25 \text{ sec}$$

TABLE 2.5
Probability of k messages in the system

	$\rho = 0.2$				
	$m = 1$	$m = 2$	$m = 3$	$m = 4$	$m = 5$
$k = 0$:	0.800000	0.666667	0.547945	0.449102	0.367816
$k = 1$:	0.160000	0.266667	0.328767	0.359281	0.367816
$k = 2$:	0.032000	0.053333	0.098630	0.143713	0.183908
$k = 3$:	0.006400	0.010667	0.019726	0.038323	0.061303
$k = 4$:	0.001280	0.002133	0.003945	0.007665	0.015326
$k = 5$:	0.000256	0.000427	0.000789	0.001533	0.003065
$k = 6$:	0.000051	0.000085	0.000158	0.000307	0.000613
$k = 7$:	0.000010	0.000017	0.000032	0.000061	0.000123
$k = 8$:	0.000002	0.000003	0.000006	0.000012	0.000025
$k = 9$:	0.000000	0.000001	0.000001	0.000002	0.000005
$k = 10$:	0.000000	0.000000	0.000000	0.000000	0.000001

	$\rho = 0.5$				
	$m = 1$	$m = 2$	$m = 3$	$m = 4$	$m = 5$
$k = 0$:	0.500000	0.333333	0.210526	0.130435	0.080100
$k = 1$:	0.250000	0.333333	0.315789	0.260870	0.200250
$k = 2$:	0.125000	0.166667	0.236842	0.260870	0.250313
$k = 3$:	0.062500	0.083333	0.118421	0.173913	0.208594
$k = 4$:	0.031250	0.041667	0.059211	0.086957	0.130371
$k = 5$:	0.015625	0.020833	0.029605	0.043478	0.065186
$k = 6$:	0.007813	0.010417	0.014803	0.021739	0.032593
$k = 7$:	0.003906	0.005208	0.007401	0.010870	0.016296
$k = 8$:	0.001953	0.002604	0.003701	0.005435	0.008148
$k = 9$:	0.000977	0.001302	0.001850	0.002717	0.004074
$k = 10$:	0.000488	0.000651	0.000925	0.001359	0.002037

	$\rho = 0.9$				
	$m = 1$	$m = 2$	$m = 3$	$m = 4$	$m = 5$
$k = 0$:	0.100000	0.052632	0.024907	0.011256	0.004959
$k = 1$:	0.090000	0.094737	0.067248	0.040522	0.022313
$k = 2$:	0.081000	0.085263	0.090785	0.072940	0.050205
$k = 3$:	0.072900	0.076737	0.081706	0.087528	0.075308
$k = 4$:	0.065610	0.069063	0.073535	0.078775	0.084721
$k = 5$:	0.059049	0.062157	0.066182	0.070898	0.076249
$k = 6$:	0.053144	0.055941	0.059564	0.063808	0.068624
$k = 7$:	0.047830	0.050347	0.053607	0.057427	0.061762
$k = 8$:	0.043047	0.045312	0.048247	0.051684	0.055586
$k = 9$:	0.038742	0.040781	0.043422	0.046516	0.050027
$k = 10$:	0.034868	0.036703	0.039080	0.041864	0.045024

(b) Suppose this time is unacceptably large. One possibility is to get a second 9600 bps channel and put 12 terminals on each channel. What is the total time in the system now?

   The system is still an M/M/1 queue, but the arrival rate, and hence the utilization, have been halved. Thus U is now 0.417 and T is given by

$$T = \frac{0.208 \text{ sec}}{0.583} = 0.357 \text{ sec}$$

TABLE 2.6

Average queue length for m queues

ρ	$m = 1$	$m = 2$	$m = 3$	$m = 4$	$m = 5$
0.05	0.003	0.000	0.000	0.000	0.000
0.10	0.011	0.002	0.000	0.000	0.000
0.15	0.026	0.007	0.002	0.001	0.000
0.20	0.050	0.017	0.006	0.002	0.001
0.25	0.083	0.033	0.015	0.007	0.003
0.30	0.129	0.059	0.030	0.016	0.009
0.35	0.188	0.098	0.055	0.032	0.020
0.40	0.267	0.152	0.094	0.060	0.040
0.45	0.368	0.229	0.152	0.105	0.074
0.50	0.500	0.333	0.237	0.174	0.130
0.55	0.672	0.477	0.358	0.277	0.218
0.60	0.900	0.675	0.532	0.431	0.354
0.65	1.207	0.951	0.782	0.658	0.562
0.70	1.633	1.345	1.149	1.000	0.882
0.75	2.250	1.929	1.703	1.528	1.385
0.80	3.200	2.844	2.589	2.386	2.216
0.85	4.817	4.426	4.139	3.906	3.709
0.90	8.100	7.674	7.354	7.090	6.862

TABLE 2.7

Total time in the system for m queues

	(Service time = 1)				
ρ	$m = 1$	$m = 2$	$m = 3$	$m = 4$	$m = 5$
0.05	1.053	1.003	1.000	1.000	1.000
0.10	1.111	1.010	1.001	1.000	1.000
0.15	1.176	1.023	1.004	1.001	1.000
0.20	1.250	1.042	1.010	1.003	1.001
0.25	1.333	1.067	1.020	1.007	1.003
0.30	1.429	1.099	1.033	1.013	1.006
0.35	1.538	1.140	1.053	1.023	1.011
0.40	1.667	1.190	1.078	1.038	1.020
0.45	1.818	1.254	1.113	1.058	1.033
0.50	2.000	1.333	1.158	1.087	1.052
0.55	2.222	1.434	1.217	1.126	1.079
0.60	2.500	1.563	1.296	1.179	1.118
0.65	2.857	1.732	1.401	1.253	1.173
0.70	3.333	1.961	1.547	1.357	1.252
0.75	4.000	2.286	1.757	1.509	1.369
0.80	5.000	2.778	2.079	1.746	1.554
0.85	6.667	3.604	2.623	2.149	1.873
0.90	10.000	5.263	3.724	2.969	2.525

(c) Suppose we get a second channel and use the two channels as an M/M/2 queue (i.e., we allow messages to use whichever channel is free). What is the total time now?

Now use the M/M/2 formula. U is 0.833; the effective utilization is $U/2 = 0.417$. The service time is unaltered

$$T = \frac{T_s}{1 - (U/2)^2} = 0.252 \text{ sec}$$

(d) Suppose we get a second channel and multiplex the two channels together into a single 19,200 bps channel and offer it the full load. What is the total time now?

We are back to the original M/M/1 queue, but now both the service time and utilization are halved. Thus,

$$T = \frac{T_s}{1 - U} = \frac{0.104 \text{ sec}}{0.583} = 0.178 \text{ sec}$$

Careful inspection shows that all three approaches significantly reduce delay by reducing the utilization. The greatest improvement, however, is made by doubling the line speed; unfortunately, this is also the most expensive as it involves obtaining both a second channel and a multiplexor. Indeed, all of these alternatives involve trading higher cost for lower delay.

2.3.3 The Effect of Message Length Distribution: The M/G/1 Queue

Thus far, it has been assumed that message lengths are exponentially distributed. This is seldom the case in reality. While many networks do carry messages of varying lengths, few have lengths that are exponentially distributed. In most data networks, messages are packetized and each packet is transmitted separately. Thus, our delay analysis should really be done in terms of the lengths of the packets, not the messages. Indeed, by packetizing the messages, their average length is reduced, which correspondingly reduces the service time. This is one of the main reasons for packetizing the messages in the first place, as shown in the following example.

Example 2.5. Consider two systems that both use a 9600 bps line to transmit messages. One produces 300 character (2400 bit) messages at the rate of 3 per second. The other produces 150 character (1200 bit) messages at the rate of 6 per second. In both systems the messages arrive independently and have exponentially distributed lengths. Compare the average time in the system for messages in each of these systems.

The traffic intensity in both systems is the same

$$(3)(2400) = (6)(1200) = 7200 \text{ bps}$$

The service rate (9600 bps) is also the same. Thus, the utilizations of both systems is also the same

$$U = 7200/9600 = 0.75$$

The service time, however, is different. In the first system it is 2400/9600 = 0.25 sec while in the second it is 0.125 sec. This means the average time in the system, which is $T_s/(1 - U)$, is halved in the second system relative to the first.

Packetization seeks to take advantage of this by decreasing the average length of messages. Of course, it may be argued that by breaking one long message into many short ones that the user now wait for many messages instead of one; in particular, for all the packets in a message to arrive. This is not entirely true, however. In some cases, the packets can be sent in parallel over multiple lower speed channels, getting the same effect as using a single higher speed channel. Also, arriving packets may require further processing inside the destination node. Each packet in a message can be processed while the next arrives. This is known as pipelining. Finally, the user may be able to make use of the beginning of the message even before the whole message arrives. For example, if the message is a screen of information, the user can begin reading the message (at a rate of 60 char/sec) while the remainder of the message arrives (at a rate of 1200 char/sec).

Packetization not only affects the average message length, it also affects the message length distribution. If the original message length is larger than the maximum packet size (which is often the case), a message is broken into several packets. All but the last has this fixed maximum length. The last packet is shorter. Another scenario is that messages to a host computer (e.g., enquiries) are short and the responses are longer and packetized as above. There may be several applications, each with its own characteristic message length. There may be many applications, which in aggregate have normally distributed messages. We now examine how the distribution of message lengths affects delays in the network.

A queueing system with independent arrivals and general message lengths is referred to as a M/G/1 queue, the G standing for general. It can be shown [Sch77] [Kle75] [All78] that the average waiting time in such a system is given by the Pollaczek-Khinchine Formula

$$T_w = \frac{A E[s^2]}{2(1 - U)} \tag{2.67}$$

where $E[s^2]$ is the second moment of the service time distribution. Here, A is the arrival rate in msg/sec and s^2 has units of $(\text{sec}/\text{msg})^2$. Recall that the second moment is defined as

$$E[s^2] = \sum_{j=1}^{M} P_j s_j^2 \tag{2.68}$$

where s_j is the service time for messages of type j, and P_j is the probability of a message being of type j. The moments of a probability distribution describe it completely. The first moment

$$E[s] = \sum_{j=1}^{M} P_j s_j \tag{2.69}$$

is just the average value. The second moment is a measure of variation about the mean and is closely related to the variance of the distribution, which was discussed earlier. In particular, the variance, V, is given by

$$V = E[s^2] - (E[s])^2 \tag{2.70}$$

Note in particular, that the square of the mean is not usually equal to the second moment. In fact, the only time they are equal is when the distribution is a constant, c. In this case $E[s] = c$, $E[s^2] = c^2$ and $V = 0$.

Example 2.6. We are given three systems. All have 56 Kbps lines. All have 50 percent utilizations. All have average message lengths of 1400 bits. The first has exponentially distributed message lengths. The second has constant (1400 bit) message lengths. In the third, half the messages are 400 bits long and the other half are 2400 bits long. Compare the waiting times in the three systems.

If the utilization is 50 percent ($U = 0.5$), the line speed is 56 Kbps and the average message length is 1400 bits, the arrival rate A, must be

$$A = \frac{(56000)(0.5)}{1400} = 20 \text{ msg/sec}$$

For an exponentially distributed random variable which is continuous, the sum in the definition of $E[s^2]$ above is replaced by an integral which, in this case, evaluates to $2E[s]^2$, that is, for an exponential distribution, the second moment is twice the square of the first moment (the average), and the variance is equal to the mean squared. The mean service time here is

$$E[s] = \frac{1400 \text{ bits/msg}}{56000 \text{ bits/sec}} = 0.025 \text{ sec/msg}$$

and thus,

$$E[s^2] = 0.00125 \ \frac{\text{sec}^2}{\text{msg}}$$

The waiting time is then given by

$$T_w = \frac{AE[s^2]}{2(1-U)} = \frac{20 \text{ msg/sec } (0.00125) \text{ (sec/msg)}^2}{(2)(1-0.5)} = 0.025 \text{ sec/msg}$$

Therefore, $T_w = T_s$ in this case. This is not a surprise, since the message length is exponential, this is an M/M/1 queue and we already know that for an M/M/1 queue

$$T_w = \frac{UT_s}{1-U}$$

For $U = 0.5$, which is the case here, this also gives $T_w = T_s$. More generally, we find that for an exponential distribution, since $E[s^2]$ is $2(E[s])^2$ and $E[s]$ is $1/D$, that the two expressions are equivalent for any utilization. This is reassuring.

In the second case, there is only one message length in the distribution, and the second moment is defined by a single term,

$$E[s^2] \ = \ (E[s])^2 \ = \ 0.000625 \text{ (sec/msg)}^2$$

and

$$T_w = 0.0125 \text{ sec/msg}$$

In the third case,

$$E[s^2] = (0.5) \left(\frac{400}{56000} \right)^2 + (0.5) \left(\frac{2400}{56000} \right)^2 = (0.000969) \text{ (sec/msg)}^2$$

and

$$T_w = 0.01938$$

Thus, when the message length is a constant (known as an M/D/1 queue), the waiting time is half what it is when the message length is exponential. In the case where there are two message lengths of very different sizes, the waiting time is intermediate between the other two cases. In general, when there are many message lengths, we expect results like in the third case. In fact, if the message lengths are not too different, we expect results closer to the constant length than they are to the exponential length. It is possible to create pathological message length distributions where the variance is larger than the mean (e.g., a very large number of messages with a very small length, and a small number of messages with a very large length) and to obtain results worse than for an exponential message length distribution, but these rarely arise in practice. The exponential message length is often studied because it not only gives rise to a model which can be easily analyzed, but also because it gives conservative results when the actual message length distribution is not known.

2.3.4 Systems with Priorities

It is seen that there is a basic tradeoff between utilization and delay. The higher the utilization, the more cost effectively the links are used, but the higher the delay. It is, of course, desirable to have both high cost effectiveness and low delay. Surprisingly, it is sometimes possible to achieve both. If there are some types of messages which do not require low delay, it is possible to give them a lower priority than the other messages, and have them served only when there are no higher priority messages waiting. Conversely, if a small fraction of the overall traffic requires exceptionally low delay, it is possible to give such messages a very high priority. Thus, there may be several priority classes.

There are two types of priority—preemptive and nonpreemptive. In preemptive priority, an arriving high priority message interrupts a low priority message in service. Thus, the high priority messages never "see" lower priority messages. Their delay can be analyzed based on a system with no lower priority messages. The analysis of delay for the lower priority messages, on the other hand, is more complex, as is the management of the actual system. Service of the interrupted message must start over from the beginning. This wastes server time and also presents the system with the problem of recognizing and discarding interrupted messages. For this reason, preemption is not widely used. Therefore we concentrate on the case of nonpreemptive priority, where a message in service is permitted to finish before starting service on another message. The analysis here follows that in [Sch77].

Suppose there are two priority classes with arrival rates A_1 and A_2, respectively, and departure rates D_1 and D_2, respectively. The high priority (priority 1) messages have only to wait for the message in service (if any) and any other high priority messages already in the queue. Thus,

$$E(W_1) = E(T_0) + E(T_1) \tag{2.71}$$

where $E(T_0)$ is the average time to complete the message currently in service and $E(T_1)$ is the time to serve the priority 1 messages already in the queue.

$$E(T_1) = E(m_1)/D_1 \tag{2.72}$$

where m_1 is the average number of priority 1 messages in the queue. Let $E(n_1)$ be the average number of priority 1 messages in the system, both waiting and in service. By Little's Formula,

$$E(n_1) = A_1 E(TT_1) \tag{2.73}$$

where TT_1 is the total time spent by the waiting messages (service and waiting).

$$E(TT_1) = E(W_1) + \frac{1}{D_1} \tag{2.74}$$

So,

$$E(n_1) = A_1 E(W_1) + \frac{A_1}{D_1} \tag{2.75}$$

But the average number of messages in service is A_1/D_1 since this is the fraction of time the server spends serving messages of class 1. So,

$$E(m_1) = A_1 E(W_1) \tag{2.76}$$

and since each such message is served on average for time $1/D_1$,

$$E(T_1) = \frac{A_1}{D_1} E(W_1) = U_1 E(W_1) \tag{2.77}$$

where $U_1 = A_1/D_1$. So,

$$E(W_1) = E(T_0) + U_1 E(W_1) \tag{2.78}$$

$$= \frac{E(T_0)}{1 - U_1} \tag{2.79}$$

As before,

$$E(T_0) = \frac{A}{2} E[s^2] \tag{2.80}$$

where A is the total arrival rate and $E[s^2]$ is the second moment of the service time of all messages.

This is the same expression already found for the M/G/1 queue, except that in the denominator we only consider utilization due to priority 1 messages.

A priority 2 message must wait for the message in service (if any), all messages in the queue when it arrives, and any priority 1 messages which arrive while the priority 2 message waits for service. Thus,

$$E(W_2) = E(T_0) + E(T_1) + E(T_2) + E(T_1') \tag{2.81}$$

where $E(T_2)$ is the time spent waiting for priority 2 messages already in the queue when the message arrived, and $E(T_1')$ is the time spent waiting for priority 1 messages which arrive while the message is waiting. Using the same arguments used before to determine $E(T_1)$,

$$E(T_2) = U_2 E(W_2) \tag{2.82}$$

and

$$E(T_1') = U_1 E(W_2) \tag{2.83}$$

So,

$$E(W_2) = \frac{E(T_0)}{(1 - U_1)((1 - (U_1 + U_2))} = \frac{E(T_0)}{(1 - U_1)(1 - U)} \tag{2.84}$$

Where U is the total utilization $(U_1 + U_2)$. For more than 2 priorities, the same type of analysis gives us

$$E(W_k) = \frac{E(T_0)}{(1 - U[k-1])(1 - U[k])} \tag{2.85}$$

where $U[k]$ is defined as

$$U[k] = \sum_{j=1}^{k} U_j \tag{2.86}$$

(i.e., it is the utilization due to messages of priority 1 through k).

Therefore, in exchange for not having to "pay" for utilization due to higher priority messages, each priority pays twice for messages of lower priority. In most realistic situations, only the lowest priority messages suffer significantly because the factor of $1 - U[k]$ in the denominator is really significant only when $U[k]$ is close to 1 and this can only happen for the largest values of k.

The total time in the system is, as always, the waiting time plus the service time. In this case, it is assumed that different priority messages may have different service times, so,

$$E(TT_k) = E(W_k) + \frac{1}{D_k} \tag{2.87}$$

Example 2.7. A 9600 bps (1200 char/sec) line carries batch and interactive traffic. The interactive traffic is comprised of 120 character messages arriving at the rate of 4 per second. The batch traffic is comprised of 1920 character messages arriving at the rate of 0.25 per second.

(a) If interactive traffic is given priority over batch, what is the average time in the system for each type of traffic?

$A_1 = 4$; $D_1 = 10$ because a 1200 char/sec line can serve 10 120 character messages per second; $U_1 = .4$; $A_2 = 0.25$; $D_2 = 0.625$; $U_2 = 0.4$; $A = A_1 + A_2 = 4.25$; $U = 0.8$

$$E(T_0) = \frac{A}{2} E[s^2] = 2.125 \left(\frac{4}{4.25} \left(\frac{120}{1200} \right)^2 + \frac{0.25}{4.25} \left(\frac{1920}{1200} \right)^2 \right) = 0.3439$$

$$E(W_1) = \frac{E(T_0)}{1 - U_1} = 0.5732$$

$$E(TT_1) = E(W_1) + \frac{1}{D_1} = 0.5732 + 0.1 = 0.6732$$

$$E(W_2) = \frac{E(T_0)}{(1 - U[1])(1 - U[2])} = \frac{0.3439}{(0.6)(0.2)} = 2.866$$

$$E(TT_2) = 2.866 + 1.6 = 4.466$$

(b) What is the average time in the system if no priorities are used?

$$E(W) = \frac{E(T_0)}{(1 - U)} = \frac{0.3499}{0.2} = 1.749$$

$$E(TT_1) = 1.849 \qquad E(TT_2) = 3.349$$

2.3.5 Networks of Queues

Networks are actually comprised of many queues, with the output of one queue feeding into another as input. For example, in Fig. 2.12 there is a network of three nodes and three links. Suppose that r_{12}, the requirement from node 1 to node 2, goes over the $(1, 2)$ link and, similarly, that r_{23} goes over the $(2, 3)$ link. Now suppose that r_{13} splits evenly, half of it going over the $(1, 3)$ link and half of it following the path $(1, 2), (2, 3)$.

This can be modeled as a network of queues in a number of different ways, depending on how the different types of traffic arrive and are treated in the competition for the outgoing links from each node. This topic is discussed more fully in Chapter 6 when routing in networks is discussed. Here, make the assumption that all traffic arrives independently to the network from outside and all traffic simply joins a single queue for an appropriate outgoing link toward its destination. Thus, all r_{12} traffic joins the queue for the $(1, 2)$ link. Half the r_{13} traffic joins the queue for the $(1, 3)$ link and half joins the queue for the $(1, 2)$ link. The r_{13} traffic arriving at node 2 joins the queue for the $(2, 3)$ link along with r_{23} traffic.

Thus, some of the arrivals to the queue for the $(2, 3)$ link are departures from the $(1, 2)$ link. Strictly speaking, such arrivals are no longer independent of one another.

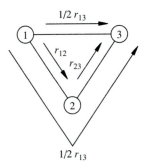

FIGURE 2.12

If there is a large number of arrivals of r_{13} traffic at node 1 over a short period of time, they queue up at node 1 (half in the $(1, 2)$ queue and half in the $(1, 3)$ queue). The r_{13} traffic traversing link $(1, 2)$, however, arrives at node 2 neatly spaced out at the speed of the link. In fact, if link $(1, 2)$ operates at the same speed as link $(2, 3)$ and if the queue for link $(2, 3)$, and link $(2, 3)$ itself are empty, the arriving r_{13} traffic does not have to wait at all at node 2.

The analysis of situations where there are networks of queues with dependent arrivals is extremely complex. It is an important area and forms the basis for the study of flow control in networks. However, it is beyond the scope of the current discussion. It is often studied via simulation of the network involved.

In fact, this dependence is most pronounced when a very small number of nodes and requirements are involved. In many realistic networks, where the number of nodes is more than ten and the number of requirements involved is likewise more than a few, if it has been observed (by measuring the response of the network directly) that a reasonably accurate analysis can be obtained without taking this dependency into account. Also, processing delays in the nodes, which involve traffic from many queues competing for common resources in the node, and also involving communication processing competing with other tasks going on in the node, tend to make the outgoing traffic streams to each queue more random.

It is common practice to assume that all arrivals to an outgoing queue for a link, both arrivals coming from outside the network and arrivals from other queues within the network, are all independent of one another. This makes the analysis of delays in such networks tractable. In fact, it is an extension of the analysis seen thus far.

Consider the simplest case here. Suppose that all messages arriving to the network are of the same type, with message lengths exponentially distributed. Suppose further that these messages arrive independently of one another and that messages leaving one queue and joining another also form an independent stream, joining other messages arriving from outside the network. This latter assumption is most reasonable, as mentioned before, when the number of streams involved is not very small. Also, it is most reasonable when traffic with a choice of path (like R_{13} in Fig. 2.12) makes its path decision randomly. Note that it is not necessary for the traffic to split evenly, just that it make its decision to take a particular link randomly with some prespecified probability. Such networks are called **Jackson type networks**, after Jackson [Jac57] who discovered many of the properties of such networks of queues. In fact, many of these results hold, even when some of the assumptions are relaxed.

The analysis of such networks proceeds in a manner similar to the one we have been using, modeling the system as being in one of a given set of states. Now the states include the number of messages in each of the queues. Figure 2.13 shows part of the state transition diagram for the network shown in Fig. 2.12. The actual state transition diagram has an infinite number of states; the queues can in theory become arbitrarily large. Also, for the sake of clarity, only some of the transitions are shown even for small queue sizes.

The state ijk represents the situation where there are i messages in the "system" for link 1 (i.e., in service or in the queue for link 1), j messages in the system for link 2 and k messages in the system for link 3. Thus, the 000 state corresponds to the

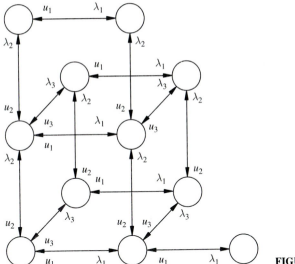

FIGURE 2.13

entire system being idle and the 021 state corresponds to no messages in the system for link 1, 2 messages in the system for link 2 (one in the queue and one in service), and 1 message in the system for link 3 (in service). This system is memoryless. We do not care how it got into the state it is in or what kind of messages are waiting in each queue. There is, of course, a conservation of messages of each type, that is, if r_{13} traffic enters the system at node 1, and chooses to take the 1–2–3 path, it appears as traffic at node 2 as well. This, however, is accounted for by the arrival rate at node 2 and not explicitly taken into account in the state of the system.

In Fig. 2.12, suppose the rates shown are in messages per second, arriving at each node from outside the system. The arrival rate to link $(1, 3)$ is 1 (half of the r_{13} traffic). The arrival rate to link $(1, 2)$ is 2 (the r_{12} traffic plus half of the r_{13} traffic). Finally, the arrival rate to link $(2, 3)$ is 3 (the r_{23} traffic plus half of the r_{13} traffic). We refer to the arrival rate to queue i as λ_i and the departure rate from each queue as μ_i.

We can now compute the probabilities of being in each state using the same techniques we have been using all along. We again set up balance equations on the flow coming into and leaving each node in the state diagram. These balance equations take into account all incoming and outgoing links from one state to another. For example,

$$(\lambda_1 + \lambda_2 + \lambda_3)P_{000} = \mu_1 P_{100} + \mu_2 P_{010} + \mu_3 P_{001} \tag{2.88}$$

for the balance equation around the 000 state, and

$$(\lambda_1 + \lambda_2 + \lambda_3 + \mu_2 + \mu_3)P_{011} = \mu_1 P_{111} + \mu_2 P_{021} + \mu_3 P_{012} + \lambda_2 P_{001} + \lambda_3 P_{010} \tag{2.89}$$

In each case, the balance equation is formed by drawing a circle around the given node in the state diagram and recording the incoming and outgoing edges. In fact, it is possible to form a balance equation by drawing a closed curve around any set of states and recording the edges separating that set of states from the remaining ones. This is exactly what has been done all along, but previously, the graphs involved were simpler. Not all balance equations are independent, however. In general, some may be sums of others. A set of balance equations formed around different individual nodes, however, are always independent.

The general case for a single state is

$$(\lambda_1 + \lambda_2 + \lambda_3 + \mu_1 + \mu_2 + \mu_3)P_{i,j,k}$$

$$= \mu_1 P_{i+1,j,k} + \mu_2 P_{i,j+1,k} + \mu_3 P_{i,j,k+1}$$

$$+ \lambda_1 P_{i-1,j,k} + \lambda_2 P_{i,j-1,k} + \lambda_3 P_{i,j,k-1} \quad (2.90)$$

This type of analysis can be further generalized to networks with an arbitrarily large number of queues. Jackson [Jac57] showed that a solution to these balance equations is

$$P_{i,j,k} = U_1^i U_2^j U_3^k P_{000} \quad (2.91)$$

where

$$U_m = \frac{\lambda_m}{\rho_m}$$

for $m = 1, 2, 3$, and

$$P_{000} = \frac{1}{(1 - U_1)(1 - U_2)(1 - U_3)} \quad (2.92)$$

This result generalizes to any number, n, of queues

$$P_{i_1, i_2, \ldots i_n} = \prod_{k=1}^{n} U_k^{i_k} P_{000\ldots 0} \quad (2.93)$$

and

$$P_{000\ldots 0} = \frac{1}{\prod_{i=1}^{n} 1 - U_i} \quad (2.94)$$

This type of solution is known as a **product form solution** and can be seen to be a generalization of the results obtained earlier for single queues. It is possible to relax some of the earlier assumptions and still obtain product form solutions. The interested reader is referred to [Kle75].

2.4   SYSTEMS WITH LOSS

Not all systems deal with congestion by allowing messages to wait. Most notably, traditional analog telephone systems simply block calls from entering the system if no capacity is available for them. The basic mechanism is different than in systems that permit waiting. Here, a call arrives and requires a fixed amount of capacity (enough to handle a conversation). If the capacity is available (i.e., if a line is available), it is dedicated to the call for its duration. If not, the caller is given a busy circuit indicator and is said to be "lost" (i.e., that call is not carried). Nobody waits.

Here, the measure of performance is the percentage of calls that are lost. This is referred to in the literature as the grade of service (GOS). In reality, such calls are not permanently lost. The caller can try again at some later time. It is even possible that the call may be carried immediately by another system as is the case when calls offered to lines in a private telephone network overflow into the public network. In this case, the issue is not one of performance (The caller doesn't know the difference) but rather one of trading off cost between the private network and the public network. For the moment, we assume that blocked calls disappear; other alternatives are examined later.

The behavior just discussed is not intrinsic to telephone networks. Circuit switched data networks (and, to some extent, virtual circuit networks as well) dedicate capacity (both line and buffer) to users; and when this capacity is not available they block calls from entering the system. Conversely, modern digital telephone networks can statistically multiplex calls or even packetize them and introduce delays in exchange for lower blocking. Also, some voice networks can store the called number until a line is available and then place the call; this is called camp-on and it also trades higher delay for lower loss.

The basic techniques for the analysis of systems with loss are the same as those used to analyze queues. In fact, these systems are simply finite queues (i.e., queues where after the number in the system reaches a given limit the queue overflows and no further arrivals are permitted). If, as before, it is assumed that arrivals are independent and message lengths or, in telephone terminology, holding times, are exponentially distributed, the system is Markovian. We can again draw a state transition diagram, set up balance equations, and find the state probabilities. Given this, we can then find loss probabilities, or even analyze systems with both loss and delay.

2.4.1 The Erlang-B Function

Consider a system with calls arriving at rate A and holding time H. The departure rate, D, is simply the inverse of H. We define the quantity A/D, which is the quantity U used earlier, as the call intensity and its units as Erlangs. To keep closer to the notation used in the analysis of telephone systems, we refer to this quantity as E. For example, if A is 4 calls/min. and H is 3 min. then the call intensity is 12 Erlangs. Note that call intensity is dimensionless. An Erlang can also be thought of as the amount of calls that occupy a line full time or, equivalently, as the average number of calls in progress (assuming none are blocked).

Some telephone literature measures call intensity in CCS, which are hundred call seconds per hour. Thus,

$$1 \text{ Erlang} = 36 \text{ CCS}$$

Call intensity is measured in Erlangs throughout this book.

Figure 2.14 illustrates the simplest case of a finite queue—a single server and no buffer. If a message arrives and the server is idle, the message immediately goes into service. If it arrives when the server is busy, the message is lost. There is no

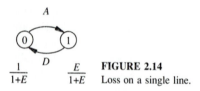

FIGURE 2.14
Loss on a single line.

queueing. It is a single balance equation:

$$AP_0 = DP_1 \tag{2.95}$$

Thus

$$P_1 = EP_0 \tag{2.96}$$

Since these are the only 2 states the system can be in,

$$P_0 + P_1 = 1 \tag{2.97}$$

So,

$$P_0(1 + E) = 1 \tag{2.98}$$

$$P_0 = \frac{1}{(1 + E)} \tag{2.99}$$

and

$$P_1 = \frac{E}{(1 + E)} \tag{2.100}$$

Since a message (or call) is lost if it arrives when the server is busy and since messages arrive independently of one another (and everything else), the probability of loss (or blocking) is simply P_1. This is denoted by

$$B(E, 1) = \frac{E}{(1 + E)} \tag{2.101}$$

In this notation, $B(E, m)$ is the probability of blocking when E Erlangs of traffic is offered to m channels. We say that $EB(E, m)$ Erlangs are lost and $E(1 - B(E, m))$ Erlangs are carried. This is illustrated in Fig. 2.15.

The state transitions for the more general case for m trunks and no queueing is shown in Fig. 2.16. We now have m balance equations of the form

$$AP_{k-1} = kDP_k \quad \forall k = 1, 2, \ldots, m \tag{2.102}$$

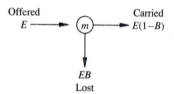

FIGURE 2.15
Offered, carried, and lost load.

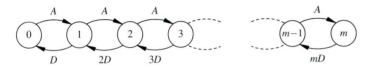

FIGURE 2.16
Loss with m lines.

from which

$$P_k = \frac{E^k}{k!} P_0 \tag{2.103}$$

Since all the P_k must sum to 1,

$$P_0 = \sum_{k=0}^{m} \frac{E^k}{k!} \tag{2.104}$$

Again, the probability of loss is the probability of a message arriving when all servers are busy, which is P_m. Thus we have

$$B(E, m) = \frac{\frac{E^m}{m!}}{\sum_{k=0}^{m} \frac{E^k}{k!}} \tag{2.105}$$

Note that this is indeed a probability. All the terms in the numerator and denominator are positive, so B must be positive, and since the numerator is one of the terms in the denominator, B can be no greater than 1. Note also that $B(E, 0)$ is 1 and $B(0, m)$ is 0. ($B(0, 0)$ is 1.)

This formula is somewhat cumbersome to calculate by hand and is even numerically difficult to evaluate with a computer because it involves calculating very large numbers. Fortunately, it can be rearranged algebraically into a much more convenient form which allows us to iteratively compute B, as follows

$$B(E, 0) = 1 \tag{2.106}$$

$$B(E, k) = \frac{E B(E, k - 1)}{E B(E, k - 1) + k} \forall k = 1, 2, \ldots, m \tag{2.107}$$

Example 2.8. There are 20 people in an office. During the course of an 8 hour day, they each attempt to make an average of 16 telephone calls which last an average of 3 minutes each. There are three telephone lines in the office. What is the probability of loss (the probability that a call be attempted when all three lines are busy)?
The call intensity is

$$\frac{(20)(16)(3)\text{min}}{(8)(60)\text{min}} = 2 \text{ Erlangs}$$

$$B(E, m) = B(2, 3) = \frac{\frac{2^3}{3!}}{\frac{2^0}{0!} + \frac{2^1}{1!} + \frac{2^2}{2!} + \frac{2^3}{3!}}$$

$$\frac{8/6}{1/1 + 2/1 + 4/2 + 8/6} = \frac{4}{19} = 0.2105$$

We can also obtain this result iteratively,

$$B(2, 0) = 1$$

$$B(2, 1) = \frac{2B(2, 0)}{2B(2, 0) + 1} = \frac{2}{3}$$

$$B(2, 2) = \frac{(2)(2/3)}{(2)(2/3) + 2} = \frac{2}{5}$$

$$B(2, 3) = \frac{(2)(2/5)}{(2)(2/5) + 3} = \frac{4}{19}$$

Another advantage to the iterative approach is that it can be used to find m when E as well as a desired value of $B(E, m)$ is known.

Example 2.9. What is the smallest number of lines which allow an offered load of 0.5 Erlangs to suffer no worse than a 2 percent loss?

$$B(0.5, 0) = 1$$

$$B(0.5, 1) = 1/3$$

$$B(0.5, 2) = 1/13$$

$$B(0.5, 3) = 1/79$$

So the answer is 3.

Figure 2.17 shows the values of $B(E, m)$ for various E and $m = 1, 2, 5$, and 10. Table 2.8 gives values of $B(E, m)$ for various values of E and m. Table 2.9 gives values of the line utilization for various values of E and m. We make the following observations from these figures and tables:

1. If the offered load increases and the number of lines remains the same, the probability of loss increases.
2. If the offered load remains the same and the number of lines increases, the probability of loss decreases.
3. The more load that is offered to a group of lines, the higher the utilization of those lines.
4. As the offered load gets much larger than the number of lines, both the probability of loss and the line utilization approach 1.
5. If both the offered load and the number of lines are increased by the same factor, the probability of loss decreases and the line utilization increases. For example, if 9 Erlangs are offered to 10 lines, the probability of loss is 0.168 and the utilization is 0.749. If 90 Erlangs is offered to 100 lines, however, the probability of loss is

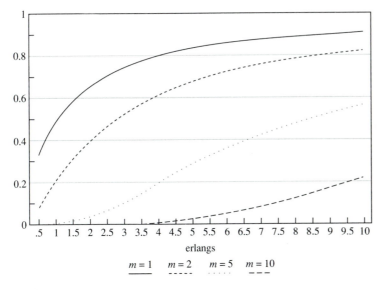

erlangs

$m = 1$ $m = 2$ $m = 5$ $m = 10$

FIGURE 2.17

0.027 and the utilization is 0.876. This is one of the many examples of economy of scale, which is a very important principle in network design. On the other hand, this effect becomes less pronounced as the number of lines gets bigger. The effect is larger going from 5 lines to 10 lines than it is going from 10 to 20. This is an example of another general principle—that economy of scale itself is most important in its earliest stages.

6. For a given number of lines, m, as the offered load E increases, we find three regions. First, when E is much smaller than m, the probability of loss is essentially 0; then, as E approaches m, the probability of loss increases quickly, and finally, as E passes m, the probability of loss is approximated by $(E - m)/E$. If we are interested in maintaining a prespecified limit on the probability of loss, typically in the range of a few percent, we find therefore that there is a critical region of E, for any given m, where the probability of loss is "interesting" (i.e., where we need to calculate it). Below this region, we can assume it is 0 and above it we can assume it is $(E - m)/E$. Often, this is sufficient for basic design calculations. The actual values bounding this region vary as m increases.

2.4.2 Systems with Both Loss and Delay (M/M/m/q)

The techniques described can also be used to analyze systems with both loss and delay. In the nomenclature these are referred to as M/M/m/q systems; systems with independent arrivals, exponential service times, m servers, and maximum queue length of q.

TABLE 2.8
Erlang-B loss

erl	$m = 1$	$m = 2$	$m = 3$	$m = 4$	$m = 5$	$m = 6$	$m = 7$	$m = 8$	$m = 9$	$m = 10$
0.5	0.333	0.077	0.013	0.002	0.000	0.000	0.000	0.000	0.000	0.000
1.0	0.500	0.200	0.063	0.015	0.003	0.001	0.000	0.000	0.000	0.000
1.5	0.600	0.310	0.134	0.048	0.014	0.004	0.001	0.000	0.000	0.000
2.0	0.667	0.400	0.211	0.095	0.037	0.012	0.003	0.001	0.000	0.000
2.5	0.714	0.472	0.282	0.150	0.070	0.028	0.010	0.003	0.001	0.000
3.0	0.750	0.529	0.346	0.206	0.110	0.052	0.022	0.008	0.003	0.001
3.5	0.778	0.576	0.402	0.260	0.154	0.082	0.040	0.017	0.007	0.002
4.0	0.800	0.615	0.451	0.311	0.199	0.117	0.063	0.030	0.013	0.005
4.5	0.818	0.648	0.493	0.357	0.243	0.154	0.090	0.048	0.024	0.010
5.0	0.833	0.676	0.530	0.398	0.285	0.192	0.121	0.070	0.037	0.018
5.5	0.846	0.699	0.562	0.436	0.324	0.229	0.153	0.095	0.055	0.029
6.0	0.857	0.720	0.590	0.470	0.360	0.265	0.185	0.122	0.075	0.043
6.5	0.867	0.738	0.615	0.500	0.394	0.299	0.217	0.150	0.098	0.060
7.0	0.875	0.754	0.638	0.527	0.425	0.331	0.249	0.179	0.122	0.079
7.5	0.882	0.768	0.658	0.552	0.453	0.362	0.279	0.207	0.147	0.100
8.0	0.889	0.780	0.675	0.575	0.479	0.390	0.308	0.236	0.173	0.122
8.5	0.895	0.792	0.692	0.595	0.503	0.416	0.336	0.263	0.199	0.145
9.0	0.900	0.802	0.706	0.614	0.525	0.441	0.362	0.289	0.224	0.168
9.5	0.905	0.811	0.720	0.631	0.545	0.463	0.386	0.314	0.249	0.191
10.0	0.909	0.820	0.732	0.647	0.564	0.485	0.409	0.338	0.273	0.215

erl	$m = 10$	$m = 20$	$m = 30$	$m = 40$	$m = 50$	$m = 60$	$m = 70$	$m = 80$	$m = 90$	$m = 100$
5	0.018	0.000	0.000	0.000	0.000	0.000	0.000	0.000	0.000	0.000
10	0.215	0.002	0.000	0.000	0.000	0.000	0.000	0.000	0.000	0.000
15	0.410	0.046	0.000	0.000	0.000	0.000	0.000	0.000	0.000	0.000
20	0.538	0.159	0.008	0.000	0.000	0.000	0.000	0.000	0.000	0.000
25	0.622	0.280	0.053	0.001	0.000	0.000	0.000	0.000	0.000	0.000
30	0.681	0.380	0.132	0.014	0.000	0.000	0.000	0.000	0.000	0.000
35	0.725	0.459	0.220	0.054	0.003	0.000	0.000	0.000	0.000	0.000
40	0.758	0.521	0.299	0.116	0.019	0.001	0.000	0.000	0.000	0.000
45	0.784	0.571	0.367	0.185	0.054	0.005	0.000	0.000	0.000	0.000
50	0.805	0.612	0.425	0.250	0.105	0.022	0.001	0.000	0.000	0.000
55	0.822	0.646	0.473	0.308	0.161	0.053	0.007	0.000	0.000	0.000
60	0.837	0.674	0.515	0.360	0.216	0.096	0.024	0.002	0.000	0.000
65	0.849	0.699	0.550	0.406	0.267	0.144	0.052	0.009	0.001	0.000
70	0.859	0.720	0.581	0.445	0.314	0.192	0.090	0.025	0.003	0.000
75	0.869	0.738	0.608	0.480	0.356	0.237	0.131	0.051	0.011	0.001
80	0.877	0.754	0.632	0.511	0.393	0.279	0.173	0.084	0.026	0.004
85	0.884	0.768	0.653	0.539	0.427	0.317	0.213	0.121	0.050	0.012
90	0.890	0.781	0.672	0.564	0.457	0.352	0.251	0.158	0.080	0.027
95	0.896	0.792	0.689	0.586	0.484	0.384	0.287	0.195	0.112	0.049
100	0.901	0.802	0.704	0.606	0.509	0.413	0.320	0.229	0.146	0.076

TABLE 2.9
Line utilizations

erl	$m = 1$	$m = 2$	$m = 3$	$m = 4$	$m = 5$	$m = 6$	$m = 7$	$m = 8$	$m = 9$	$m = 10$
0.5	0.333	0.231	0.165	0.125	0.100	0.083	0.071	0.062	0.056	0.050
1.0	0.500	0.400	0.313	0.246	0.199	0.167	0.143	0.125	0.111	0.100
1.5	0.600	0.517	0.433	0.357	0.296	0.249	0.214	0.187	0.167	0.150
2.0	0.667	0.600	0.526	0.452	0.385	0.329	0.285	0.250	0.222	0.200
2.5	0.714	0.660	0.598	0.531	0.465	0.405	0.354	0.312	0.278	0.250
3.0	0.750	0.706	0.654	0.595	0.534	0.474	0.419	0.372	0.332	0.300
3.5	0.778	0.741	0.698	0.647	0.592	0.535	0.480	0.430	0.386	0.349
4.0	0.800	0.769	0.732	0.689	0.641	0.589	0.536	0.485	0.439	0.398
4.5	0.818	0.792	0.761	0.724	0.681	0.634	0.585	0.535	0.488	0.445
5.0	0.833	0.811	0.784	0.752	0.715	0.673	0.628	0.581	0.535	0.491
5.5	0.846	0.827	0.803	0.776	0.744	0.707	0.666	0.622	0.578	0.534
6.0	0.857	0.840	0.820	0.796	0.768	0.735	0.699	0.659	0.617	0.574
6.5	0.867	0.852	0.834	0.813	0.788	0.759	0.727	0.691	0.652	0.611
7.0	0.875	0.862	0.846	0.827	0.805	0.780	0.751	0.719	0.683	0.645
7.5	0.882	0.870	0.856	0.840	0.820	0.798	0.772	0.743	0.711	0.675
8.0	0.889	0.878	0.865	0.851	0.834	0.814	0.791	0.764	0.735	0.703
8.5	0.895	0.885	0.874	0.860	0.845	0.827	0.807	0.783	0.757	0.727
9.0	0.900	0.891	0.881	0.869	0.855	0.839	0.821	0.800	0.776	0.749
9.5	0.905	0.897	0.887	0.877	0.864	0.850	0.833	0.814	0.793	0.768
10.0	0.909	0.902	0.893	0.883	0.872	0.859	0.844	0.827	0.808	0.785

erl	$m = 10$	$m = 20$	$m = 30$	$m = 40$	$m = 50$	$m = 60$	$m = 70$	$m = 80$	$m = 90$	$m = 100$
5	0.491	0.250	0.167	0.125	0.100	0.083	0.071	0.063	0.056	0.050
10	0.785	0.499	0.333	0.250	0.200	0.167	0.143	0.125	0.111	0.100
15	0.884	0.716	0.500	0.375	0.300	0.250	0.214	0.188	0.167	0.150
20	0.924	0.841	0.661	0.500	0.400	0.333	0.286	0.250	0.222	0.200
25	0.944	0.900	0.789	0.624	0.500	0.417	0.357	0.313	0.278	0.250
30	0.956	0.930	0.868	0.739	0.600	0.500	0.429	0.375	0.333	0.300
35	0.964	0.947	0.910	0.828	0.698	0.583	0.500	0.437	0.389	0.350
40	0.969	0.957	0.934	0.884	0.785	0.666	0.571	0.500	0.444	0.400
45	0.973	0.965	0.949	0.917	0.851	0.746	0.643	0.562	0.500	0.450
50	0.976	0.970	0.959	0.938	0.895	0.815	0.713	0.625	0.556	0.500
55	0.979	0.974	0.965	0.951	0.923	0.868	0.780	0.687	0.611	0.550
60	0.981	0.977	0.970	0.960	0.941	0.904	0.837	0.748	0.667	0.600
65	0.983	0.979	0.974	0.966	0.952	0.927	0.880	0.805	0.722	0.650
70	0.984	0.981	0.977	0.971	0.961	0.943	0.910	0.853	0.775	0.700
75	0.985	0.983	0.979	0.974	0.967	0.954	0.931	0.890	0.824	0.749
80	0.986	0.984	0.981	0.977	0.971	0.962	0.945	0.916	0.866	0.797
85	0.987	0.985	0.983	0.979	0.975	0.967	0.955	0.934	0.897	0.840
90	0.988	0.986	0.984	0.981	0.977	0.971	0.962	0.947	0.920	0.876
95	0.989	0.987	0.985	0.983	0.980	0.975	0.968	0.956	0.937	0.904
100	0.989	0.988	0.986	0.984	0.981	0.978	0.972	0.963	0.949	0.924

Suppose, as is depicted in Fig. 2.18, that there are two servers and two spaces in the queue (i.e., $m = q = 2$). If a message arrives at a time when there is an idle server, it immediately goes into service. If it arrives when there is no available server but there is an empty space in the queue, it waits. Otherwise, the message is lost.

As before, U is defined as the arrival rate, A, divided by the departure rate, D. Note that U is the traffic intensity in Erlangs. Here are balance equations:

$$AP_0 = DP_1 \qquad P_1 = UP_0$$

$$AP_1 = 2DP_2 \qquad P_2 = \frac{U^2}{2}P_0$$

$$AP_2 = 2DP_3 \qquad P_3 = \frac{U^3}{4}P_0$$

$$AP_3 = 2DP_4 \qquad P_4 = \frac{U^4}{8}P_0$$

Again, P_0 is found by summing the P_k,

$$1 = P_0\left(1 + U + \frac{U^2}{2} + \frac{U^3}{4} + \frac{U^4}{8}\right)$$

So,

$$S = 1 + U + \frac{U^2}{2} + \frac{U^3}{4} + \frac{U^4}{8}$$

$$P_0 = \frac{1}{S}$$

In a more general case, if there are m servers and q places in the queue we have

$$P_k = \frac{U^k}{k!}P_0 \qquad \forall k = 1, 2, \ldots, m \tag{2.108}$$

$$P_{m+k} = \frac{U^k}{m^k}P_m = \frac{U^m}{m!}\frac{U^k}{m^k}P_0 \qquad \forall k = 1, 2, \ldots, q \tag{2.109}$$

$$S = \sum_{k=0}^{m}\frac{U^k}{k!} + \frac{U^m}{m!}\sum_{k=1}^{q}\frac{U^q}{m^q} \tag{2.110}$$

$$P_0 = 1/S \tag{2.111}$$

Given these state probabilities, and since messages are lost if they arrive when the system (all servers and the queue) is full, the probability of loss is

$$B = P_{m+q} \tag{2.112}$$

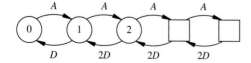

FIGURE 2.18
A system with loss and delay.

Since messages wait if they arrive when all servers are busy and the queue is not full, the probability of waiting is

$$P_w = \sum_{k=0}^{q-1} P_{m+k} \qquad (2.113)$$

The average queue size is simply the sum, over all states of the size of the queue, times the probability of being in the state, and only the sum over states where the queue is non-zero.

$$Q = \sum_{k=1}^{q} k P_{m+k} \qquad (2.114)$$

By Little's Formula, the average waiting time is

$$T_w = \frac{Q}{A} \qquad (2.115)$$

The average time in the system is the waiting time plus the service time.

$$T = T_w + T_s = \frac{Q}{A} + \frac{1}{D} \qquad (2.116)$$

Example 2.10. Suppose in the system shown in Fig. 2.18 that $A = D = 1$.

(a) What is P_0?

$$U = \frac{A}{D} = 1$$

$$P_0 = 1/(1 + U + \frac{U^2}{2} + \frac{U^3}{4} + \frac{U^4}{8}) = 0.3478$$

(b) What is the probability of loss?

$$B = P_4 = \frac{U^4}{8} P_0 = 0.0435$$

(c) What is the average waiting time?
The average queue length is

$$Q = \sum_{k=1}^{q} k P_{m+k} = P_3 + 2P_4 = 0.087 + 0.087 = 0.174$$

$$T_w = Q/A = .174$$

(d) How does the probability of loss compare with a system of two servers and no queue, and with a system of four servers and no queue?

$$B(E, m) = \frac{\dfrac{U^m}{m!}}{\sum_{k=0}^{m} \dfrac{U^k}{k!}}$$

$$B(1, 2) = 0.2$$

$$B(1, 4) = 0.0153$$

So, the probability of loss is between these two values. This is as expected since the presence of the queue saves some messages from being lost but having two places in the queue does not help as much as having two more servers.

Table 2.10 shows various statistics for a traffic intensity of 3 Erlangs, various numbers of servers (m), and various maximum lengths (q). We make the following observations from examining this table:

1. If $q = 0$, we have a system with loss and no queueing. The results are the same as for the Erlang-B function.

2. If $A/D \geq m$, the queue size and the time in the system grows with q; the servers cannot keep up with the arrivals.

3. If $A/D < m$, as q increases the queue size and time in the system converges to the values for an M/M/m queue and the probability of loss goes to 0.

2.4.3   Other Models of Retry Behavior

2.4.3.1   ERLANG-C FUNCTION.   Until now, it is assumed that blocked calls disappear, that is, that callers do not retry. They either go away or are served in some other way. We now consider some of the other possibilities. One possibility is that calls arriving when no server is available wait until a server becomes free (i.e., they camp on). This is exactly the case of an M/M/m/q system as q goes to infinity. It can be approximated using the previous formulas with a very large q, but the computation becomes tedious. A better choice is to recognize that this is the M/M/m system which was analyzed before. If the calls really wait, there is no blocking. This model, however, is also used as a worst case approximation to the behavior of systems where callers retry very quickly. In this case, we are interested in the probability of an arriving call not being served immediately. This is simply the probability that a call arrives when the servers are all busy and this probability is given by the sum of there being m or more messages in the system. We recall from Section 2.3.2 that

$$\frac{1}{P_0} = \sum_{k=0}^{m-1} \frac{U^k}{k!} + \frac{U^m}{m!(1 - U/m)} \tag{2.117}$$

The last term corresponds to the states with all the servers busy. Therefore, C, the probability that all servers are busy is

$$C = \frac{\frac{U^m}{m!(1-U/m)}}{\sum_{k=0}^{m-1} \frac{U^k}{k!} + \frac{U^m}{m!(1-U/m)}} \tag{2.118}$$

C is known as the Erlang-C function and it is closely related to the Erlang-B function. Recall that U is the traffic intensity in Erlangs. The above expression for C can be

TABLE 2.10

M/M/m/q statistics

Arrival rate = 3.000, departure rate = 1.000

(a) Queue Size

	$m = 1$	$m = 2$	$m = 3$	$m = 4$	$m = 5$
$q = 0$:	0.000	0.000	0.000	0.000	0.000
$q = 1$:	0.692	0.443	0.257	0.134	0.062
$q = 2$:	1.575	1.064	0.614	0.304	0.131
$q = 3$:	2.529	1.789	1.019	0.477	0.192
$q = 4$:	3.511	2.584	1.452	0.638	0.239
$q = 5$:	4.504	3.431	1.901	0.783	0.275
$q = 6$:	5.502	4.316	2.363	0.910	0.300
$q = 7$:	6.501	5.230	2.831	1.019	0.318
$q = 8$:	7.500	6.167	3.306	1.112	0.330
$q = 9$:	8.500	7.121	3.785	1.189	0.338
$q = 10$:	9.500	8.087	4.267	1.254	0.344
$q = 15$:	14.500	13.016	6.708	1.439	0.353
$q = 20$:	19.500	18.003	9.175	1.501	0.354
$q = 25$:	24.500	23.000	11.653	1.520	0.354
$q = 30$:	29.500	28.000	14.139	1.526	0.354
$q = 35$:	34.500	33.000	16.628	1.528	0.354
$q = 40$:	39.500	38.000	19.119	1.528	0.354
$q = 45$:	44.500	43.000	21.613	1.528	0.354
$q = 50$:	49.500	48.000	24.107	1.528	0.354

(b) Time in System

	$m = 1$	$m = 2$	$m = 3$	$m = 4$	$m = 5$
$q = 0$:	1.000	1.000	1.000	1.000	1.000
$q = 1$:	1.231	1.148	1.086	1.045	1.021
$q = 2$:	1.525	1.355	1.205	1.101	1.044
$q = 3$:	1.843	1.596	1.340	1.159	1.064
$q = 4$:	2.170	1.861	1.484	1.213	1.080
$q = 5$:	2.501	2.144	1.634	1.261	1.092
$q = 6$:	2.834	2.439	1.788	1.303	1.100
$q = 7$:	3.167	2.743	1.944	1.340	1.106
$q = 8$:	3.500	3.056	2.102	1.371	1.110
$q = 9$:	3.833	3.374	2.262	1.396	1.113
$q = 10$:	4.167	3.696	2.422	1.418	1.115
$q = 15$:	5.833	5.339	3.236	1.480	1.118
$q = 20$:	7.500	7.001	4.058	1.500	1.118
$q = 25$:	9.167	8.667	4.884	1.507	1.118
$q = 30$:	10.833	10.333	5.713	1.509	1.118
$q = 35$:	12.500	12.000	6.543	1.509	1.118
$q = 40$:	14.167	13.667	7.373	1.509	1.118
$q = 45$:	15.833	15.333	8.204	1.509	1.118
$q = 50$:	17.500	17.000	9.036	1.509	1.118

TABLE 2.10

M/M/m/q statistics (continued)

Arrival rate = 3.000, departure rate = 1.000

(c) Probability of Waiting

	$m = 1$	$m = 2$	$m = 3$	$m = 4$	$m = 5$
$q = 0$:	0.000	0.000	0.000	0.000	0.000
$q = 1$:	0.231	0.295	0.257	0.179	0.103
$q = 2$:	0.300	0.443	0.409	0.284	0.159
$q = 3$:	0.322	0.527	0.509	0.351	0.191
$q = 4$:	0.330	0.577	0.581	0.396	0.209
$q = 5$:	0.332	0.609	0.634	0.427	0.220
$q = 6$:	0.333	0.629	0.675	0.449	0.227
$q = 7$:	0.333	0.642	0.708	0.465	0.230
$q = 8$:	0.333	0.650	0.735	0.477	0.233
$q = 9$:	0.333	0.656	0.757	0.485	0.234
$q = 10$:	0.333	0.659	0.776	0.491	0.235
$q = 15$:	0.333	0.666	0.839	0.505	0.236
$q = 20$:	0.333	0.667	0.874	0.508	0.236
$q = 25$:	0.333	0.667	0.896	0.509	0.236
$q = 30$:	0.333	0.667	0.912	0.509	0.236
$q = 35$:	0.333	0.667	0.924	0.509	0.236
$q = 40$:	0.333	0.667	0.933	0.509	0.236
$q = 45$:	0.333	0.667	0.940	0.509	0.236
$q = 50$:	0.333	0.667	0.945	0.509	0.236

(d) Probability of Loss

	$m = 1$	$m = 2$	$m = 3$	$m = 4$	$m = 5$
$q = 0$:	0.750	0.529	0.346	0.206	0.110
$q = 1$:	0.692	0.443	0.257	0.134	0.062
$q = 2$:	0.675	0.399	0.205	0.091	0.036
$q = 3$:	0.669	0.374	0.170	0.064	0.021
$q = 4$:	0.668	0.360	0.145	0.046	0.012
$q = 5$:	0.667	0.350	0.127	0.033	0.007
$q = 6$:	0.667	0.345	0.113	0.024	0.004
$q = 7$:	0.667	0.341	0.101	0.018	0.003
$q = 8$:	0.667	0.338	0.092	0.013	0.002
$q = 9$:	0.667	0.337	0.084	0.010	0.001
$q = 10$:	0.667	0.335	0.078	0.007	0.001
$q = 15$:	0.667	0.334	0.056	0.002	0.000
$q = 20$:	0.667	0.333	0.044	0.000	0.000
$q = 25$:	0.667	0.333	0.036	0.000	0.000
$q = 30$:	0.667	0.333	0.030	0.000	0.000
$q = 35$:	0.667	0.333	0.026	0.000	0.000
$q = 40$:	0.667	0.333	0.023	0.000	0.000
$q = 45$:	0.667	0.333	0.021	0.000	0.000
$q = 50$:	0.667	0.333	0.019	0.000	0.000

rearranged to give

$$C(U, m) = \frac{B(U, m)}{1 - (U/m)(1 - B(U, m))}$$ (2.119)

where $B(U, m)$ is the Erlang-B function. Thus, the Erlang-C function can be computed once the value of the Erlang-B function is known for the same U and m.

Since the denominator is less than 1, it is immediately understood that C is always greater than B (i.e., the "loss" probability is greater when blocked calls wait). In fact, this is the most conservative assumption that can be made about retries. Note also that the Erlang-C function only makes sense for $U < m$; if $U \geq m$, then the queue would overflow; the m servers cannot keep up with the traffic intensity of U.

2.4.3.2 POISSON (MOLINA) FUNCTION. Another assumption about retries is that a blocked call stays in the system as long as it would have if it were served. If a server becomes available while it is still in the system, it occupies the server for its remaining time in the system. This gives rise to the Poisson Formula, also known as the Molina Formula, for computing the probability of loss. This, seemingly odd assumption, which is known as the lost-calls-held assumption, is widely used in telephony because it gives rise to simple engineering rules. It was adopted within the Bell System, and it gives results intermediate between lost calls cleared (Erlang-B) and lost calls delayed (Erlang-C).

The Poisson Formula is related to the Poisson probability distribution which arises from independent arrivals. The binomial distribution has already been studied. If there are n sources of traffic and each source generates a call with probability p in a given interval of time, then the probability that k calls are generated in that interval of time is

$$P_k = \frac{n!}{k!(n - k)!} p^k (1 - p)^{n-k}$$ (2.120)

and the average number of calls generated is np. Suppose now that p is the probability that a call is generated in the time it takes to serve an average call (holding time). Then np is the number of calls which arrive during a holding time and is just the traffic intensity, U, which we have been discussing all along. Suppose further that n gets very large and that p correspondingly small so that U remains the same. As n becomes infinite,

$$\frac{n!}{k!(n - k)!} p^k \quad \rightarrow \quad \frac{U^k}{k!}$$

and,

$$(1 - p)^{n-k} \rightarrow e^{-U}$$

Thus,

$$P_k = e^{-U} \frac{U^k}{k!}$$ (2.121)

This is indeed a probability distribution since

$$\sum_{k=0}^{\infty} \frac{U^k}{k!} = e^U \tag{2.122}$$

So the sum of all the P_k is 1.

This assumes that calls arrive independently from an infinite population of sources. Thus, the probability of k calls arriving within a single holding time is P_k, as defined earlier. In fact, for any finite value of the time interval, t, since p varies linearly with the t, we have

$$P(k, t) = e^{-Ut} \frac{(Ut)^k}{k!} \tag{2.123}$$

where t is measured in units of holding times. For example, if the holding time is 3 minutes and the length of the interval is 12 minutes, then t is 4.

The lost-calls-held assumption and Little's Formula tell us that the number of messages in the system is equal to the number of arrivals during a holding time. Thus, the number of calls in the system, both in service and waiting, follows a Poisson distribution. Another way of seeing this is to note that the lost-calls-held assumption gives rise to an M/M/∞ queueing system (although only m of the servers are real) and for such a system

$$(P_0)^{-1} = \sum_{k=0}^{\infty} \frac{U^k}{k!} = e^U \tag{2.124}$$

and P_k is a Poisson distribution as defined above.

All servers are busy if k, the number in the system, is greater than or equal to m, the number of servers. Thus, a call is blocked with probability equal to

$$P = \sum_{k=m}^{\infty} P_k = 1 - \sum_{k=0}^{m-1} P_k = 1 - \sum_{k=0}^{m-1} e^{-U} \frac{U^k}{k!} \tag{2.125}$$

This is the Poisson Formula for loss, given the traffic intensity, U, and the number of servers, m.

Example 2.11. Suppose traffic with intensity 2 Erlangs is offered to 3 lines. What is the probability of loss predicted by the Erlang-B, Erlang-C and Poisson formulas?

$$U = 2; m = 3$$

$$\text{Erlang-B} = 0.2105$$

$$\text{Erlang-C} = \frac{0.2105}{1 - 0.667(1 - 0.2105)} = 0.4444$$

$$\text{Poisson} = 1 - (P_0 + P_1 + P_2)$$

$$= 1 - e^{-2}(1 + 2 + 2)$$

$$= 0.3231$$

Thus we see that, as expected, the Poisson Formula gives results intermediate between the Erlang-B and Erlang-C formulas.

2.4.3.3 EXTENDED ERLANG-B METHOD.

Another possibility is to assume that callers retry with probability p and that the retry traffic simply joins the original offered load. This is illustrated in Fig. 2.19. The entire traffic stream is assumed to be a Poisson process and the Erlang-B function is then used to determine the probability of blocking, which in this case is different from the probability of loss. The probability of loss is equal to the probability of a call being blocked and not retrying. Note that a call can be blocked more than once and continue to retry. It is only lost if, after being blocked on some particular try, it then chooses not to try again. Thus, if B is the probability of being blocked on any given try and p is the probability of retrying, then L, the probability of loss, is given by

$$L = B(1-p) + BpB(1-p) + (Bp)^2 B(1-p) + \cdots + (Bp)^k B(1-p) + \cdots \quad (2.126)$$

The first term corresponds to the case where the call is blocked once and chooses not to retry; the next term corresponds to the case where the call is blocked, retries, is blocked again and then does not retry, etc. Thus,

$$L = B(1-p) \sum_{k=0}^{\infty} (Bp)^k = \frac{B(1-p)}{1 - Bp} \quad (2.127)$$

Note that if p is one (i.e., if blocked calls always retry), then L is zero and the call will eventually be carried.

Thus, there are two figures of merit, B and L. B is the probability of a call being blocked on its first (or any) try. It represents the fraction of time the system fails to carry a call when the caller originally wanted it placed. L is the probability of the system not carrying the call at all. B is usually the figure of merit used and is closer to all the other models presented so far.

To use this approach, determine the total offered load, which includes both original offered load and retries. This is done iteratively. Given m servers and an

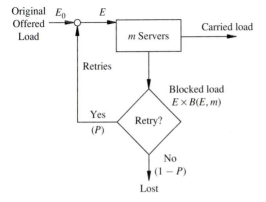

FIGURE 2.19
Extended Erlang-B model.

offered load of E_0 Erlangs, the Erlang-B function is used to determine the amount of blocked (overflow) traffic, A_0. Thus,

$$A_0 = E_0 B(E_0, m)$$

Then form the next estimate of the offered load, E_1, as the sum of the original offered load and the blocked traffic that retries. Thus,

$$E_1 = E_0 + pA_0$$

Now form a new estimate of the blocked traffic,

$$A_1 = E_1 B(E_1, m)$$

and a new estimate of the offered load

$$E_2 = E_0 + pA_1$$

In general,

$$A_k = E_k B(E_k, m)$$

and

$$E_{k+1} = E_0 + pA_k$$

Note that pA_k is added to E_0, not E_k. Because E_{k+1} is somewhat larger than E_k, A_{k+1} is somewhat larger than A_k, but as this iteration is continued and the increments in A_k and E_k get smaller and smaller and the process converges, provided that the carried load remains below m. The only time this does not happen is if the retrial probability is 1 and the original offered load is greater than or equal to m, in which case we certainly want to increase m. From the point of view of the analysis, we would terminate the iteration and return a value of 1 for the blocking probability. There are several ways of testing for convergence, one of which is to test if

$$\frac{E_{k+1} - E_k}{E_k} < \delta$$

where δ is a prespecified tolerance (say, 1 percent). If so, then the process has converged.

> **Example 2.12.** Suppose 0.2 Erlangs of traffic is offered to a single line and that 50 percent of the blocked calls retry. What are the probabilities of blocking and loss predicted by the Extended Erlang-B Method?
>
> For a single line, the Erlang-B function gives
>
> $$B(E, 1) = \frac{E}{1 + E}$$
>
> Compute successive iterative estimates of E_k (offered load), B_k (probability of blocking), A_k (overflow traffic), and R_k (retrying traffic). All calculations are carried out to three decimal places
>
> $$E_0 = 0.2 \quad B_0 = 0.167 \quad A_0 = E_0 B_0 = 0.033 \quad R_0 = 0.5 A_0 = 0.017$$

$$E_1 = E_0 + 0.5A_0 = 0.217$$

$$B_1 = 0.178A_1 = 0.038R_1 = 0.019$$

$$E_2 = 0.219B_2 = 0.180A2 = 0.039R2 = 0.019$$

$$E_3 = 0.219 = E_2B_3 = B_2$$

The iteration has converged to within the tolerance of the calculations. The effect of retries has raised the blocking probability from 16.7 percent to 17.8 percent. The probability of loss is

$$L = \frac{B(1-p)}{1-Bp} = 0.099$$

and the carried load is

$$C = E_0(1-L) = 0.180$$

This is also obtainable from

$$C = E_3(1-B_3) = 0.180$$

When the iteration converges, these two values should be (about) the same, to within our tolerance.

Table 2.11 gives a comparison among the Erlang-B, Erlang-C, Poisson, and Extended Erlang-B functions. The table shows the probability of blocking when 10 Erlangs are offered to various numbers of servers. As expected, the Erlang-B and Erlang-C functions bound the blocking probabilities from below and above. The columns marked EEB-50 and EEB-100 are the Extended Erlang-B formula for 50% and 100% retries, respectively. As can be seen, even with 100% retries the Extended Erlang-B function predicts less blocking than the Poisson function does in almost all situations.

TABLE 2.11
Comparison of blocking based on retry behavior
Offered load = 10 Erlangs

m	Erlang-B	Erlang-C	Poisson	EEB-50	EEB-100
10	.215	—	.542	.289	—
11	.163	.682	.417	.215	.459
12	.120	.449	.303	.152	.250
13	.084	.285	.208	.102	.142
14	.057	.174	.136	.066	.082
15	.036	.102	.083	.041	.047
16	.022	.057	.049	.024	.026
17	.013	.031	.027	.014	.014
18	.007	.016	.014	.007	.008
19	.004	.008	.007	.004	.004
20	.002	.004	.003	.002	.002
21	.001	.002	.002	.001	.001
22	.000	.001	.001	.000	.000
23	.000	.000	.000	.000	.000

We also note that the differences among these formulae are most pronounced for high blocking probability.

Note also that if we are designing to meet a specific GOS (grade of service, blocking probability) then there is not that great a difference in the results based on the formula used. Thus, for example, to design for 3% GOS, the number of lines would be 16, 18, 17, 16, 16 for Erlang-B, Erlang-C, Poisson, EEB-50, and EEB-100, respectively. To design for a 1% GOS, the numbers would be 18, 19, 19, 18, and 18, respectively. Even at 10% GOS the numbers do not vary that much; 13, 16, 15, 14, and 14, respectively. The conclusion is that we can use whichever model we prefer until the very end of the design process and then determine the network configuration in detail.

2.4.4   Non-Random Traffic

All of the analysis thus far has been carried out under the assumption that arrivals are independent of one another. This is not always realistic, but it is usually very difficult, both to characterize the nature of dependency among arrivals and also to analyze loss and delay in the presence of dependencies. In this section we briefly examine two cases which are comparatively tractable.

Random (Poisson) traffic has the characteristic that the variance of the arrival process is equal to its mean. The variance to mean ratio (VMR) is defined as

$$VMR = \frac{\text{variance}}{\text{mean}} \tag{2.128}$$

Thus, random traffic has a VMR of 1. Traffic with $VMR > 1$ is referred to as rough, and traffic with $VMR < 1$ is called smooth. The statistics of carried and overflow traffic have been worked out [Sys60]; and it was found that the carried traffic is smooth while the overflow traffic is rough (see Fig. 2.20). The precise expressions for the mean and variance of the carried and overflow traffic are:

$$E_c = E(1 - B(E, m)) \tag{2.129}$$

$$V_c = E_c(1 - L_c) \tag{2.130}$$

$$E_o = E B(E, m) \tag{2.131}$$

$$V_o = E_o \left(1 - E_o + \frac{E}{m + 1 + E_o - E} \right) \tag{2.132}$$

where E is the offered load (Poisson)

m is the number of servers

E_c and V_c are the mean and variance of the carried load

E_o and V_o are the mean and variance of the overflow

L_c is the load carried by the last server and is given by

$$L_c = E(B(E, m - 1) - B(E, m))$$

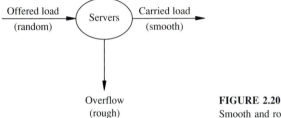

Offered load
(random)

Servers

Carried load
(smooth)

Overflow
(rough)

FIGURE 2.20
Smooth and rough traffic.

Example 2.13. 2 Erlangs of Poisson traffic is offered to three lines.

(a) What are the mean and variance of the carried and overflow traffic?

$$B(E, m) = B(2, 3) = 0.211$$

$$B(2, 2) = 0.4$$

$$E_c = 2(1 - 0.211) = 1.578$$

$$E_o = 2(0.211) = 0.422$$

$$L_c = 2(0.4 - 0.211) = 0.378$$

$$V_c = 1.578(1 - 0.378) = 0.982$$

$$V_o = 0.422x \left(1 - 0.422 + \frac{2}{3 + 1 + 0.422 - 2}\right) = 0.592$$

(b) What are the $VMRs$ for the carried and overflow traffic?

$$\text{carried load} = \frac{0.982}{1.578} = 0.622 = (1 - L_c)$$

$$\text{overflow} = \frac{0.592}{0.422} = 1.404$$

2.4.4.1 LOSS IN SYSTEMS WITH NON-RANDOM TRAFFIC. The smoother the arriving traffic pattern is, the lower the probability of loss. In the extreme case where the arrival pattern is totally uniform, there would be no loss at all if the traffic intensity were less than the number of servers. Conversely, if the traffic pattern is rough, the probability of blocking for a given traffic intensity and number of servers is higher than it would be for an equal (average) amount of random traffic.

Example 2.14. The problem of analyzing loss with non-random traffic arises when the load carried by, or overflowing from one set of lines is then offered to other sets of lines in a network. This situation is depicted in Fig. 2.21. E_1 Erlangs of random traffic arrives at node s and is destined to node d. It is first offered to the group of m_1 lines connecting s to b. E_3 Erlangs of traffic are carried and the remaining load is blocked. The E_3 Erlangs reach node b and are joined by E_2 Erlangs of random traffic, also destined for d. Together, they form a non-random, smooth load, offered to a group of m_2 lines connecting b to d. Of these, E_6 are lost and the remaining E_4 Erlangs reach d. The load

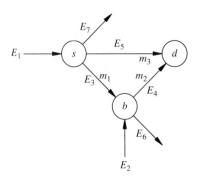

FIGURE 2.21
Loads in a network.

originally blocked at s is offered to a group of m_3 lines directly connecting s and d. Of this load, E_5 Erlangs are carried and reach d. The remaining E_7 Erlangs are lost. Now, estimate the probabilities of blocking in each of these situations and the amount of traffic which reaches d.

The approach taken is to model arriving traffic with mean intensity E and variance V as another distribution for which we can analyze the blocking. Although two distributions with the same mean and variance are not the same distribution, modeling load, by taking both its mean and variance into account, is better than just modeling the traffic by its mean (which is what we would do if we assume it was random.)

It is possible to model traffic with VMR less than 1 as a binomial distribution. When the binomial distribution was first looked at it was found that a binomial distribution with n sources and probability p of an arrival from a source has a mean arrival rate of np and a variance of $np(1 - p)$. Thus, the VMR is $(1 - p)$, which is less than 1. In the limit, as $n \to \infty$ and $p \to 0$, we have a Poisson distribution with a VMR of

$$1 - p = 1 - 0 = 1$$

which is the VMR expected in this case.

The probability of loss for a binomial distribution is the probability of more than m arrivals (in 1 service time), that is,

$$B = \sum_{m+1} n P_k \qquad P_k = 1 - \sum_{k=0} m P_k \tag{2.133}$$

where P_k, as in the binomial distribution examined before, is given by

$$P_k = \frac{n!}{k!(n - k)!} p^k (1 - p)^{n-k} \tag{2.134}$$

For VMR's greater than 1, there is also a negative binomial distribution, which has no simple physical analogy like the binomial distribution, but nevertheless extends the model to rough traffic. The negative binomial distribution has the distribution

$$P_k = \frac{(k + n - 1)!}{(n - 1)!} p^k (1 - p)^n \tag{2.135}$$

Note that unlike the binomial distribution, here P_k has a non-zero value for all $k > 0$. In a manner similar to that above, the probability of blocking can be estimated as the probability of more than k arrivals.

Given m servers, the probability of blocking is then just the probability that there are $m + 1$ or more arrivals. Thus,

$$B = \sum_{k+1}^{n} P_k \qquad (2.136)$$

where P_k is either a binomial or negative binomial distribution, as appropriate. The values of p and n are set to match the mean and variance of the offered load. Specifically, we have already seen that the mean, u, variance, v, and variance to mean ratio VMR of a binomially distributed random variable are given by

$$u = np \qquad v = np(1 - p) \qquad VMR = 1 - p \qquad (2.137)$$

For the negative binomial,

$$u = \frac{np}{1 - p} \qquad v = \frac{np}{(1 - p)^2} \qquad VMR = \frac{1}{1 - p} \qquad (2.138)$$

Thus, given the offered load with mean u and variance v, first form the VMR to choose whether to use the binomial or negative binomial distribution and then find the n and p which match u and v best.

Note that as n gets large and p gets small, both of these distributions approach the Poisson distribution (and one another) from opposite sides and as expected, the VMR approaches 1. Therefore, we have a model of arrivals spanning all possible VMR's. It is a Poisson model for $VMR = 1$. It becomes more and more "binomial" as the VMR decreases, and becomes more and more "negative binomial" as the VMR increases.

Another model, actually more widely used, assumes that when the VMR is less than 1, there is a finite population of sources, and the arrival rate is proportional to the number of idle sources. This leads to the Engset formula [Min74]

$$B = \frac{\frac{n!}{m!(n-m)!} b^m}{\sum_{k=0}^{m} \frac{n!}{m!(n-m)!} b^k} \qquad (2.139)$$

where b is the offered traffic per idle source and is given by:

$$b = \frac{p}{1 - h(1 - B)} \qquad (2.140)$$

Note that B and b must be evaluated iteratively starting with an estimate for b (say $b = p$). Note also that $B = 0$ for $m \geq n$.

Similarly, if $VMR < 1$, the arrival rate can be assumed to be proportional to the number of busy sources. This gives rise to the formula (sometimes known as the negative binomial loss formula):

$$B = \frac{\frac{n!}{m!(n-m)!} b^m}{\sum_{k=0}^{m} \frac{n!}{m!(n-m)!} b^k} \qquad (2.141)$$

This formula too must be evaluated iteratively. Unlike the Erlang formulas, this formula cannot be used incrementally to easily determine an appropriate value m during a design. Instead, it must be evaluated over and over again for different m. Also, the variances of the carried and overflow traffic are very difficult to evaluate precisely. For all these reasons, these formulas are not widely used. Instead, the Wilkenson Equivalent Random Theory [Wil56], described in the next section, is more widely used.

2.4.4.2 THE WILKENSON EQUIVALENT RANDOM METHOD. This method outlined in [Wil56] proposed that instead of dealing directly with the analysis of non-random overflow traffic, that such traffic be modeled by an equivalent amount of random traffic offered to an appropriate group of lines. Look at Fig. 2.20 again, this time from a different perspective, and think of any offered load, with mean E and variance v as overflow from a random load, $E[e]$, being offered to a group of m lines. E_e is called the equivalent random load and m_e the equivalent lines. If this overflow traffic is offered to another group of m lines, the probability of loss can be computed using the Erlang-B function.

Thus, we have an offered load with mean E and variance v, which may or may not be overflow from other groups of lines. Now, find the mean and variance of the overflow when this load is offered to a group of m lines. First determine E_e and m_e, using the methods described as follows, and then use the Erlang-B function to compute the mean, E_1 of the overflow traffic, that is,

$$E_1 = EB(E_e, m_e + m)$$

This approach relies on the observation that offering traffic to m_1 lines, and then offering the overflow to m_2 lines is the same as offering the original load to $m_1 + m_2$ lines. This is valid because we can think of the traffic being offered first to the group of m_1 lines and then, if necessary, to the rest. While this point of view affects the utilization of the lines in the individual groups, it does not affect the blocking probability for the traffic because from the point of view of the traffic, the lines are identical.

It is not immediately clear that just because we are matching the mean and variance of the offered load with another distribution, which has the same mean and variance, that it is possible to accurately estimate blocking probabilities. Wilkenson verified that this approach is reasonable, both via analysis and extensive experimentation, which are reported in the previous reference. The method has been used extensively over the past 50 years and holds up well in practice.

This method allows us to cascade and merge traffic sources in networks. Since means and variances add for independent sources, we can create equivalent streams of traffic by adding the means and variances of the individual sources. Therefore, we have traffic offered first to a group of lines. The overflow from this group, along with random traffic present, attempts to use other groups. An example of this is shown in Fig. 2.21. Additional examples follow this discussion.

Note that this method is meant to deal with overflow traffic, where the variance is greater than the mean and all the computations are set up for it. Thus, if we add carried traffic to random or overflow traffic, and the variance of the resultant process is less than the mean, the following formulas do not work. If we are willing to do a conservative analysis, we can treat such traffic as random.

A method to determine $E[e]$ and $m[e]$ was developed by [Rap64]. Begin with a load with mean E and variance v. We compute the VMR, z, as v/E. Note that z must be no less than 1. Rapp's Approximation then gives

$$E[e] = v + 3z(z - 1) \tag{2.142}$$

$$m[e] = E[e]\frac{E + z}{E + z - 1} - E - 1 \tag{2.143}$$

Notice that for random traffic that $v = E$, $z = 1$, $E_e = E$ and $m_[e] = 0$, all as expected. Also note that $m[e]$ is, in general, non-integer. The Erlang-B function can be extended to the case of non-integer m; it then becomes a gamma function, which can be evaluated [FK78] using the approximation:

$$S = 0.711093(1 + 0.415775/E)^f + 0.278518(1 + 2.29428/E)^f$$
$$+ 0.010389(1 + 6.28995/E)^f \tag{2.144}$$

$$B(E, f) = \frac{1}{S} \tag{2.145}$$

where f is a fraction between 0 and 1. Note that when $f = 0$, it yields 1, and when $f = 1$, it yields $E/(1 + E)$, as expected. This gives $B(E, m)$ for $0 < m < 1$. We can then find $B(E, m)$ for any m using the recursive expression discussed before,

$$B(E, m) = \frac{EB(E, m)}{EB(E, m) + m + 1} \tag{2.146}$$

which holds for (positive) non-integer m as well as integers.

An alternative to this approach is to compute an integer estimate for $m[e]$ and then adjust $E[e]$ accordingly. Specifically,

$$m[e]' = \lfloor m[e] \rfloor \tag{2.147}$$

$$E[e]' = \frac{(m'_e + E + 1)(z + E - 1)}{(z + E)} \tag{2.148}$$

For many situations of interest, z is close to 1 and the above approximations are fine. In fact, if z is really close to 1, we can ignore the fact that the traffic is non-random and simply use the Erlang-B function. If z is greater than 1.6 (which it rarely is in cases of interest), we can get a somewhat better estimate of $E[e]$ from:

$$p = \frac{z}{2(E + z)} \tag{2.149}$$

$$c = \left(\frac{3(z + 6)(z - 1.5)}{20E}\right)^p \tag{2.150}$$

$$E_e = v + (2 + c)z(z - 1) \qquad\qquad (2.151)$$

This is the same as Rapp's approximation with $2 + c$ in place of 3 for $z > 1.6$. The calculation of $m[e]$ is the same as before.

> **Example 2.15.** 5 Erlangs of random traffic is offered to a group of three lines (see Figure 2.22). 9 Erlangs of random traffic is offered to another group of four lines. The overflow from both groups is offered to a group of x lines. We wish to determine x so that no more than 1 percent of the original offered load is blocked by the final group of lines.
>
> Using the Erlang-B function the mean of the overflows from the first two groups is found to be 2.65 and 5.52 Erlangs, respectively. From the formula for variance, the overflows are 3.67 and 7.62. Since the overflows from the two groups are independent, the means and variances can be added to determine the mean and variance of the load offered to the final group. Thus, we have an offered load with mean 8.17 and variance 11.29. z is 1.38. From Rapp's approximation, we get $E[e] = 12.88$ and $m[e] = 5.21$. (These numbers are obtained using an automated implementation which used more than two significant figures; if the calculation is performed manually, the results are 12.87 and 5.24, which are also reasonable approximations.) If we choose to round down, these numbers become 12.69 and 5.00 respectively.
>
> All of these approximations are reasonable. If 12.88 Erlangs go to 5.21 trunks, the mean and variance of the overflow traffic are 8.17 and 11.39, respectively. If 12.69 Erlangs of traffic go to 5 trunks, the mean and variance of the overflow are 8.17 and 11.28, respectively.
>
> We now want to determine the smallest value of x such that at most 1 percent of the offered load is lost. There are several ways of interpreting this requirement. We can insist that at most 1 percent of the original load (i.e., 1 percent of the 14 Erlangs of original offered traffic) is lost. Or, we can require that at most 1 percent of the offered load from each of the two original loads (considered separately) be lost. Or, we can insist that at most 1 percent of the load offered to the final group is lost. Each of these alternatives is reasonable. The last is most conservative. The first is most liberal. The second is probably fairest.
>
> We begin with the first alternative and require that the overflow from the final group of x lines be no more than 0.14 Erlangs, which is a probability of blocking of 0.011. Using the Erlang-B function, we find that if 12.69 Erlangs are offered to 20 trunks, the probability of loss is 0.015 percent, and is 0.009 if it is offered to 21 trunks. Therefore, we require 21 trunks, including the 5 trunks in the equivalent group. So, x is 16 lines.
>
> What if we had ignored the variance of the overflow traffic and treated the overflow as random traffic? We could then offer 8.17 Erlangs to x trunks, requiring that at most

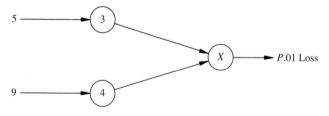

FIGURE 2.22

0.14 Erlangs be blocked, which is a probability of 0.017. Using the Erlang-B function, we need 15 trunks. Thus, Wilkenson's method requires one trunk more. This is typical. The Wilkenson method is more conservative, but there is rarely a great difference in outcome.

Suppose we provide separate groups of trunks for each of the requirements. The 5 Erlang requirement then needs 11 lines, and the 9 Erlang requirement needs 17 trunks. Thus, a total of 28 trunks is required to handle the two requirements separately.

Compare this with allowing the two requirements to share the final group of 16 lines. A total of 23 lines (16+3+4) are required. This is a significant savings. There is a catch, however, to make use of this approach, we have to get all the overflow traffic to one place in order to offer it to the final group. This approach makes the most sense if the original two requirements come from locations that are near one another and are destined for someplace far away. In this case we can transport the overflow from one location to another, over a large but inexpensive group of lines, with very low loss. The combined load then can be offered to the final group of lines.

EXERCISES

2.1. Give an example of a distribution where the variance is larger than the mean.

2.2. Verify that the sum of the $P(n, k, p)$ from 0 to n is 1 for $n = 5$ and $p = 1/3$.

2.3. Verify that the variance in the number of heads in three flips is indeed 0.75 using the definition of V and the probabilities of 0, 1, 2, and three heads.

2.4. (*a*) What is the probability of a message having a length greater than 1000 characters, in a system with exponential message lengths with a mean of 200 characters?

(*b*) What is the probability of the length being between 100 and 300 characters long?

2.5. A 19200 bps (2400 char/sec) line has nine terminals on it. Each terminal generates 10 messages per minute. 4 terminals generate 100 character messages, three terminals generate 400 character messages, and two terminals generate 1600 character messages.

(*a*) What is the utilization of the line?

(*b*) What is the average service time?

(*c*) What is the average waiting time?

(*d*) What is the average time in the system?

(*e*) Suppose that the messages are packetized before they are transmitted, and that the packetization process takes no time and adds nothing to the length of the messages. Answer (*a*) through (*d*) above again where the times are now for packets instead of messages.

(*f*) Repeat (*e*) assuming that the packetization takes 25 ms and adds a 10 character header to each packet. (The time in the system now includes packetization time.)

2.6. There are three classes of messages in a system:

batch (low priority); 1000 characters; 10 msg/sec

interactive (medium priority); 100 characters; 100 msg/sec

control (high priority); 20 characters; 10 msg/sec

(*a*) What is the average time in the system for each type of message? What is the average time in the system overall?

(*b*) What is the average time in the system if no priorities are used? Why is this different than in part (*a*)?

2.7. Verify Equation 2.56. The proof follows along the same lines as for the M/M/1 queue.

2.8. Suppose that when an M/M/2 system is empty an arrival is always given to server 1. Draw the state transition diagram, distinguishing between state 1,0 (where server 1 is busy and server 2 is idle) and state 0,1 (where server 1 is idle and server 2 is busy. Set up the balance equations, and find the probabilities of being in states 1,0 and 0,1. Show that these probabilities sum to P_1 and that none of the other state probabilities have changed.

2.9. Find n and p as functions of u and v for the binomial and negative binomial distributions.

2.10. A small bank has two tellers, one fast and one slow. The fast teller serves one customer every 2 minutes. The slow teller serves one customer every 5 minutes. The service times of both tellers are exponentially distributed. There is only enough room in the bank for four customers at any given time, including the ones being served. Customers enter the bank and take service from the fastest, idle teller currently available. If neither teller is idle and there is room to wait, they wait and take service from the first teller who becomes idle. If there is no room to wait, they leave. Customers arrive independently at a rate of one customer every 4 minutes.

(*a*) Draw a state transition diagram for this system.

(*b*) What is the probability a customer leaves without being served?

(*c*) What is the average service time for customers who are served?

(*d*) What is the average waiting time for customers who are served?

BIBLIOGRAPHY

[All78] Allen, A. O.: *Probability, Statistics and Queueing Theory*, Academic Press, New York, 1978.

[BG87] Bertsekas, D., and R. Gallager: *Data Networks*, Prentice-Hall, Englewood Cliffs, NJ, 1987.

[Coo72] Cooper, R. B.: *Introduction to Queueing Theory*, Macmillan, New York, 1972.

[Fel57] Feller, W.: *An Introduction to Probability Theory and Its Applications*, John Wiley, New York, 1957.

[FK78] Farmer, R. F., and I. Kaufman: "On the numerical evaluation of some basic traffic formulae," *Networks*, 8:153–186, 1978.

[Jac57] Jackson, J. R.: "Networks of waiting lines," *Oper. Res.*, 5:518–521, 1957.

[Kle67] Kleinrock, L.: *Communications Nets: Stochastic Message Flow and Delay*, Dover, New York, 1964.

[Kle75] Kleinrock, L.: *Queueing Systems, Volume 1: Theory*, Wiley-Interscience, New York, 1975.

[Lit61] Little, J.: A proof of the queueing formula 1 = w, *Oper. Res.*, 18:172–174, 1961.

[Liu90] Liu, H.: "Telecommunication Network Design with Bulk Facilities: The T1 Problem," unpublished paper, 1990.

[Min74] Mina, R.: *Introduction to Teletraffic Engineering*, Telephony Publishing Co., Chicago, 1974.

[Pap65] Papoulis, A.: *Probability, Random Variables and Stochastic Processes*, McGraw-Hill, New York, 1965.

[Rap64] Rapp, Y.: "Planning of junction networks in a multi-exchange area," *Ericson Technics*, 1:???, 1964.

[Sch77] Schwartz, M.: *Computer Communication Network Design and Analysis*, Prentice-Hall, Englewood Cliffs, NJ, 1977.

[Sys60] Syski, R.: *Introduction to Congestion Theory in Telephone Systems*, Oliver and Boyd, London, 1960.

[Wil56] Wilkenson, R. I.: "Theories for toll traffic engineering in the USA," *BSTJ*, 35:421–514, 1956.

CHAPTER 3

MODELING NETWORKS AS GRAPHS

3.1 GRAPH TERMINOLOGY

In this section the basic terminology is introduced for describing networks, graphs, and their properties. Although graph theory is a well-established discipline, there are, unfortunately, several different accepted terms for many of its basic concepts. Also, yet other terms are often used to describe networks as graphs. The terminology given here is just one of a number of possible accepted sets of terms and is used consistently throughout the remainder of this book.

A **graph**, G, is defined by its vertex set, V, and its edge set, E. The vertices are more commonly called **nodes** and these represent locations (e.g., sources of traffic or sites of communications equipment). The edges are also called **links**, and these represent communication facilities. This is expressed:

$$G = (V, E)$$

Figure 3.1 shows an example of a graph.

While in theory, V can be empty or infinite, V is usually a non-empty finite set, that is,

$$V = \{v_i \mid i = 1, 2, \ldots, N\}$$

where N is the number of nodes. Similarly, E is denoted by

$$E = \{e_j \mid j = 1, 2, \ldots, M\}$$

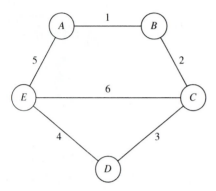

FIGURE 3.1
A simple graph.

A link, e_j, corresponds to a connection between a pair of nodes. This is sometimes noted by a link, e_j, between nodes i and k as

$$e_j = (v_i, v_k)$$

or more simply as

$$e_j = (i, k)$$

A link is **incident on a node** if the node is one of its endpoints. Nodes i and k are **adjacent** if there exists a link, (i, k), between them. Such nodes are also referred to as **neighbors**. The **degree** of a node is the number of links incident on the node or, equivalently, the number of neighbors it has. These notions are equivalent for ordinary graphs. However, for graphs with more than one link between the same pair of nodes there can be some confusion. In this case, degree is defined in terms of the number of links.

Sometimes the link is bidirectional. In this case the order of the nodes in the link does not matter. Other times, the order does matter. If the order of the nodes does matter, the link is referred to as an **arc** and denoted by

$$a_j = [v_i, v_k]$$

or more simply as

$$a_j = [i, k]$$

k is said to be **outwardly adjacent** to i if an arc $[i, k]$ exists and the **out-degree** of i is the number of such arcs. The notions **inward adjacency** and **in-degree** are defined in a similar way.

A graph is called a **network** if the links and nodes associated with it have properties (e.g., length, capacity, type, etc.). Networks are used to model problems in communications, and the specific properties of the nodes and links relate to the specific problem at hand.

The difference between links and arcs is important, both in terms of how they model networks and how they function within algorithms, so the difference should always be clear. Graphically, links are lines connecting pairs of nodes. Arcs are lines with an arrow at one end, indicating the direction of the arc.

A graph with links is called an **undirected graph** while one with arcs is called a **directed graph**. Some directed graphs also include undirected links. Usually, graphs are assumed to be undirected, or that the difference does not matter.

It is possible for there to be more than one link between the same pair of nodes. This can, for example, correspond to multiple communication channels between two switches. Such links are referred to as **parallel links**. A graph with parallel links is called a **multigraph**.

It is also possible for there to be a link between a node and itself. These links are called **self loops**. This is relatively rare, but might result from treating two nodes as one in modeling a network, or may arise during the course of an algorithm which merges nodes. Figure 3.2 illustrates a graph with parallel links and self loops. A graph without parallel links or self loops is called a **simple graph**. Simple graphs are easier to represent and manipulate, so it is assumed the graphs dealt with here are simple. Exceptions to this rule are indicated when they occur.

A **path** in a network is a sequence of links which begins at some node, s, and ends at some node, t. Such a path is also referred to as an **s,t-path**. Note that the order of the links in the path matters. A path can be directed or undirected depending on whether its components are links or arcs. A path is said to be simple if a node appears no more than once in the path. Note that a simple path in a simple graph can be specified by the sequence of nodes it contains since a unique sequence of links is specified by such a node sequence.

If s is the same as t, the path is called a **cycle**, and if an intermediate node appears no more than once, the cycle is said to be a **simple cycle**. A simple cycle in a simple graph can also be specified by a node sequence.

Example 3.1. In Fig. 3.1, links 1, 2, and 3 form a simple A-D path; links 1, 2, 6, and 5 form a simple cycle; and links 4, 3, 6, and 5 form a non-simple E-A path.

In a communications network a channel is said to be **full duplex** if it can be used in both directions at the same time. It is said to be **half duplex** if it can be used in only one direction at a time. Occasionally, as with satellite networks, a channel

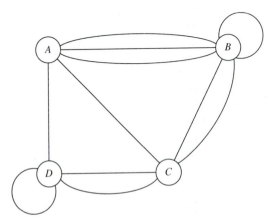

FIGURE 3.2
Graph with parallel edges and self loops.

can have capacity in only one direction. Clearly, if network channels have capacity in only one direction, they are modeled as arcs. Half duplex links are usually modeled as pairs of arcs. Full duplex links can be modeled as links with two capacities (two separate properties), or as a pair of oppositely directed arcs between the same pair of nodes. This is often an important decision which can have an effect on both the simplicity and effectiveness of the algorithm used to solve the problem. It is also possible, of course, to convert from one form to another if necessary, but it's best to try to avoid this as it complicates implementation and increases runtime and storage requirements.

A graph is **connected** if there is at least one path between every pair of nodes. The sets of nodes with paths to one another are **connected components**, or more simply, **components**. The links between these nodes are also part of the components. A connected graph has a single component.

Notice that if there is a path from i to j there is also a path from j to i. Also, if there is a path from i to j and a path from j to k there is also a path from i to k. By definition, there is always a path from i to i. Thus, components form equivalence classes over the set of nodes in a graph. Each node is a member of exactly one component. Also, each link is a member of exactly one component. Therefore, the component structure of a graph can be described as a partition on its node set or link set.

A directed graph with a directed path *from* every node *to* every other node is called **strongly connected**. Connectivity in directed graphs is not symmetric. There may be a directed path from i to j without there being one from j to i. A set of nodes with directed paths from any one node to any other is called a **strongly connected component**. Note that again a node is part of exactly one strongly connected component but that an arc can be part of at *most* one. Specifically, some arcs may not be part of any strongly connected component.

Example 3.2. Consider the directed graph in Fig. 3.3. The strongly connected components are defined by the nodal partition

$$\{ABCD\} \; \{EFG\} \; \{H\} \; \{I\} \; \{J\}$$

The arcs (A, H), (D, I), (I, J) and (J, G) are not part of any strongly connected component. Considered as an undirected graph (i.e., treating the arcs as undirected links), the graph has a single component, and is a connected graph.

Given a graph, $G = (V, E)$, H is a **subgraph** of G if $H = (V', E')$, where V' is a subset of V and E' is a subset of E. These subsets may or may not be proper.

A **tree** is a graph without cycles. A **spanning tree** is a connected graph without cycles. Such a graph is refered to more simply as a tree. If the graph is not necessarily connected, it is referred to it as a **forest**. We generally speak of trees in undirected graphs.

In directed graphs, there is an analogous structure, called an **arborescence**. An arborescence is a directed graph which forms paths from one node (called the **root of**

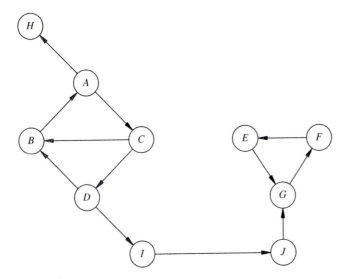

FIGURE 3.3
Directed graph.

the arborescence) to all other nodes; alternatively, the paths might be from all other nodes to the root. An arborescence would be a tree if it were undirected.

Spanning trees have many interesting properties which make them useful in designing communications networks. First, they are minimally connected, that is, they are connected graphs but no subset of the edges in a tree forms a connected graph. Thus, if the objective were to simply design a connected network of minimum cost, a tree would be the optimal solution. Closely related to this is the fact that there is exactly one path between every pair of nodes in a tree. This makes routing a trivial problem in trees and greatly simplifies the communications equipment involved.

Note that if a graph has N nodes, any tree spanning the nodes has exactly $N - 1$ edges. Indeed, any forest with k components contains exactly $N - k$ edges. This can be seen by noting that a graph with N nodes and no edges has N components, and each edge added connects two previously unconnected components thereby reducing the number of components by one.

A set of edges whose removal disconnects a graph (or, more generally, increases the number of its components) is called a **disconnecting set**. A disconnecting set which partitions the set of nodes into two sets, X and Y, is called a **cutset** or sometimes an **XY-cutset**. The most concern is with minimal cutsets (i.e., cutsets which are not subsets of other cutsets). In a tree, any edge is a minimal cutset. A minimal set of nodes whose removal partitions the remaining nodes into two sets is called a **cut**. Again, the interest is usually with minimal cuts.

Example 3.3. Figure 3.4 shows an undirected graph. The sets of links

$$\{(A, C), (B, D)\}$$

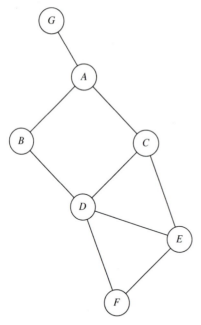

FIGURE 3.4
Cutsets, cuts, trees.

and

$$\{(C, E), (D, E), (E, F)\}$$

are examples of minimal cutsets. The latter set is an example of the fact that the set of all links incident on any node is a cutset separating that node from all the others.

The set of nodes $\{C, D\}$ is a cut. The node A by itself is also a cut. A single node whose removal disconnects the graph is called an **articulation point**.

The set of links

$$\{(A, B), (A, C), (A, G), (C, D), (C, E), (E, F)\}$$

is a tree. Any subset of this set, including the entire set and the empty set, is a forest.

3.2 REPRESENTATION OF NETWORKS

Our purpose here is to implement algorithms for the design and analysis of networks modeled as graphs. To do so, these graphs must be represented in a computer. This involves the area of **data structures**. There are entire texts devoted to this topic and the interested reader can find more information in [Knu73], [Smi87], [Sta80]. Here, are the basic issues and alternatives and as specific algorithms are presented, more detail is given.

The goal is to use the simplest data structures that do the job efficiently, not to squeeze out a small amount of runtime or storage efficiency by implementing a complex and highly specialized structure. By keeping the structure simple, the implementation of the algorithms is easier and more reliable.

Another goal is to implement complete network design tools that include many algorithms working on the same network. Each algorithm requires a representation of the network. Ideally, all the algorithms would use the same representation and no conversion would be necessary. In practice, some conversion is necessary. However, by sticking to simple structures, the amount of conversion can be minimized.

3.2.1 External Representation

External representation is what the user sees in inputs and outputs. It is more a database than a data structure and the network is resident on disk rather than in memory. The principal goal is to represent information rather than relationships, between the data objects. Space is not as critical a constraint for external representation as it is for internal representation.

To begin, choose an input file format which contains the data as clear text (i.e., it is not a database with embedded pointers or cross referencing, and it contains no embedded formatting information except carriage returns and linefeeds). This enhances the portability of the files and facilitates retrieving data from them.

There are many possible approaches to organizing the actual file. One successful approach is to choose a simple, record-oriented structure. Each node, link, requirement, device, and parameter has a record associated with it giving information about the object. In some cases, there may be several records associated with an object, but the number is fixed. The format of these records is columnar but flexible.

The information appears in columns, and all records of a particular type in a particular input file have the same columns in the same order. However, the number of columns in each record, the order of the columns, and their width can vary from one input file to another. The records in a particular input file can include columns which the tool ignores.

This flexibility allows the tool to make use of files which were originally created for other purposes, without having to modify the file. Thus, it is easy to obtain machine-readable data from external sources. Even reports from existing software systems may be used as input sources. It is also easy to use text editors and spreadsheets in data preparation and editing, and to read the files and check them for completeness and consistency. By minimizing the difference between input and output formats, it is possible to chain processes together and to interface network design tools to other systems.

In order for the tool to accept input in this flexible form, each part of the tool must also have a "dictionary" that identifies the columns it uses. This dictionary lists the column headers that the tool uses, and allows the tool to extract required information, ignoring data in columns not listed in the dictionary. For example, the section of the file containing node information might have a column header of

```
NODE_ID  CITY_NAME____    STATE LATA   STREET_ADDRESS____
```

Each record in the node table has its data under these headers. The dictionary for the tool might include only the NODE_ID and LATA in its dictionary. The tool looks for these headers and extracts the needed data from the columns beneath them, ignoring data in other columns. In another file, this data can be in different columns, the fields can be in a different order, and the headers even be of different widths. (The underscore characters are used to extend the fields.) The tool simply keys off the header.

Even with this flexibility, it is sometimes necessary to accommodate variations in the input format. Requirements data can be available from several sources in different formats. In such cases, take what is available and convert it to a single uniform format at the earliest possible time. This process can be thought of as external to the network design tool itself. It is usually better to use a good text editor with macro capability and global search and replace functions to do this job rather than building the capabilities into the tool itself. Think of the input to the network design tool as a file with a single format of records for each type of object.

In this representation, each location has a brief identifier, typically between three and six characters in length. These identifiers are used to denote endpoints of links as well as sources and destinations of traffic. The selection of this set of location identifiers is a key to the organization of the database. Note that, unlike node numbers, these identifiers need not change if nodes are added or deleted. Thus, it is possible to edit one part of the database without forcing changes in other parts.

It is possible to map one location set into another. Sources and destinations of traffic can then be identified in existing files by names associated with a specific application. All such names would be converted to a unique location ID. In doing so, important decisions have to be made as to when to consolidate multiple names to a single ID. If there are ten terminals in a room, all attached to the same terminal control unit, they can safely be considered as a single location as long as there is no interest in designing the part of the network connecting the terminals to the terminal control unit. If there are 1000 terminals inside a 50 story office building, however, a decision to consolidate them into a single location prevents us from carefully considering the possibility of placing several nodes inside the building and partitioning the terminals among them.

At the other extreme, if a WATS (Wide Area Telephone Service) network is designed where 800 numbers are used to access whole areas of the country, it may be possible to consolidate all traffic in a telephone exchange, or even an entire area code, into a single location. This dramatically reduces the amount of data required, but it presupposes that there is a great deal already known about the final design.

Another important example is that some sources and destinations of traffic can be completely filtered out and not considered at all in the network design. Thus, in designing a corporate telephone network for a company with all its facilities in the U.S., overseas traffic can simply be ignored. Doing so only reduces the size of the database slightly but it can dramatically reduce the overall effort in maintaining it because there is no need to be concerned with any overseas tariffs.

Note that although we speak of a single location set, it is not necessary that all data be present for all locations. Thus, each floor of a building can be a location, for the sake of exploring the partitioning of terminals among nodes in the building, but at the same time we can choose to keep tariff information on the basis of area code and exchange, which is how it is available from an external source. In this scheme, the cost of a link between two nodes in the same building is approximated or even ignored so that the tariff database could be more easily maintained.

Such decisions are difficult and they often profoundly affect the network design. The amount of data involved is large and the consolidation often involves manual effort. Nevertheless, these decisions have to be made in order to proceed with the design. In practice, the work is done in stages, postponing important decisions as long as possible. If a big mistake is made, it may be necessary to back up and recompress.

The external representation must accommodate all user design decisions. At one extreme, the user could specify the entire design, entering all the nodes, links, and requirements. The tool is then used in an analysis mode to check the feasibility of the network and its performance. Alternatively, the user might wish to specify that a particular location is a mandatory site for a backbone switch or that some other site is forbidden to be a switch site.

Such input is easily represented as type information at each location. However, it is important to distinguish between inputs and outputs. Thus, a user may specify a location as a mandatory switch site. A design algorithm may then designate another site as a switch site. If the algorithm were rerun with different parameters, it might redesignate the second site but it should not redesignate the first. Thus, the database should distinguish between the categories: mandatory, forbidden, current, and not current with respect to the switch site property.

It is important to distinguish between locations and devices. There may be several switches at the same physical location. One possibility is to treat each device as a logical location. The advantage of doing this is that it then becomes clearest which links terminate on which device. This approach, however, requires that the design add and delete locations as devices are added and deleted. An alternative is to simply associate a count of the number of devices with each location.

Which approach we take is a function of how detailed a model of the design we want to maintain. If there is to be a detailed node configuration, it is important to keep track of exactly what links and nodes home on each device. If, on the other hand, only a topological optimization is done, the simpler approach of counting devices suffices.

Similarly, it is possible to keep track of each physical channel or more simply just keep track of logical links. Here again, the level of detail in the representation should match the type of design desired.

A hybrid approach is also possible. Thus, there can be a field containing the count of the number of devices or physical channels, and still record individual devices and channels as separate locations and links if the details are significant in specific instances.

The consequences of selecting each alternative are examined in more detail in later chapters. At this point, simply note that there are important decisions to be made. These decisions are, unfortunately, very difficult to change as they impact the implementation of the network design tools themselves. In practice, many tools are implemented with extremely general external network representations, even to the point of using full scale database management systems for the implementation. This, of course, leads to great flexibility and wide applicability of the tools. Often, this decision, however, also makes the tools much more cumbersome to implement and use. Such tools require more hardware and more running time than they would have had a simpler external representation been chosen.

3.2.2 Internal Representation

The internal representation of the network is aimed at making the network design and analysis algorithms easy to implement and efficient both in terms of their memory requirements and running time. It is possible to have several representations although it is preferable not to have too many because the conversion code complicates implementation. Also, the alternative representations take up space unless dynamic storage allocation is used (which further complicates implementation).

Here we are concerned with the relationships among data objects as well as the ability to compress data to increase efficiency. Readability is not a concern as the network is not edited in the internal representation.

3.2.2.1 NODE AND LINK NUMBERS. Nodes are identified by numbers; conceptually, each node has a number associated with it. This number may be its position in an alphabetical list of node ID's, an index into an array of node properties, or a pointer to a record describing the node. In all cases, this number provides immediate access to information about the node. Note that the node ID itself, while valuable in the external representation, requires some sort of table lookup to actually access any information about the node.

Thus, the first step in creating the internal representation is to associate a number with each node. This is done each time the network is read into memory and the conversion is carried out over the entire database. Thus, unlike the situation with the external representation, there is no concern about the fact that node numbers change as nodes are added and deleted.

There can also be several sets of node numbers depending on the part of the design process currently being developed. For example, many processes (e.g., routing) work only on the backbone nodes. It is possible to then map from the complete node set to the backbone node set and work entirely with backbone node numbering during these processes. Another example is that once the node set has been partitioned into regions, it is possible to work with one region at a time when designing local networks within each region. This can save a considerable amount of time and memory. It is important, however, to be clear which node numbers are being used in each process and, except for processes whose function is to convert from one numbering to another, to work with only one set of numbers at a time.

In a similar way, numbers can be associated with the links and information about the links accessed via these numbers. The endpoints of the links are node numbers. If the node numbering is changed, as described above, the link endpoints must also be converted. This is generally done by converting once, while the network is being read into memory, from node ID's to node numbers, using the complete node set. Then, as subsets of the nodes are extracted, the conversion of link endpoints can be done directly using a table mapping the full node set into the subset.

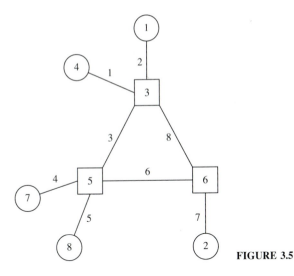

FIGURE 3.5

Example 3.4. Consider the network shown in Fig. 3.5. Two types of nodes are shown in the figure. **Backbone nodes** are nodes which act as via points for traffic, that is, they may carry traffic which neither originates or terminates at them. **Local nodes** are nodes which only handle their own traffic. The squares are backbone nodes and the circles are local nodes homed to (i.e., associated with) the backbone nodes. Node and link numbers are shown. The mapping for the backbone nodes and links is shown in Table 3.1. In this case, only some of the original nodes map into backbone nodes. The other nodes map into 0, a value recognizable as not being a backbone node number. (If we started numbering nodes at 0, we would map such nodes into −1.) This mapping can then be used to convert the links into a backbone link set with endpoints as backbone node numbers. The logic for doing so is simply to pass through the original list of links and extract the links whose endpoints are backbone nodes, and replace the endpoints by backbone node numbers.

Example 3.5. A second type of mapping is also possible. Consider the network shown in Fig. 3.5 again. Suppose now that it is desired to convert a traffic matrix on the original

TABLE 3.1
Backbone link mapping

Node #	Backbone node #	Backbone link endpoints	Original link #
1	0	1 2	3
2	0	2 3	6
3	1	1 3	8
4	0		
5	2		
6	3		
7	0		
8	0		

TABLE 3.2

Homing nodes to backbone notes

Node #	Home
1	1
2	3
3	1
4	1
5	2
6	3
7	2
8	2

node set into a backbone traffic matrix. Use the mapping in Table 3.2 to create the backbone traffic matrix. Table 3.2 gives the home of each node (i.e., the node number of the backbone node to which the node is homed). Backbone nodes home to themselves. Note that this mapping is identical to the one in Table 3.1, except that non-backbone nodes show their home rather than a 0. Indeed, the second mapping could serve both purposes by putting a flag (e.g., a minus sign) on the backbone nodes.

Suppose there is a traffic matrix (i.e., an array containing the point-to-point requirements between each pair of nodes)

$$T = \{t_{ij} \mid i = 1, 2, ..., 8; \quad j = 1, 2, ..., 8\}$$

on the original node set. The new matrix, T_B, on the backbone nodes is

$$T_B = \{b_{km} \mid k = 1, 2, 3; \quad m = 1, 2, 3\}$$

where

$$b_{km} = \Sigma\{t_{ij} \mid H_i = k \quad \text{and} \quad H_j = m\}$$

3.2.2.2 ADJACENCY AND INCIDENCE RELATIONS.

Many algorithms make use of data structures that indicate which nodes are adjacent to others and which links are incident on each node. There are several common ways of representing this information. One of the simplest ways is, for a network with N nodes, to create an N by N boolean matrix with 1s where nodes are adjacent and 0s elsewhere. Similarly, for a network with N nodes and M links, it is possible to create an N by M matrix with 1s in rows i and j of column k if undirected link k connects nodes i and j. If k is an arc directed from i to j, it is possible to place a $+1$ in row i and a -1 in row j of column k of the matrix. Table 3.3 gives the adjacency and incidence matrices for the network in Fig. 3.5.

Notice that for an undirected graph the adjacency matrix is symmetric but that the incidence matrix is not. Indeed, the incidence matrix is not even square unless (as is the case in this example) the number of links happens to equal the number of nodes.

TABLE 3.3
√ **Adjacency and incidence matrices**

	Node		Link 1 2 3 4 5 6 7 8
	0 0 1 0 0 0 0 0		1 0 1 0 0 0 0 0 0
N	0 0 0 0 0 1 0 0	N	2 0 0 0 0 0 0 1 0
o	1 0 0 1 1 1 0 0	o	3 1 1 1 0 0 0 0 1
d	0 0 1 0 0 0 0 0	d	4 1 0 0 0 0 0 0 0
e	0 0 1 0 0 1 1 1	e	5 0 0 1 1 1 1 0 0
	0 1 1 0 1 0 0 0		6 0 0 0 0 0 1 1 1
	0 0 0 0 1 0 0 0		7 0 0 0 1 0 0 0 0
	0 0 0 0 1 0 0 0		8 0 0 0 0 1 0 0 0
	(a) Adjacency Matrix		(b) Incidence Matrix

A matrix is said to be **sparse** if the majority of its entries are 0s. More specifically, it is sparse if the number of non-zero entries is of a lower order than the size of the matrix. The notion of order is discussed more fully in the following section. For the moment, think of something being of order N, denoted by $O(N)$, if it is proportional to N.

An N by N matrix is said to be sparse if the number of non-zero entries is of order less than N^2. Adjacency matrices have exactly two 1s for each link in the √ network. Thus, the adjacency matrix is as sparse as the network itself. The incidence √ matrix, on the other hand, contains exactly two 1s in each column and therefore is always sparse.

Note that the incidence matrix contains more information than the adjacency √ matrix as it tells us not only that a link is incident on a node but also which link is incident on it. Also, an incidence matrix can represent networks with multiple links √ between the same pair of nodes, by allotting multiple columns to the multiple links.

The degree of a node can be computed by adding up the number of 1s in the row of the incidence or adjacency matrix corresponding to that node. In directed networks, the inward (or outward) degree can similarly be computed by summing the number of +1s (or −1s). It is sometimes useful to store the degree of each node, explicitly, to avoid having to recalculate it.

It is possible to represent adjacency and incidence information more concisely than with explicit matrices by maintaining a list of adjacent nodes or incident links.

Example 3.6. Consider the network shown in Fig. 3.6a. The data structure in Fig. 3.6b represents the adjacency and incidence relationships for this network. There is a vector, First_adj, which gives the link number of the "first" link incident on each node. In some cases, first may have a significance. Here it is simply the link with the smallest link number. The links can be numbered based on their lengths; in that case, first would be shortest. In other cases, first might mean nothing at all and can simply be a function of how the adjacency relation was set up.

The endpoints of each link are given by the vectors End1 and End2. Each link, k, has two properties, Next_adj1[k] and Next_adj2[k]. These properties are

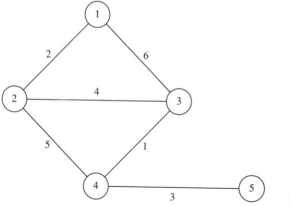

FIGURE 3.6*a*

represented as vectors indexed on the link numbers. Next_adj1[k] is the number of the next link incident on End1[k]; similarly, Next_adj2[k] is the number of the next link incident on End2[k]. Thus, First_adj and the two Next_adjs form a linked list of all the links incident on a node. To traverse this linked list, it must be known whether to follow Next_adj1 or Next_adj2. This is done by keeping track of the node whose adjacency we are following, and as each link, *k*, is reached we check if this node is End1[k] or End2[k].

Unlike the explicit matrix representations, this representation requires a total amount of storage of $O(N + M)$ for a network with N nodes and M links.

A slightly different representation, which also requires order $N + M$ storage, is shown in Figs. 3.6*c* and 3.6*d*. This is a more explicit linked representation where the adjacency can be traversed by following a chain of pointers without checking end-points. This format is easier to work with in languages with built-in list manipulation operations (e.g., LISP). It is essentially equivalent to the data structure given in Fig. 3.6*b*.

NODE	FIRST ADJ
1	2
2	2
3	1
4	1
5	3

FIGURE 3.6*b* √

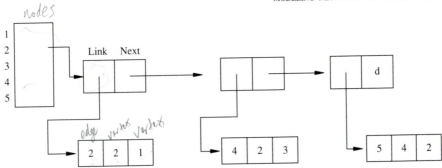

√FIGURE 3.6c

Each representation has advantages and disadvantages. The amount of space re-quired has already been mentioned. Rows of the boolean matrices can be represented as bit vectors with W bits-per-word, where W is the wordsize of the computer (or a virtual wordsize supported by the programming language being used). This com-pression by a factor of W might outweigh the difference in the order of complexity, especially for small, dense networks.

An important special case is where N is not greater than W. In this case, operations on the entire adjacency set might be carried out as single instructions, dramatically simplifying the implementation of an algorithm. While it is not good to limit an algorithm's applicability to cases where N is not greater than the wordsize of the computer, it should be noted that some languages, such as APL, have built-in operators that work on vectors of arbitrary length. And there are other languages, (e.g., LISP, SCHEME, C++) that allow you to define your own operators.

This also has an impact on the efficiency of the algorithms. Traversing a row of a matrix is an $O(N)$ (or M) operation (i.e., it requires effort proportional to N) if it must be done one element at a time. Considered as a vector operation, it is of $O(1)$. Traversing an explicit list with d (e.g., the degree of a node) elements is of $O(d)$. In many cases the latter is considerably faster, especially when N is large.

Link	End 1	End 2	Next adj. 1	Next adj. 2
1	3	4	4	3
2	2	1	4	6
3	4	5	5	0
4	2	3	5	6
5	4	2	0	0
6	1	3	0	0

√ FIGURE 3.6d

But, as mentioned above, there are special cases, where the $O(1)$ vector computation is realistic. As parallel computers come into wider use, such situations will become more prevalent.

3.2.2.3 MUTABILITY.

A **mutable** data structure is one whose size or shape changes. Conversely, **immutable** data structures do not change but are easier to implement, more efficient, and take up less space. It is best, therefore, to work with immutable data structures wherever possible (i.e., unless the algorithm requires mutability).

> **Example 3.7.** In a routing algorithm, the network itself does not change, that is, we are not adding or deleting nodes or links. Thus, it is possible to use a matrix that gives the distance between each pair of nodes. The value of the "distance" may in fact be the value of our current estimate of delay, and this estimate may change during the course of the algorithm. If the occurs, we would then change the values inside the matrix. The distance matrix is an immutable data structure; its shape does not change although the values inside it do. In contrast to this, if the algorithm selects the backbone nodes and needs to keep track of the distances between the backbone nodes selected thus far, the structure would have to be mutable because the number of nodes is changing. In this case, a matrix would not be as good a choice because it would have to be reallocated every time the number of nodes changed.

Generally, linked lists work best for mutable structures because they do not require sequential memory and it is possible to allocate and free memory as required. However, linked structures do not provide random access as well as sequential structures do, so there is a definite tradeoff. However, if algorithms are used that work with sequential access to data, the linked structures are attractive.

In making such decisions, we should also examine how the shape of a data object is changing. If nodes and links are only added, with no deleting, for example, and if it is affordable to allocate the maximum sized object, then a sequential structure might be used even though the object is mutable. In the previous example if there was enough space to store a matrix on all nodes, it would work well. Alternatively, we might be willing to limit the maximum number of backbone nodes and allocate a matrix of this size, mapping the full nodeset into the current backbone nodeset, and reusing positions in the mapping vector as nodes entered and left the backbone nodeset.

In making these decisions, the key issue is whether mutability is required or not. If a data object is local to a specific algorithm, the decision is easily made by looking at how that algorithm uses the object. If, however, the data is used more globally, this decision may be harder to make as several algorithms may be involved.

3.2.2.4 NODE AND LINK PROPERTIES.

If the number of nodes and links is fixed, or at least limited to a prespecified maximum, each property is stored in a vector with length equal to the number of nodes or links. This has the advantage of simplicity and also allows individual properties to pass as arguments to functions, thereby increasing the locality of data. This "vertical" view of properties is illustrated in Fig. 3.7a for a set of link properties.

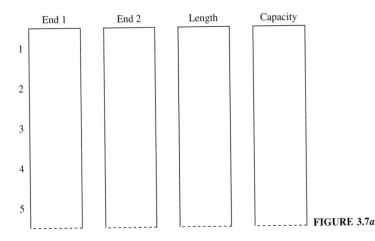

FIGURE 3.7*a*

Alternatively, a "horizontal" view can be taken, where each node or link can be treated as a record, containing all its properties. This view is illustrated in Fig. 3.7*b*. In this case, each node or link is a record which can be treated as a single entity. This view of the properties is closer to the external representation of the network and is most useful for functions relating to input and output. Also, it is possible, by forming a linked list of nodes or links, to deal with problems where the number of nodes or links varies. Generally, the vertical view is more useful for algorithms and is the view which is usually taken.

3.3   COMPUTATIONAL COMPLEXITY

It is advantageous to be able to compare the runtime and storage requirements of different algorithms without dealing with the details of their implementation. This allows the selection of an algorithm from among a set of choices, without having to implement all of them. It is also advantageous to be able to predict the running time of an algorithm on different size problems without having to run it. This allows us to determine if the runtime is excessive without actually having to implement the algorithm.

	End 1	End 2	Length	Capacity
1				

	End 1	End 2	Length	Capacity
2				

FIGURE 3.7*b*

3.3.1 Order of Complexity

Both of these goals are achieved by determining the **order of complexity** of the runtime (or storage). A function, $f(N)$, is of order $g(N)$, denoted $f(N)$ is of $O(g(N))$ if

$$\lim_{N \to \infty} \frac{f(N)}{g(N)} \leq c$$

for some constant $c \geq 0$.

The idea of this is that $g(N)$ is a much simpler function than $f(N)$ and that many functions, $f(N)$ map into the same $g(N)$. $g(N)$ captures the basic behavior of $f(N)$ without any of the detail. Most important, it is usually possible to estimate the order of complexity of an algorithm without actually implementing it.

> **Example 3.8.** We might find, by looking at the specific code for an algorithm and running it on a set of problems, that its runtime is, on average,
>
> $$f(N) = 1312 + 286N + 82N^2 \text{ milliseconds}$$
>
> when run on a network with N nodes. Alternatively, we might determine, by examining the basic structure of the algorithm itself; that is, by examining a description in pseudocode or even in English, that the algorithm is $O(N^2)$.

Note that the order of complexity is not unique. For example, in the previous example, if any higher order function is chosen (e.g., N^3) the limit goes to 0. Given a function, $f(N)$, of the above form, it is easy to determine its order of complexity; simply take the dominant term.

Complexity order ignores constants. Thus, $1000N$ is $O(N)$ while $3N^2$ is $O(N^2)$. We consider the latter function to be of a higher order of complexity and usually consider algorithms with the latter complexity to be inferior to those with the former. This is usually appropriate because most of the concern is with an algorithm's performance on large problems, and for sufficiently large N, an algorithm with a lower order of complexity always outperforms an algorithm with a higher order of complexity. Also, while constants differ among algorithms they seldom differ enough to overwhelm a difference in complexity order once N becomes large.

The running time of an algorithm varies from one problem to another even when the problems are the same "size" (e.g., when the networks involved have the same number of nodes). Sometimes, the rate at which an algorithm converges to a solution is a function of the specific data values. It is possible that the order of complexity may vary as well.

Ideally, it is best to obtain the average complexity of an algorithm. This is usually very difficult to do. First, it is not clear what to average over. Second, few significant algorithms lend themselves to such analysis. Therefore, worst case complexity analyses are usually used.

Most frequently, the "size" of a problem is given in terms of the number of nodes. It is also possible that size may be a function of the number of links or requirements. In analyzing an algorithm's complexity, try to work with as few variables as possible, considering only the dominant ones.

An important special case of complexity order is when the order is a constant (i.e., the running time is not a function of anything). In this case, $f(N)$ is $O(1)$.

All that has been said about runtime complexity also holds for storage complexity. Therefore, we can speak of an internode distance matrix as being $O(N^2)$, where N is the number of nodes in the network.

3.3.2   Pseudocode

To analyze algorithms, or even to discuss them seriously, a way of describing them is required. The description should be clear, concise, unambiguous, and sufficiently detailed to permit analysis. An implementation in a high level language such as LISP or APL is unambiguous and sufficiently detailed. It is not that concise, however, and only clear to those who know the language well. Most books use pseudocode descriptions, sacrificing some detail and unambiguity for conciseness and ease of reading; this will be done here as well. Source code, when it is available, provides an additional level of detail and specification. Some of the basic elements of the pseudocode used are described below. The remaining elements are standard and (hopefully) self explanatory.

global constants - Global constants are used to numerically designate quantities which do not change and which are available to all algorithms. They are designated by upper case letters; that is, N is used for the maximum number of nodes, L is used for the maximum number of links and R is used for the maximum number of requirements. Global constants are used, for example, for declaring the size of arrays. We also assume the existence of the boolean constants TRUE and FALSE.

variables - Variables are denoted by lower case character strings. Type information is declared only when important and not clear from context. The size of arrays is declared symbolically using global constants. Thus,

```
dcl link_cap[L] , req_mag[N,N]
```

declares that link capacity is an array on the links and requirement magnitudes are given between pairs of nodes. It is not necessary to mention if these quantities are integer or real as this does not matter for the definition or analysis of the algorithms. Assume the convention that properties of nodes, links, and requirements are prefixed by n_, l_, and r_, respectively. We try to use the same names throughout the book to denote major variables (such as link length).

It is also possible to declare that a variable is a list, that is,

```
dcl edges[list]
```

or

```
dcl n_adj[N,list-of-nodes]
```

This latter declaration says that n_adj is an array, each element of which is a list of nodes. It is important to understand what each variable stands for and, in the case where the variable is a structure, what its structure is. Specifically, it is important to understand what an array is indexed on and what kind of items comprise a list.

If it is important in understanding or analyzing an algorithm, the physical structure will then be described. Thus, using the previous example we might declare:

```
dcl n_first_adj[N], l_next_adj[L]
```

to indicate that the linked list holding the adjacency information is represented as two arrays where each node points at an adjacent link, and each link points at the next adjacent link. The pointers in this case are array indices.

assignment - The assignment operator is

```
<-
```

For conciseness, multiple assignments are allowed for in both of the following ways:

```
x <- y <- z
```

means that the value of z is assigned to both x and y.

```
x , y <- a , b
```

means the value of a is assigned to x and the value of b is assigned to y.

When the meaning is clear, the assignment to an entire structure is also permitted. Thus,

```
dcl v[N], x[N]

v <- 0
```

assigns 0 to each element of the vector v and

```
v <- x
```

copies x into v, element by element.

iterators - An iterator is a language element which allows the repeat of a block of code. The most commonly used iterator, present in one form or another in all higher level languages, is:

```
for n = 1 to nn
        . . .
        . . .
```

This says that the indented code is executed nn times, incrementing the loop index as the execution proceeds. It also shows that the loop index is a node number. The scope of the loop is indicated by the indentation, which is adhered to strictly. Note that nn is the actual number of nodes while *N* is the maximum possible number of nodes.

The above iteration assumes that the nodes are represented as indices in arrays. This is not necessarily true. Another iterator which is less restrictive and often clearer is:

```
for each ( node, nodeset )
    ...
    ...
```

This is read: do for each node in the nodeset. It says to execute the indented code once for each node in the nodeset, and it allows the nodeset to be an array, linked list, or any other appropriate structure. It is assumed that the nodes are considered in their natural order within the nodeset. A special case of for each is

```
for each ( node , nn )
```

where nn is an integer. This means that the nodes are in fact indices, and we go through them in order from 1 to nn. This form of **for each** is equivalent to the **for i = 1 to n** above but calls attention to the fact that the integers iterated over are in fact node indices.

Other iterators are permitted, such as

```
while (condition)
    ...
    ...
```

alternators - The standard alternator is if-then-else

```
if ( condition is true )
    ...
    ...
else
    ...
    ...
```

The else is optional.

functions - A function, f, with two arguments, x and y, which returns an index would be defined as follows:

```
index <- f( n , v )
dcl v[n]
...
...
return( i )
```

The first line of the function shows that the function returns an index and that it takes two arguments. The types of the arguments are given in declarations inside the function. We usually do not bother to declare scalars. Thus, no mention is made of the (obvious) fact that n is an integer; this is apparent from the fact that n is used as the length of the array v. Also, no mention is made if the elements of v are integers or real numbers. Such things are declared only when it is important to do so.

The **return** statement shows what value is returned by the function. Some functions, sometimes referred to as procedures, do not return any value. This is indicated by,

```
void <- f( x )
```

Functions ordinarily take their arguments by value, that is, the function is not permitted to change the value of the argument in the calling routine. In the case where an argument is to be passed by reference, permitting the function to change its value globally, the name of the argument is prefixed by an asterisk (*). Functions are allowed to return more than one value as their result.

> **Example 3.9.** The definition of a function, f, with two arguments a and b, the first passed by value and the second passed by reference, and returning two numbers looks like this:
>
> ```
> ( number , number ) <- f( a , *b )
>     . . .
>     . . .
>     return( x , y )
> ```
>
> Functions may be recursive (i.e., they may call themselves).
> We allow a function, f2, to be defined inside another function, f1. f2 is callable only from inside f1 and has access to local variables defined in f1. Thus, a call to f1(4), with f1 defined as follows

```
number <- f1( x )
    number <- f2( y )
        return( x + y + z )
    z = 3
    return( f2(6) )
```

would return the value 13.

3.3.3   Analysis of Complexity

In some cases, the analysis of the complexity of an algorithm is straight forward. In particular, if the algorithm is comprised entirely of iteration, the complexity is apparent.

Example 3.10. Suppose an algorithm has the structure:

```
for each ( node nn )
    ...
    ...
for each ( node1 , nn )
    for each (node2, nn )
       ...
       ...
```

This is clearly $O(nn^2)$, where nn is the number of nodes.

Almost as simple is the case:

```
for each ( node , nn )
    if ( f(node) > 0 )
       ...
    ...
```

Here, it is tempting to say the complexity is $O(nn\ P(f(node) > 0))$ where $P(e)$ is the probability that event e happens. Rarely, however, are such probabilities known and so it is usually assumed that this is $O(nn)$ in the worst case.

The interesting cases occur when recursion is involved as in

```
x <- f(y)
if ( y <= 1 )
    x <- y
else
    x <- y * f(y-1)
```

This is, of course, the factorial function. The complexity of this function is $O(y)$ but this is not obvious from examination of the function. The complexity of such functions is analyzed using **recurrence relations**. Allowing $C(n)$ to be the complexity of the above function for $y = n$, it is possible to write

$$C(n) = C(n-1) + 1 \tag{3.1}$$

and

$$C(1) = 1 \tag{3.2}$$

It is easy to see that $C(1) = 1$ since there is no explicit iteration in this function and so it takes a constant amount of time to execute except for the recursion. Thus, what is really being measured is the number of times the function is called. Therefore, the function

$$C(n) = n \tag{3.3}$$

satisfies this relation. A complete discussion of recurrence relations is beyond the scope of this book. An excellent discussion can be found in [Knu73], but we will discuss one important class briefly. It is often possible to solve simple recurrence relations by inspection. And indeed, a program which gives rise to a very complex recurrence relation is itself very complex and hence suspect.

Often, it is possible to develop an algorithm to solve a problem by recursively breaking it down into several smaller problems. This is called *divide and conquer*. Divide and conquer algorithms result in recursive programs whose complexity can be analyzed via recurrence relations.

The simplest example is the case where the problem is broken down by removing one variable at a time. The factorial algorithm is an example of this. Start with an algorithm, $a(n)$, which acts on a set of n variables. $a(n)$ calls $b(n)$ which selects one of the variables, does some calculation and then removes it from the problem. $a(n-1)$ is then recursively called to act on the remaining $n-1$ variables. Note that a and b can in reality, have many arguments; the focus here is only on the number of variables because this relates directly to the discussion of complexity. Proceeding as above, the expression can be written:

$$C(n) = C(n-1) + C_b(n) \tag{3.4}$$

where $C_b(n)$ is the complexity of b when called to act on n variables. Unlike a, assume that b is an iterative function whose complexity is known. This is often the case in practice. In the factorial examples, $C_b(n)$ is a constant. Often b iterates over n one or more times and $C_b(n)$ is a polynomial in n. In this case, $C(n)$ turns out to be a polynomial one order higher as shown in the example.

> **Example 3.11.** (Selection Sort) The following algorithm sorts a vector of n numbers in ascending order.
>
> ```
> void <- sort ( n , *numbers )
> dcl numbers[n]
> while ( n > 0 )
>     j <- find_index_of_max ( n , numbers )
>     exchange ( numbers , j , n )
>     sort ( n-1 , *numbers )
> ```
>
> where **find_index_of_max** returns the index of the largest number in the vector and **exchange** swaps the numbers in positions **j** and **n**. Note that **numbers** is passed by reference and thus, that **sort** actually rearranges **numbers**. Also note that the recursive call to **sort** only acts on the first **n-1** elements of **numbers**; the largest number, now in its correct position at the end of the vector, is removed from further consideration.
>
> If **find_index_of_max** simply loops through numbers, its complexity is $O(n)$. **exchange** is $O(1)$ and thus dominated by **find_index_of_max**; there is no need to consider **exchange** further in the analysis of the complexity of **sort**. Thus, **find_index_of_max** plays the role of the function b in this case, and $C_b(n) = n$. Therefore:
>
> $$C(n) = C(n-1) + n$$
> $$C(n-1) = C(n-2) + n - 1$$
> $$\cdots$$
> $$C(3) = C(2) + 3$$
> $$C(2) = C(1) + 2$$
> $$C(1) = 1$$

We can now work from bottom to top substituting the known value of $C(1)$ in the expression for $C(2)$, the now known value of $C(2)$ into the expression for $C(3)$, etc. Doing so, we find

$$C(n) = 1 + 2 + 3 + \cdots + n$$
$$= n(n+1)/2$$

which is $O(n^2)$. Similarly, if $C_b(n)$ is $O(n^p)$, we have

$$C(n) = 1^p + 2^p + \cdots + n^p$$

We can see that this sum is $O(n^{p+1})$ by the following arguments. First, the sum is clearly no larger than $O(n^{p+1})$, since there are n terms, each of which is no larger than n^p. Also, since the last $n/2$ terms are $\geq n/2$, and all the terms are positive, the sum is greater than $n^{p+1}/2^p$, which is $O(n^{p+1})$. The sum, then, is thus bounded both from above and below by the same order and so must be $O(n^{p+1})$. Therefore, in this case, the order of complexity of the overall function is n, raised one power higher than $C_b(n)$.

Another common situation is that we recursively divide the problem in half until we are left with a trivial problem. An example can be drawn, again, from sorting.

Example 3.12. (Merge Sort) Again, a vector of numbers is to be sorted into ascending order.

```
void <- sort ( 1 , n , *numbers )
dcl numbers[n]
if ( n > 1 )
   h = floor( n / 2 )
   merge( sort(1, h, *numbers) , sort( h+1, n, *numbers ) )
```

where `floor (x)` returns the largest integer less than or equal to x, and `merge` takes two sorted lists as input and returns a single sorted list as output. `merge` makes a single pass through its two input lists and hence is $O(m_1 + m_2)$, where m_1 and m_2 are the lengths of the two input lists. So, in this case $C_b(n)$ is $O(n)$ and it is possible to obtain the complexity of the overall procedure from the recurrence

$$C(n) = 2C(n/2) + n \qquad n > 1$$
$$C(n) = 1 \qquad n = 1$$

Back substituting as above produces:

$$C(2) = 2C(1) + 2 = 4$$

$$C(4) = 2C(2) + 4 = 12$$

$$C(8) = 2C(4) + 8 = 32$$

By inspection, we find that

$$C(2^k) = (k+1)2^k$$

TABLE 3.4
Orders of complexity

C_b	$C(n)$
$\log(n)$	n
n	$n \log n$
$n \log n$	$n(\log n)^2$
n^2	n^2
n^3	n^3

or, alternatively, that

$$C(n) = n(\log_2 n + 1)$$

for n a power of 2. Thus, this algorithm is $O(n \log n)$.

Table 3.4 lists the complexity order of algorithms whose run times obey the recurrence relation

$$C(n) = 2C(n/2) + C_b(n) \quad \text{for various common } C_b(n)$$

It is apparent that once $C_b(n)$ exceeds $n \log n$, the $C_b(n)$ dominates the recurrence and $C(n)$ is of the same order as $C_b(n)$.

Sometimes, an algorithm creates two or more subproblems with $n - 1$ variables from a problem with n variables. This gives rise to the recurrence

$$C(n) = kC(n - 1)$$

$$C(1) = 1$$

which has the solution

$$C(n) = k^n$$

In this case, the run time grows very quickly with n. It grows so quickly that the algorithm is not practical except for very small values of n. Such an algorithm is said to have *exponential complexity*. The other algorithms mentioned all have *polynomial complexity*. Generally, we consider a problem intractable if we can only find algorithms with exponential complexity for its solution.

EXERCISES

3.1. How many different ways can a specific cycle be specified by a sequence of nodes?

3.2. A complete graph is defined as a graph with an edge between every pair of nodes. How many edges are there in a complete graph with N nodes? How many arcs are there in a complete directed graph?

3.3. What is the largest possible number of edges in a graph that is not connected?

3.4. Estimate the number of paths between a given pair of nodes in a complete graph with N nodes.

3.5. (a) What must be done to convert external node identifiers (used in representing the network on disk) to internal node numbers (used to represent the network in memory)?

(b) Based on your answer in (a), does it make sense to maintain a sorted list of the node identifiers?

3.6. Given a network with N nodes and M links, keep track of which nodes are adjacent.

(a) Compare the use of bit vectors and explicit lists for this purpose, giving required memory in terms of M, N, and the number of bits, W, per word.

(b) Compare these approaches based on the effort required to add or delete nodes or links.

3.7. Compare the horizontal and vertical approaches to network representation based on the amount of effort required to add and delete nodes, links, and properties.

3.8. C, the product of two N by N matrices A and B, is defined by

$$c_{ij} = \sum_{k=1}^{N} a_{ik}b_{kj}$$

(a) Suppose an N by N matrix is represented as an N by N array. What is the complexity of matrix multiplication?

(b) It is also possible to represent a matrix by explicit lists of its non-zero elements, organized by rows, by columns, or simply as vectors of indices and values. This is particularly attractive for sparse matrices. Describe such a representation and the order of complexity of matrix multiplication using it.

3.9. Sets can be represented as linked lists of their elements. A linked list can be stored with its elements in order (i.e., alphabetical order) or not. It is sometimes necessary to be able to merge sets. Sometimes it is apparent that the elements of different sets are disjoint; other times this is not the case.

(a) What is the order of complexity of forming the union of two sets, stored as unordered linked lists, if the sets have N and M elements, respectively? Consider both the cases where the sets are known to be disjoint and when they are not.

(b) What is the order of complexity of forming the union of two sets, stored as ordered linked lists if the sets have N and M elements, respectively? Consider both the cases where the sets are known to be disjoint and when they are not.

3.10. Figure P3.1 shows a graph which is know as a grid with N rows and M columns.

(a) How many nodes are there in the grid?

(b) How many edges are there in the grid?

(c) How many paths are there between nodes 11 and 34?

(d) How many paths are there between nodes 11 and NM?

(e) If $N = 2$ and $M = 3$, how many trees are there in the grid?

(f) If $N = 2$ and general M, estimate how many trees are there in the grid?

(g) If $N = 2$ and $M = 4$, how many cycles are there in the grid?

(h) If $N = 2$ and general M, how many cycles are there in the grid?

3.11. Suppose a computer executes one million instructions per second and a person is willing to devote one minute to running an algorithm whose running time grows with N, the number of nodes in his/her network. How large can N be if their algorithm is of the

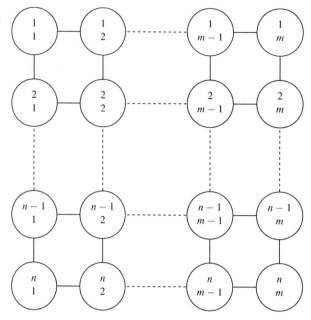

FIGURE P3.1

following orders? (Assume the constant in the order is C (i.e., that an algorithm of $O(f(N))$ has a running time of $Cf(N)$).

(a) $O(N)$

(b) $O(N^2)$

(c) $O(N^3)$

(d) $O(N \log N)$

(e) $O(2^N)$

(f) $O(1.1^N)$

BIBLIOGRAPHY

[Knu73] Knuth, D. E.: *The Art of Computer Programming Volume I: Fundamental Algorithms*, Addison-Wesley, Reading, MA, 1973.

[Smi87] Smith, H.: *Data Structures*, Harcourt Brace Jovanovich, San Diego, 1987.

[Sta80] Standish, T. A.: *Data Structure Techniques*, Addison-Wesley, Reading, MA, 1980.

CHAPTER
4

FUNDAMENTAL GRAPH ALGORITHMS

4.1 FINDING TREES IN GRAPHS

We turn now to the first set of basic algorithms — those for finding trees which are used to design and analyze networks. Recall that a tree is a graph with no loops; there is only one path between any pair of nodes. Consider the model of undirected graphs where links are used in both directions in forming paths.

Trees are very useful for a number of reasons and they will be used as the basis for many algorithms and design and analysis techniques. First, they are minimal networks; they provide connectivity without any unnecessary additional links. Second, by providing a unique path between each pair of nodes, they eliminate the routing problem (i.e., deciding how traffic should flow between nodes). This simplifies both the network and its design. However, since trees are minimally connected they are also minimally reliable and robust. This is why actual networks are usually more highly connected. Nevertheless, the design of a network often starts with a tree.

4.1.1 Tree Traversals

Given a tree, we may wish to visit all its nodes. This is called a tree traversal. In doing so, all the edges in the tree are traversed twice, once in each direction. There are several ways of traversing a tree. Begin by identifying a node in the tree as the root. The traversal is carried out relative to this node. Sometimes there is a logical

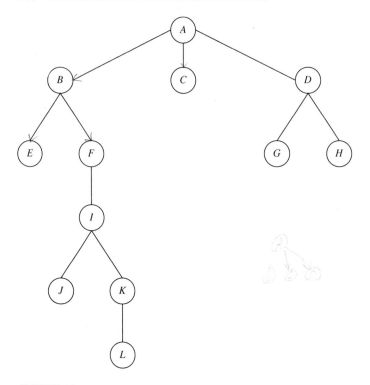

FIGURE 4.1
Tree traversal.

reason for selecting the root (e.g., it is the central computer site). Other times, the root is selected arbitrarily.

Suppose node A in Fig. 4.1 is the root of the tree shown. It is possible to begin by visiting A and then visit the neighbors of A, nodes B, C, and D, in that order. Next, visit their neighbors, nodes E, F, G, and H. Continue to visit the neighbors of these nodes in the order the nodes themselves were visited. In this case, the traversal is finished by visiting nodes I, J, K, and L. This is called a **breadth first search**. A breadth first search has the property that nodes closest to the root are visited first. The search fans out in all directions at once. This is sometimes useful and is easy to implement.

An algorithm that visits all the nodes of a tree is called a tree traversal algorithm. The following algorithm, `BfsTree`, implements a breadth first search of a tree. (By convention, we will capitalize the names of functions to distinguish them from variable names.) `BfsTree` makes use of an adjacency list, `n_adj_list`, which lists the nodes adjacent to each node. For the sake of simplicity, we assume that the tree is a directed tree outward from the root and, thus, that `n_adj_list` includes only nodes adjacent to a node but farther from the root than that node is. It is left as an exercise for the reader to modify this algorithm for the case where the links are undirected.

```
✓ void <- BfsTree( n , root , n_adj_list ):
      dcl n_adj_list[ n , list ]
          scan_queue[ queue ]

      InitializeQueue( scan_queue )
      Enqueue( root , scan_queue )
      while ( NotEmpty( scan_queue ) )
          node <- Dequeue( scan_queue )
          Visit( node )
          for each( neighbor , n_adj_list[node] )
              Enqueue( neighbor , scan_queue )
```

Visit is a routine which does whatever it is instructed to do to each node.

This algorithm is implemented using a queue. Queues functions in a first-in first-✓ out (FIFO) fashion; new items are added to the rear of the queue and removed from the front. The procedures InitializeQueue, Enqueue, Dequeue, and NotEmpty work on queues. InitializeQueue sets up an empty queue. Enqueue and Dequeue add an element at the end of a queue and remove an element from the front of a queue, respectively. NotEmpty returns TRUE or FALSE depending on whether or not the queue is empty. The algorithms for these functions and other primitives which manipulate elementary data structures are given in Appendix A.

n_adj_list is an array of lists. n_adj_list[n] is the list of nodes adjacent to node n. for_each(element , list) is, as mentioned in the preceding chapter, a control structure which loops over the members of list and executes the code inside the loop using each element in list in turn. Assume n_adj_list is already set up by the time BfsTree is called.

In a similar manner, it is possible to define a **depth first search**. Again, start at the root. Now, however, proceed by visiting an unvisited neighbor of the node just visited. Again, assume that the tree is comprised of links directed away from the root.

Example 4.1. Returning to the graph in Figure 4.1, we might visit B after A. Next, we might visit E, a neighbor of B, the most recently visited node. E has no unvisited neighbors, so we backtrack to B and visit F. We proceed, visiting I, J, K (having backtracked to I) and L. Then we backtrack all the way to A and proceed, visiting C, D, G and H. Thus, the entire traversal is:

$$A, B, E, F, I, J, K, L, C, D, G, H$$

Note that the traversal order is not unique. We chose to explore the neighbors of a node from left to right. Other orders are possible, for example:

$$A, B, F, I, J, K, L, E, D, H, G, C$$

The actual order of the traversal is dependent on the specifics of the algorithm. The same is true for breadth first search. Examining the algorithm BfsTree, it is apparent that the order is a function of the order of neighbors in n_adj_list.

The following algorithm, `DfsTree`, implements a depth first search of a tree:

```
void <- DfsTree( n , root , n_adj_list ):
     dcl n_adj_list[ n , list ]

     Visit( root )
     for each( neighbor , n_adj_list[node] )
          DfsTree( n , neighbor , n_adj_list )
```

Depth first search is done with the aid of a stack, which functions in a last-in, first-out (LIFO) fashion; we insert and delete items from the top of the stack. In this case, we simply have `DfsTree` recursively call itself, implicitly using the system stack, that is, the stack which the system uses to hold function calls and arguments.

Both of these traversals are **preorder traversals** (i.e., they visit a node and then visit its successors). It is sometimes necessary to visit the nodes in **postorder**, where a node is visited after its successors. It is possible, of course, to form a preorder list and then reverse it. Alternatively, it is possible to alter the order of search directly, as in:

```
void <- PostorderDfsTree( n , root , n_adj_list ):
     dcl n_adj_list[ n , list ]

     for each( neighbor , n_adj_list[node] )
          PostorderDfsTree( neighbor , n_adj_list )
     Visit( root )
```

4.1.2   Connected Components in Undirected Graphs

It is possible to generalize the notion of a traversal to an undirected graph, simply by keeping track of which nodes were visited and not visiting them again. Thus:

```
void <- Dfs( n , root , n_adj_list ):
     dcl n_adj_list[ n , list ]
          visited[n]

     void <- DfsLoop( node )
          if( not( visited[node] )
               visited[node] <- TRUE
               Visit(node)
               for each( neighbor , n_adj_list[node] )
                    DfsLoop( neighbor )

     visited <- FALSE
     DfsLoop( root )
```

Notice that the statement

```
visited <- FALSE
```

initializes the entire array visited to FALSE. Also, notice that the local procedure DfsLoop is defined inside Dfs and therefore DfsLoop has access to visited and n_adj_list. (Note: It is easiest to read pseudocode for functions like Dfs above by first reading the body of the main function and then going back to read the body of embedded functions like DfsLoop.)

Note that during the course of the traversal we implicitly examine all the edges in the graph, once from each end. In particular, for each edge, (i, j), in the graph, j is an element in n_adj_list[i] and i is an element in n_adj_list[j]. Actually, it is possible to put the edges themselves into the adjacency lists and then look up the node at the other end of the edge using a function

```
node <- OtherEnd( node1 , edge )
```

which returns the end of edge which is not node1. This complicates the implementation only slightly. Viewed in this way, it is clear that the complexity of all these traversal algorithms is $O(E)$, where E is the number of edges in the graph.

It is now possible to find the connected components of an arbitrary (undirected) graph by traversing each component. We will label each node with a component number as we proceed. The variable ncomponents keeps track of what component we are currently up to.

```
void <- LabelComponents( n , n_adj_list )
     dcl n_component_number[n] , n_adj_list[n,list]

     void <- Visit(node)
          n_component_number[node] <- ncomponents

     n_component_number <- 0
     ncomponents <- 0
     for each( node , node_set )
          if( n_component_number[node] = 0 )
               ncomponents += 1
               Dfs( node , n_adj_list )
```

Notice that here we actually define a function Visit to set the component number of visited nodes. This function is local to LabelComponents and can only be called from inside that function. Dfs, on the other hand, is defined outside and, hence, can be called from anywhere.

Note that in running a BFS or DFS traversal of an arbitrary graph, the edges which actually result in a node being visited (i.e., the edges connecting a node to a previously unvisited neighbor) form a tree or, if the graph is not connected, a forest. This observation will prove useful later.

Figure 4.2 shows a graph with four components. Assuming the loop on the node set goes through the nodes in alphabetical order, the components are numbered in the

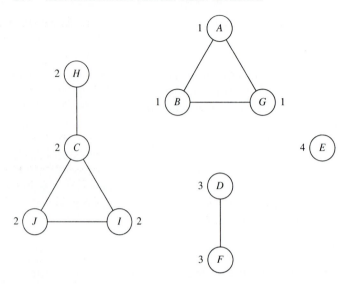

FIGURE 4.2
Components.

order of the node with the "lowest" letter and the component numbers shown next to the nodes in Fig. 4.2 result.

The algorithm above calls `Dfs` to examine each component once. It also examines each edge once. Thus, its complexity is of the order of the sum of the number of nodes plus the number of edges in all the components (i.e., it is $O(N + E)$).

4.1.3 Minimum Spanning Trees (MST)

It is possible to use `Dfs` to find a spanning tree in a graph, if one exists. The tree found is, in general, arbitrary. It is often useful to find the "best" tree. Thus, we may assign a length to each edge in the graph and ask for a tree of minimum total length. The "length" may in fact be distance, cost, or some measure of delay or reliability. A tree of minimum total cost is called a *minimum spanning tree.*

In general, if the graph is not connected, we may wish to find a minimum spanning forest, a set of edges of minimum total length which connects the graph as much as possible. This problem can be thought of as selecting a subgraph of the original graph which contains all nodes of the original graph and the edges selected. Initially, the graph contains n nodes, 0 edges and n components. Each time we select an edge to add to our graph, two previously unconnected components are joined. If this were not so, we would be bringing in an edge connecting two previously connected components and forming a loop. Thus, at any stage in the algorithm, the relationship

$$n = c + e \tag{4.1}$$

holds, where n is the number of nodes in the graph, e is the number of edges selected so far, and c is the number of components so far in the graph. At the end of the algorithm, e is equal to n minus the number of components in the original graph; if the original graph was connected, we would find a tree with $n-1$ edges. As explained earlier, Dfs will find a spanning forest. However, it will not, in general, find the one of minimum total length.

4.1.3.1 THE GREEDY ALGORITHM.

One possible approach to finding a tree of minimum total length is, at each stage in the algorithm, to select the shortest edge possible. This is called a **greedy algorithm**. Greedy algorithms are myopic (near-sighted). They do not look ahead at the eventual consequences of the greedy decision they make at each step. Instead, they just make the best choice for each individual selection. In general, greedy algorithms do not find the optimal solution to a problem. In fact, they may not even find a feasible solution, when one exists. They are, however, efficient and simple to implement. Hence, they are widely used. Also, they often form the basis of other more complex and effective algorithms.

Therefore, the first question to ask when considering the use of an algorithm to solve a given problem is whether the problem has some structure which guarantees that the algorithm will work well. Hopefully the algorithm would at least guarantee a feasible solution, if one exists. Ideally, it would also guarantee optimality, and give some guarantees about running time. In the case of minimum spanning trees, the problem at hand does indeed possess structure, which allows the greedy algorithm to guarantee both optimality and reasonable computational complexity.

Informally, the general form of the greedy algorithm is:

```
Start with an empty current solution, s.

While any elements remain to be considered,

    Find e, the "best" element not yet considered

    If adding e to s is feasible, add it; if not, discard it.
```

This requires the ability to:

- Compare the values of the elements to determine which is "best"
- Test a set of elements for feasibility

The notion of best is relative to the objective. If the objective is minimizing, best is smallest. If the objective is maximizing, best is largest.

Often, a value is associated with each element, and the value associated with a set is simply the sum of the values associated with the elements in that set. This is the case for the minimum spanning tree problem considered in this section. It need not, however, be the case in general. For example, we might say that instead of minimizing the total length of all the edges in the tree, the objective is to minimize the length of the longest edge in the tree. In that case, the value of an edge would be its length, and the value of a set would be the length of the longest edge in the set.

To find the best edge to add, evaluate edges in terms of how they affect the value of the set. Let $V(S)$ be the value of a set, S. Then, $v(e, S)$, the value of an element e, relative to a set, S, with value $V(S)$ is given by

$$v(e, S) = V(S \cup e) - V(S)$$

In the case of minimizing the length of the longest edge in a tree, $v(e, S)$ is 0 for any edge not longer than the longest edge already selected. Also, it is otherwise equal to the difference between the edge's length and the longest edge already selected, when this difference is positive.

In the most general case, it is possible for the value of a set to vary irregularly as elements are added to it. We might assign the value 1 to sets with an even number of elements and the value 2 to sets with an odd number of elements. This results in element values alternating between $+1$ and -1. The greedy algorithm is not of much use in this case. Assume for now that the weight of a set varies in a more reasonable manner and hence, that there is a reasonable basis for identifying the "best" element. It is important to realize, however, that as the set grows it is possible that the value of unconsidered elements may change as more elements are added to the set. To the extent that this happens, the greedy algorithm may make mistakes in its choices, and the quality of the solution we obtain may suffer.

Similarly, in the most general case, feasibility may be affected in an irregular way by the addition of elements. Thus, sets with an even number of elements might be considered feasible and those with an odd number of elements considered not feasible. The greedy algorithm, or any algorithm which adds elements one at a time, will not function in such cases. We will therefore assume the following property, which holds in most cases of interest:

Property 1: Any subset of a feasible set is feasible and, in particular, the empty set is feasible.

Assume further that the complexity of the algorithms to compute the value of a set and to test its feasibility is reasonable, in particular, that it is a polynomial in the number of nodes and edges in the graph.

We now give a more formal statement of the greedy algorithm. It is somewhat abstract in that it relies on the definition of two other functions, `Test` and `SelectBestElement`, to test feasibility and evaluate sets. We also assume the existence of a structure, `properties`, which is a list of lists and that contains all the information necessary to test and evaluate all sets. A list of lists is simply a linked list, each of whose members is itself a list. It is possible for this structure to be nested even more deeply, with lists inside of lists inside of lists. Such a structure is quite general and can be used to represent almost any type of information. It is possible to hold lengths, link types, capacities, or addresses. Each of these items may itself be a complex structure; that is, the structure may hold the costs and capacities of several alternative types of channels for each link.

In practice, it is beneficial to maintain auxiliary data structures to allow the algorithm to function more efficiently. The minimum spanning tree problem is an example of this. Here, however, for the sake of clarity, assume all computation is carried out in terms of `properties`.

```
list <- Greedy( properties )
     dcl properties[ list , list ]
         candidate_set[ list ]
         solution[ list ]

     void <- GreedyLoop( *candidate_set , *solution )
         dcl  test_set[list] , solution[list] ,
              candidate_set[list]

         element <- SelectBestElement( candidate_set )
         test_set <- Append( element , solution )
         if( Test( test_set ) )
             solution <- test_set
         candidate_set <- Delete( element , candidate_set )
         if( not( Empty( candidate_set ) ) )
             Greedy_loop( *candidate_set , *solution )


     solution <- Φ
     if( !( Empty( element_set ) ) )
         candidate_set <- ElementsOf( properties )
         GreedyLoop( *candidate_set , *solution )
     return( solution )
```

Φ is used to designate the empty set. `Append` and `Delete` are functions which, respectively, add and remove an element from a list. `ElementsOf` simply identifies the elements of a list; so initially, all elements in `properties` are candidates. There are many ways of actually implementing these functions. `Properties` might in fact be an array and functions such as `Append`, `Delete`, and `ElementsOf` might work with lists of indices. The way chosen in practice may be a function of how to best implement `Test` and `SelectBestElement`.

This description of the greedy algorithm assumes that it stops when there are no elements left to consider. In fact there may be other reasons for stopping. One is that the solution value deteriorates if more elements are added. This would be the case if all remaining elements had negative values and we were trying to find a solution of maximum value. Another reason for stopping would be if it was known that there were no elements left in the candidate set which form a feasible solution with the elements already chosen. This would be the case if it were known that an entire spanning tree had already been found. Assume the algorithm stops when it is reasonable to do so or, if not, that extraneous elements are discarded from the solution.

Suppose the solutions to a problem satisfy Property 1 and that the value of a set is simply the sum of the values of the elements in the set. Suppose further that the following property also holds:

√*Property 2*: If two sets, S_p and S_{p+1}, with p and $p + 1$ elements, respectively, are solutions, then there exists an element, e, in S_{p+1} and not in S_p such that $S_p \cup \{e\}$ is a solution.

As we will see, the edges of forests satisfy Property 2, that is, if there are two forests, one with p edges and the other with $p + 1$, it is always possible to find an edge in the larger set that can be added to the smaller set without forming a cycle.

A set of solutions satisfying these properties is called a **matroid**. A complete discussion of matroids is beyond the scope of this book. The interested reader is referred to [Law76] for more information. However, the following theorem is useful. The proof follows Lawler's.

Theorem 4.1. The greedy algorithm guarantees an optimal solution to a problem if and only if the solutions to that problem form a matroid.

Proof. (*if*) Suppose the greedy solution (which is not optimal) is $x_1, x_2, \ldots, x_j$ and the optimal solution is $y_1, y_2, \ldots, y_k$. Suppose further, without loss of generality, that the xs and ys are sorted in descending order and that we are maximizing. (If the solutions form a matroid, it will be shown that we can easily transform between minimization and maximization.

It is known that all the xs and ys are positive because the greedy algorithm stops before adding any negative elements, and that the optimal solution clearly does not contain any negative elements. (Recall that any subset of a feasible set is feasible.) Since it is assumed that the greedy solution is not optimal, the sum of the ys must be strictly larger than the sum of the xs. Thus, there must either be more ys than xs or else at least one of the ys must be larger than one of the xs.

Suppose there are more ys than xs (i.e., $k > j$ above). The set $y_1, y_2, \ldots, y_{j+1}$ is a subset of the optimal solution and is therefore a feasible set containing $j + 1$ elements. By Property 2, it must contain at least one element, y_m, which is feasible to add to the greedy solution, and since the value of y_m is positive, the greedy algorithm would have added it. So, k is not larger than j.

Suppose then that there is an element, y_m, which is larger than x_m. Since the ys are sorted in descending order, all the y_j for $j \leq m$ are larger than x_m. Thus, again by Property 2, the set $y_1, y_2, \ldots, y_m$ contains an element which is feasible to add to the set $x_1, x_2, \ldots, x_{m-1}$, and since this element is larger than x_m, the greedy algorithm would have chosen it in place of x_m. So, again there is a contradiction. It is therefore concluded that if Properties 1 and 2 hold, the greedy algorithm finds an optimal solution.

(*only if*) Suppose there is a problem where the greedy algorithm always finds the optimal solution regardless of the values of the elements. It is possible then to set the element values to anything and expect the greedy algorithm to find the optimal solution. Therefore, any feasible set, $x_1, x_2, \ldots, x_m$, is optimal by setting $v(x_j) = 1$ for x_j in this set, and setting $v(x_j) = 0$ otherwise. Since the greedy algorithm moves from one feasible solution to another by adding single elements, any subset of a feasible set must be feasible, and Property 1 must hold.

We now show that Property 2 must also hold using proof by contrapositive; that is, if Property 2 does not hold then there is a set of weights for which the greedy algorithm does not find an optimal solution. Let the optimal solution be $y_1, y_2, \ldots, y_m$ with $v(y_j) = 1$ for all elements in this set. Let the greedy solution be $x_1, x_2, \ldots, x_{m-1}$ with $v(x_j) = 1 + \epsilon$ for all elements in this set, where ϵ is less than $1/m$. Let $v(x_j) = 0$ for all other elements. The value of the optimal solution is m and the value of the greedy solution is $(m - 1)(1 + \epsilon)$, which is less than m. Because Property 2 does not hold, assume that there is no y_j in the optimal solution which can be feasibly added to the greedy solution and therefore the greedy algorithm does not find an optimal solution.

Thus, it is seen that Properties 1 and 2 are both necessary and sufficient to guarantee the optimality of the greedy algorithm. Therefore, always check to see if the solutions to a problem satisfy Properties 1 and 2. If they do, simply use a greedy algorithm to solve them. This is in fact the case for spanning trees, as will be seen.

Before leaving this section, here is one more useful result.

Theorem 4.2. If the feasible solutions to a problem form a matroid, then all maximal feasible sets contain the same number of elements.

Proof. A maximal feasible set is one which will not maintain feasibility if elements are added; it does not necessarily contain a maximum number of elements nor does it necessarily have maximum weight. Suppose that there are two different maximal sets with p and $p + 1$ elements, respectively. Again, Property 2 assures there must exist an element in the larger set which can be added to the smaller set, contradicting the assertion that the smaller set is maximal.

Less obvious, but also true (c.f. [Law76]) is the converse to this theorem, that is, if Property 1 holds and all maximal feasible sets have the same number of elements, then Property 2 must also hold.

Theorem 2 allows us to transform a minimization problem, P, to a maximization problem, P', by altering the values of the elements. Suppose that all the $v(x_j)$ in P are negative. The optimal solution to P contains a maximum number of elements, say m elements. We can create a maximization problem, P', from P by setting the value of x_j in P' to $-v(x_j)$. The element values in P' are all positive, and again, an optimal solution contains m elements. In fact, the order of the maximal solutions have been reversed; the solution with the maximum value in P' is the one with the minimum value in P.

Suppose now that we wish to find a solution with minimum value, subject to the constraint that it have a maximum number of elements. Elements would then be included even if they had positive values. It is possible to solve this problem P, as a maximization problem, P' by setting the element values in P' to be $B - v(x_j)$, where B is larger than the largest x_j. Again, the element values in P' are all positive, and the optimal solution contains m elements. The order of all the maximal feasible sets have been reversed; a set with value V in P has a value $mB - V$ in P'. A maximum in P' is then a minimum in P. This observation, together with the fact that spanning trees satisfy Properties 1 and 2, make it possible to find minimum spanning trees using a greedy algorithm.

4.1.3.2 KRUSKAL'S ALGORITHM. Kruskal's algorithm is a greedy algorithm for finding minimal spanning trees. Its validity relies on the truth of the following theorem.

√ **Theorem 4.3.** Forests satisfy Properties 1 and 2.

Recall that a forest is a set of edges which do not include a cycle. Clearly any subset of the edges in a forest (even the empty set) is a forest, so Property 1 holds.

To see that Property 2 also holds, consider the graph shown in Fig. 4.3. Suppose there is a forest, F_1, containing p edges. The forest $\{2, 4\}$ is an example with $p = 2$,

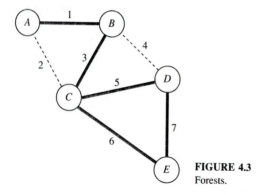

FIGURE 4.3
Forests.

and is shown with dotted lines in Fig. 4.3. Now, consider another forest, F_2, with $p + 1$ edges. There are two cases to consider:

Case 1: F_2 is incident to a node, n, which F_1 is not incident to. An example of this is the forest $\{1, 4, 6\}$, which is incident on node E. In this case, we are free to add an edge from F_2 incident on the new node to F_1. In the example, it is possible to create the forest $\{2, 4, 6\}$ by adding edge 6 to $\{2, 4\}$.

Case 2: F_2 is incident only on nodes which F_1 is incident on. An example of this is forest $\{1, 3, 5\}$. Consider S, the set of nodes F_1 is incident upon. Let there be k nodes in S. Since F_1 is a forest, each edge in F_1 reduces the number of components in S by 1 and the total number of components in S is therefore $k - p$. By similar reasoning, F_2 forms $k - (p + 1)$ components from S (one fewer component than in F_1.) Therefore, an edge exists in F_2 whose endpoints are in different components in F_1 and this edge can be added to F_1 without forming a cycle. Edge 3 is such an edge in our example (as are edges 1 and 5, coincidentally.)

We see therefore that Properties 1 and 2 both hold and that a greedy algorithm finds an optimal solution to both the minimal spanning tree and maximal spanning tree problems. Note that a spanning tree is a forest with a maximum number, $N - 1$, of edges, where N is the number of nodes in the network. We now consider the minimization problem.

Kruskal's algorithm, informally, says to sort the edges, shortest first and then include all edges which do not form cycles with the edges previously selected. Thus, a simple implementation of Kruskal's algorithm is

```
list <- kruskal_1( n , m , lengths )
    dcl lengths[m], permutation[m], solution[list]

    permutation <- VectorSort ( n , lengths )
    solution <- Φ
    for each ( edge , permutation )
        if( Test ( edge , solution ) )
            solution <- Append( edge , solution )
    return ( solution )
```

`VectorSort` takes a vector of length n as input and returns a permutation (rearrangement) of the integers 1 through n as output. The permutation gives the values of the vector in ascending order.

Example 4.2. Suppose $n = 5$ and the vector contained the values

$$31, 19, 42, 66, 27,$$

`VectorSort` would return the permutation

$$2, 5, 1, 3, 4$$

`Test` takes a list of edges as input and returns `TRUE` if they do not contain a cycle. Since `Test` is called for each edge, the efficiency of the overall algorithm depends on how efficiently `Test` is implemented. If we are willing to keep track of which component each node is in as edges are added to the tree, `Test` becomes very simple; it is only required to check if the endpoints of the edge being considered are in the same component. If they are, the edge will form a cycle. If not, it will not.

Next, consider how to maintain the component structure. There are several natural ways of doing this. One approach, due to Tarjan [KT81], is to keep a pointer to another node in the same component, and to have one node in each component, which is called the **root** of the component, point at itself. Thus, initially, each node is in a component by itself and points at itself. When an edge is added between nodes i and j, point i at j. Later in the algorithm, when an edge is added between node i in the component with root k and node j in the component with root l, is pointed k at l. Thus, we test an edge by following the pointers from its endpoints and to see if they lead to the same place. The shorter the chain of pointers, the less effort this is. To keep these chains short, Tarjan suggests collapsing the chains as they are traversed during the testing. Specifically, he suggests that a function, `FindComponent` be created as follows:

```
index <- FindComponent( node , *next )
     dcl next[]

     p = next[node]
     q = next[p]
     while( p != q )
          next[node] = q
          node = q
          p = next[node]
          q = next[p]
     return(p)
```

`FindComponent` returns the root of the component containing `node`. It also adjusts `next` which points towards the root of the component containing it. Specifically, it adjusts `next` to point one level higher. Tarjan showed that by doing this, rather than collapsing the path to the root entirely or not collapsing the path at all, that the overall

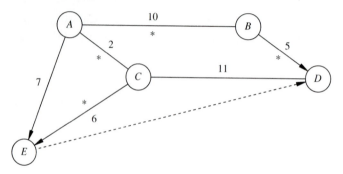

FIGURE 4.4

MST calculation.

effort in testing and updating `next` is only slightly larger than $O(n + m)$, where n is the number of nodes and m is the number of edges tested.

> **Example 4.3.** Consider the network shown in Fig. 4.4. (The ∗s in the figure are explained below.) First, sort the edges and then consider them in order, smallest first. We thus consider (A, C) first. Calling `FindComponent` for A, we find that both p and q are set to A, and `FindComponent` returns A as the root of the component containing node A. Similarly, `FindComponent` returns C as the root of the component containing node C. We therefore bring (A, C) into the tree, setting `next[A]` to C. Then consider (B, D). It too passes the test and so we add (B, D) to the tree, setting `next[B]` to D. We then consider (C, E), again accepting it and setting `next[C]` to E.
>
> Now, consider (A, E). In `FindComponent`, p is C and q is E. We therefore enter the `while` loop, setting `next[A]` to E and shortening the path from A to E, the root of its component. `Node, p,` and `q` are all then set to E and `FindComponent` returns E as the root of the component containing node A. `FindComponent` also returns E as the root of the component containing node E. Thus, both endpoints of (A, E) are in the same component and (A, E) is rejected.
>
> Next, (A, B) is considered. Calling `FindComponent` for node A, we find $p = q = E$ and `next` is not altered. Similarly, calling `FindComponent` for node B, we find $p = q = D$. We thus set `next(E)` to D. Note that we do not set `next[A]` to B, but rather, set `next` for the root of A's component to be the root of B's component.
>
> Finally, (C, D) is tested and rejected.
>
> In Fig. 4.4, the edges in the spanning tree are indicated by an "∗" next to them. The contents of `next` is indicated by arcs (directed edges) with arrows. Thus, for example, the fact that `next[B]` is D is indicated by an arrow on (B, D), pointing from B to D. We note that the arcs defined by `next` form a tree but that this tree is not in general the minimal spanning tree. Indeed, as is the case with the arc (E, D), these arcs need not even be part of the graph. Thus, `next` itself only defines the component structure as the algorithm proceeds. We make an explicit list of the edges selected for inclusion in the tree. The value of the tree defined by `next` is that it is relatively flat, that is, the paths to the roots of components are short, making `FindComponent` efficient.

It is apparent that the complexity of Kruskal's algorithm, as implemented above, is dominated by the sort of the edges, which has complexity $O(m \log m)$. If there is the possibility that the spanning tree will be found before examining most of the edges, we can improve upon this by doing a partial sort. Specifically [KS72], we can place the edges in a heap and then pop edges out, testing each edge, until a tree is formed. It is easy to know when to stop; simply keep track of the number of edges accepted and stop when $n - 1$ edges have been accepted.

We assume the reader is generally familiar with the operation of a heap. Appendix A includes functions for managing a heap, setting it up as well as pushing and popping. The important things to note here are that the effort to set up a heap on m elements is $O(m)$, the effort to find the smallest element is $O(1)$, and the effort to restore a heap after adding, deleting, or changing a value is $O(\log m)$. Thus, if we consider k edges in order to find the spanning tree, the effort in maintaining the heap √ is $O(m + k \log m)$, which is lower than $O(m \log m)$, if k is of order lower than m. k is, of course, at least $O(n)$, so if the graph is sparse, using the heap will not gain √ much. If the graph is dense, however, the savings can be considerable. Here is the final version of Kruskal's algorithm, which takes advantage of these efficiencies.

```
list <- Kruskal( n , m , lengths )
     dcl lengths[m], ends[m,2], next[n], solution[list],
         l_heap[heap]

     for each( node , n )
         next[node] <- node

     l_heap <- HeapSet ( m , lengths )
     #_accepted <- 0
     solution <- Φ

     while( ( #_accepted < n-1 ) and !( HeapEmpty(l_heap) ) )
         edge <- HeapPop ( *l_heap )
         c1 = FindComponent ( ends[edge,1] , *next )
         c2 = FindComponent ( ends[edge,2] , *next )
         if( c1 != c2 )
             next[c2] <- c1
             solution <- Append( edge , solution )
             #_accepted <- #_accepted + 1
     return( solution )
```

HeapSet defines a heap on the values given to it and returns the heap itself. Heap-Pop returns the index of the value at the top of the heap, not the value itself. This is better than returning the value since it is always possible to obtain the value from the index but not to obtain the index from the value. Notice also that HeapPop modifies the heap. HeapEmpty returns TRUE if the heap is empty. The array ends gives the endpoints of the edges.

4.1.3.3 PRIM'S ALGORITHM.
It is sometimes advantageous, especially if the network is dense, to consider an alternative approach to finding MSTs. Also, other more complex algorithms are built on top of these MST algorithms; and some of them

work better with the data structures used for the following algorithm, which is due to Prim [Pri57]. Finally, this algorithm is particularly well suited to parallel implementation because it can be implemented using vector operations. Prim's algorithm can be described informally as:

```
Start with one node in the tree and all other nodes not in the
    tree (out-of-tree)

While there are still out-of-tree nodes,

    Find the out-of-tree node which is nearest to the tree

    Bring that node into the tree and record the edge which
    connects it into the tree
```

Prim's algorithm relies on the following theorem for its validity

Theorem 4.4. A tree is an MST if and only if it includes the shortest edge in every cutset that divides the nodes into two components.

Proof. Recall that a cutset is a set of edges whose removal disconnects the graph.

(*only if*) Since a spanning tree is a connected graph, every spanning tree contains at least one edge in every cutset. Suppose a tree, T, does not contain the shortest edge, e, in some cutset but instead contains another longer edge, e'. It is possible to replace e' with e and obtain a shorter spanning tree. Hence, T is not an MST.

(*if*) Suppose T contains the shortest edge in every cutset that divides the nodes into two components but is not an MST, M. Let e be an edge in M but not in T. The removal of e from M separates the nodes into two components. Consider the cutset defined by these node sets, that is, the set of edges whose removal separates the graph into the two parts defined by these nodes. T contains e', the shortest edge in this cutset. We can replace e by e' forming a shorter tree, thus contradicting the assertion that M was an MST.

In order to implement Prim's algorithm, it is necessary to keep track of the distance from each out-of-tree node to the tree and update this distance each time a node is brought into the tree. This is straight forward. Simply keep an array, d_tree, with this information. This yields the following simple implementation for Prim's algorithm.

```
array[n] <- Prim( n , root , dist )
    dcl dist[n,n], pred[n], d_tree[n], in_tree[n]

    index <- FindMin( )
        d_min <- INFINITY
        for_each( i , n )
            if( !(in_tree[j]) and (d_tree[i] < d_min) )
                i_min <- i
                d_min <- d_tree[i]
        return( i_min )
```

```
void <- Scan( i )
     for_each(  j , n )
          if( !(in_tree[j] )
              && ( d_tree[j] > dist[i,j] ) )
              d_tree[j] <- dist[i,j]
              pred[j] <- i

d_tree <- INFINITY
pred <- -1
in_tree <- FALSE
d_tree[root] <- 0
#_in_tree <- 0

while( #_in_tree < n )
     i <- FindMin( )
     scan( i )
return(pred)
```

FindMin returns the out-of-tree node which is closest to the tree. Scan updates the distance to the tree for out-of-tree nodes.

It is seen that this algorithm is $O(n^2)$; both FindMin and Scan are $O(n)$ and each is executed n times. Comparing this with Kruskal's algorithm, it is apparent that if m, the number of edges, is $O(n^2)$, then this is faster while if m is $O(n)$ then Kruskal's is faster.

It is possible to accelerate Prim's algorithm when the graph is sparse by observing that only neighbors of i, the node just brought into the tree, need be scanned. If there is adjacency information on hand, the for loop in Scan can then can become

```
     for_each ( j , n_adj_list[i] )
```

The complexity of Scan then becomes $O(d)$, where d is the degree of node i. Thus, the total effort for Scan is reduced from $O(n^2)$ to $O(m)$.

Setting up an adjacency set for an entire graph is itself an $O(m)$ operation:

```
index[nn,list] <- SetAdj( n , m , ends )
     dcl ends[m,2], n_adj_list[n,list]

     for node = 1 to n
          n_adj_list[node] <- Φ
     for edge = 1 to m
          Append( edge , n_adj_list[ends[edge,1]] )
          Append( edge , n_adj_list[ends[edge,2]] )
```

It is also possible to accelerate FindMin if a heap is set up on the values in d_tree. We can then simply pop out the next lowest value and the total effort for popping would be $O(n \log n)$. The problem with this is that we also have to adjust the heap whenever a value in d_tree improves. This requires $O(m \log n)$ effort in the worst case as each edge potentially could result in an update and each update requires

TABLE 4.1

Prim's algorithm

Last node brought in-tree

node	init.	A	C	E	B	D
A	0	0(−)	0(−)	0(−)	0(−)	0(−)
B	100	10(A)	10(A)	10(A)	10(A)	10(A)
C	100	2(A)	2(A)	2(A)	2(A)	2(A)
D	100	100(−)	100(A)	100(A)	5(B)	5(B)
E	100	7(A)	6(C)	6(C)	6(C)	6(C)

$O(\log n)$ effort. Thus, the overall complexity of this version of Prim's algorithm is $O(m \log n)$. Experiments [KS72] indicate that careful implementations of Prim's and Kruskal's algorithms are comparably fast but that in general Prim's is preferable for dense networks, and Kruskal's is preferable for sparse networks. However, these algorithms may be part of larger, more complex procedures which work better with one of these algorithms.

> **Example 4.4.** Returning to Fig. 4.4, suppose all edges not shown have lengths of 100. Kruskal's algorithm would: select (A, C), select (B, D), select (C, E), reject (A, E) because it forms a cycle with (A, C) and (C, E) which were already selected, select (A, B), and then stop because a complete spanning tree has been found.
>
> Prim's algorithm, starting from node A, would bring A into the tree, followed by C, E, B, and D. Table 4.1 summarizes the action of Prim's algorithm, showing d_tree and pred as the algorithm proceeds. At the end of the algorithm, pred[B] is A, corresponding to (A, B) being part of the tree. Similarly, pred shows (A, C), (B, D), and (C, E) as being included in the tree. Thus, Prim's algorithm has selected the same tree as Kruskal's but has selected the links in a different order.

4.2 SHORTEST PATHS

The problem of finding shortest paths plays a central role in the design and analysis of networks. Most routing problems can be solved as shortest path problems once an appropriate "length" is assigned to each edge (or arc) in the network. Design algorithms thus seek to create networks which satisfy path length criteria.

The simplest problem in this class is to find the shortest path between a given pair of nodes. One may also wish to find the shortest path from one node to all others or, equivalently, the shortest path from all nodes to a single node. Sometimes, the shortest path between all pairs of nodes may be required. There may be constraints on the paths (e.g., a limit on the number of edges in the path). Such constraints are dealt with in later chapters as they arise.

In the following, we work with a directed graph model and assume that l_{ij}, the length of the arc between each pair of nodes, i and j, is known. These lengths need not be symmetric. In the case where an arc does not exist, assume that l_{ij} is very large (e.g., larger than n times the longest arc in the network). Note that it is

FIGURE 4.5
Nested shortest paths.

possible to also handle the case of undirected networks by simply replacing each edge by two arcs with the same lengths. Initially, assume the l_{ij} are strictly positive; later this assumption can be relaxed.

4.2.1   Dijkstra's Algorithm

All shortest path algorithms rely on the observation, illustrated in Fig. 4.5, that shortest paths nest, that is, if a node, k, is part of the shortest path from i to j, then the shortest i, j-path must be the shortest i, k-path followed by the shortest j, k-path. Thus, we can find shortest paths using the following recursion:

$$d_{ij} = \min_k (d_{ik} + d_{kj}) \tag{4.2}$$

where d_{xy} is the length of the shortest path from x to y. The difficulty with this approach is that there must be some way to start the recursion, since we do not start with any known values on the right hand side of equation 4.2. There are several ways to do this, each giving rise to a different algorithm.

One way to proceed, due to Dijkstra [Dij59], is suitable for finding the shortest path from one node, i, to all others. Begin by setting

$$d_{ii} = 0$$

and

$$d_{ij} = \infty \qquad \forall j \neq i$$

Then set

$$d_{ij} \leftarrow l_{ij} \quad \forall j \text{ adjacent to } i$$

Then, find the node, j, with the minimum d_{ij}, and use it to attempt to improve other nodes' distances (i.e., by setting)

$$d_{ik} \leftarrow \min (d_{ik}, d_{ij} + l_{jk})$$

At each stage in the process, the value of d_{ik} is the current estimate of the shortest path from i to k and is, in fact, the length of the shortest path found so far. Refer to d_{ik} as the label on node k. The process of using one node to improve other nodes' labels is referred to as scanning the node.

Proceed in this manner, finding the unscanned node with the smallest label and scanning it. Note that since it is assumed that all the l_{jk} are positive, a node cannot give another node a label smaller than its own. Therefore, once a node is scanned it need never be relabeled. So, each node need only be scanned once. If the label on a node changes, it must be rescanned.

Dijkstra's algorithm can be implemented as follows:

```
array[n] <- Dijkstra( n , root , dist )
     dcl dist[n,n], pred[n], sp_dist[n], scanned[n]

     index <- FindMin( )
          d_min <- INFINITY
          for_each ( i , n )
               if( !( scanned[j] ) && (sp_dist[i] < d_min) )
                    i_min <- i
                    d_min <- sp_dist[i]
          return( i_min )

     void <- Scan( i )
          for_each ( j , n )
               if( (sp_dist[j] > sp_dist[i]+dist[i,j] ) )
                    sp_dist[j] <- sp_dist[i] + dist[i,j]
                    pred[j] <- i

     sp_dist <- INFINITY
     pred <- -1
     scanned <- FALSE
     sp_dist[root] <- 0
     #_scanned <- 0

     while( #scanned < n )
          i <- FindMin( )
          scan( i )

     return( pred )
```

In the previous implementation, only the `pred` array was returned. This gives all the paths. The `sp_dist` array could have been returned instead, giving the lengths of the paths, or both the `pred` and `sp_dist` arrays if both values are required.

This algorithm should look familiar. It is nearly identical to Prim's MST algorithm. The only difference is that the nodes are labelled with the length of an entire path rather than the length of a single edge. Note also that it works on directed graphs, where Prim's algorithm only works on undirected graphs. Structurally, however, the algorithms are very similar. The complexity of Dijkstra's algorithm, like Prim's, is $O(N^2)$.

Again, like Prim's algorithm, Dijkstra's is most suitable for dense networks and is particularly appropriate for parallel implementation (where the scan operation can be carried out in parallel, essentially making it $O(1)$ rather than $O(N)$). It is also sometimes appropriate as a part of other more complex procedures. Its major drawbacks are that it does not take advantage of sparsity very well and it is only appropriate for networks with positive arc lengths.

Example 4.5. As an example, consider the network shown in Fig. 4.6. The objective is to find the shortest paths from A to all other nodes. Initially, A has a label of 0 and all other nodes have labels of infinity. Scanning node A, B is assigned a label of 5 and C

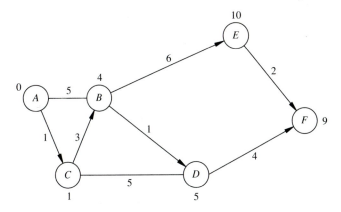

FIGURE 4.6

Shortest paths from A.

a label of 1. C, the node with the smallest label, is then scanned, and B is given a label of 4 (= 1 + 3), while D is given a label of 6. Next, B (label = 4) is scanned; D and E are given labels of 5 and 10, respectively. D is then scanned (label = 5), giving F a label of 9. E is scanned resulting in no new labels. F, the node with the largest label, need not be scanned as it can not possibly relabel any other node. Each node has been scanned at most once. Note that it was important to scan the nodes in increasing order of label because several nodes were relabeled during the course of the algorithm. Had they been scanned immediately, they would have had to be rescanned later.

Note that the edges in the paths from A to all other nodes (i.e., (A, C), (C, B), (B, D), (B, E) and (D, F)) form a tree. This is not a coincidence. It is a direct consequence of the fact that the shortest paths nest within one another. For example, if k is on the shortest path from i to j then the shortest path from i to j includes the shortest path from i to k and the shortest path from k to j.

4.2.2 Bellman's Algorithm

An alternative to Dijkstra's algorithm, due originally to Bellman [Bel58], and subsequently refined by Moore and Pape [Moo59], is to scan the nodes in the order they are labeled. This eliminates the requirement to find the smallest label, but opens up the possibility that a node might need to be scanned more than once.

In its simplest form, Bellman's algorithm keeps a queue of nodes to scan. When a node is labeled, it is added to the queue unless it is already in the queue. The queue is managed first in, first out. Thus, nodes are scanned in the order they are labeled. If a node is relabeled after it is scanned, it is put back into the queue and scanned again.

> **Example 4.6.** In the example shown in Fig. 4.6, we begin by placing A in the scan queue. Scanning A, the labels 5 and 1 are given to nodes B and C, respectively, and both B and C are placed in the scan queue (since they have received new labels and they are not already in the scan queue.) Next, we scan node B, labeling E and D with 11 and 6, respectively. D and E are also placed in the scan queue. We then scan C. B is assigned a label of 4 and is put back in the scan queue. E is scanned and F is assigned

a label of 13 and placed in the scan queue. D is scanned and F is given a label of 10; F is still in the scan queue, however, and so is not put there again. B is now scanned for a second time. It gives E and D labels of 10 and 5, respectively, and both are put back in the scan queue. F is scanned and labels no other node. E is scanned, labeling no other nodes. D is scanned, labeling F with a 9 and F is put back into the scan queue. F is scanned labeling no other node.

It is apparent that nodes B, C, D, and F are scanned twice. This is the price paid for not scanning the nodes in order. On the other hand, it was not necessary to search for the node with the smallest label.

Bellman's algorithm can be implemented as follows:

```
array[n] <- Bellman( n , root , dist )
    dcl dist[n][n], pred[n], sp_dist[n], in_queue[n],
        scan_queue[queue]

    void <- Scan( i )

            in_queue[i] <- FALSE
            for j = 1 to n
                if( sp_dist[j] > sp_dist[i]+dist[i,j] )
                    sp_dist[j] <- sp_dist[i] + dist[i,j]
                    pred[j] <- i
                    if( not( in_queue[j] ) )
                        push( scan_queue , j )
                        in_queue[j] <- TRUE

    sp_dist <- INFINITY
    pred <- -1
    in_queue <- FALSE
    initialize_queue( scan_queue )
    sp_dist[root] <- 0
    Push( scan_queue , root )
    in_queue[root] <- TRUE

    while( not( Empty(scan_queue) )
        i <- Pop( scan_queue )
        scan( i )

    return(pred)
```

A standard queue was used in the previous implementation. It is also possible to use the array in_queue to keep track of which nodes are currently in the scan queue; this saves having to duplicate copies of nodes in the queue at any given time.

As implemented above, Bellman's algorithm is a breadth-first search. The root node is scanned first resulting in paths with one hop. Then, each of the nodes at the end of these one hop paths are scanned, and paths with two hops are produced. This continues, with $k + 1$ hop paths being produced from k hop paths. We can think of the scan list as being divided into "levels" based on the number of hops in the paths labeling the nodes. Nodes placed in level k in the scan list can produce nodes at level $k + 1$. Note, however, that while a node may be relabeled several times with k hop labels, it is not rescanned until level $k + 1$. Thus, at worst, a node is scanned $n - 1$ times. The scan itself is at worst $O(n)$, where n is the number of nodes. Thus, entire algorithm is at worst $O(n^3)$. In practice, however, there is usually a strong correlation between the length of a path and the number of hops. Therefore, nodes are not usually rescanned many times.

In most practical cases, the average number of scans per node is very small, typically at most three or four, even for networks with thousands of nodes. If the average nodal degree is small, as it is in many real networks, the time it takes to find the smallest unscanned node is the dominant part of Dijkstra's algorithm. Bellman's algorithm is then in fact faster *in practice* even though its worst case complexity is worse.

Also, it is possible to improve the complexity of the Scan procedure by maintaining an adjacency list for each node. Scan then becomes $O(d)$ instead of $O(n)$, where d is the degree of the nodes being scanned. Thus, in practice, Bellman's algorithm is usually $O(E)$, where E is the number of edges in the graph.

Bellman's algorithm can be accelerated on average by making its worst case complexity even worse! As can be observed from the preceding example, node B receives a 2-hop label which is better than its 1-hop label. Node B is therefore rescanned. This is inevitable. Node B then labels nodes E and D, and node E labels node F. However, all these nodes are also rescanned. This, as it turns out, is not inevitable. If, after node B is relabeled, we place it at the front of the scan queue rather than at the end, it immediately relabels nodes E and D before they are scanned and they will not have to be rescanned. This modification, which is easily made, in practice, generally improves the performance of Bellman's algorithm. Indeed, this algorithm has been observed [RDK79] in extensive computational experiments to be the fastest, shortest path algorithm in practice. All that is required is to manage the scan list as a double ended queue (dequeue), where values may be added at either end. Simply keep track not only whether a node is currently in the scan queue, but also if it ever was before, and on the basis of this add it either at the front or the back of the queue.

Example 4.7. Interestingly, it was found [Ker81] that the worst case complexity of this modified version of Bellman's algorithm is exponential. An example of this worst case behavior is illustrated in Fig. 4.7. If the neighbors of each node are labeled in alphabetical order, then root R labels A with an 11, B with a 9, etc. A is then scanned, labeling nothing. When B is scanned, A is relabeled 10 and placed at the front of the scan list. A is scanned again, labeling nothing. C is scanned labeling A with a 9, and B with a 7. D is scanned again. B is scanned again, labeling A with an 8. This continues, with A receiving all possible labels between 4 and 11 and being scanned each time it is relabeled. A is scanned 2^{N-1} times.

This situation is, admittedly, pathological and depends not only on unusual arc lengths, but also on the unfortunate order of nodes being labeled by their neighbors. Note, in particular, how paths with a large number of hops are shorter than other

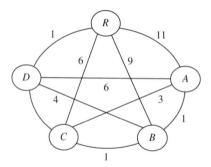

FIGURE 4.7
Pathological case for Bellman's algorithm.

paths with fewer hops. In practice, if we find such a situation occuring frequently, it is possible to revert to Dijkstra's algorithm.

In addition to improved average performance in many cases, Bellman's algorithm offers the advantage of working even when some of the arc lengths are negative. Dijkstra's algorithm relies on the fact that a node cannot give another node a label smaller than its own. This is true only if none of the arc lengths are negative. Bellman's algorithm, on the other hand, does not make this assumption and rescans nodes whenever they are relabeled. Thus, it is appropriate when negative arc lengths are present. Note, however, that if the graph contains a cycle of negative total length, that even Bellman's algorithm is not useful. In this case, the algorithm does not terminate and nodes continue to label other nodes indefinitely.

> **Example 4.8.** Consider the network shown in Fig. 4.8. A labels B, 3. B labels D, 5. D labels C, 6. C labels B, 1. The cycle B-D-C-B has a length of -2 and we continue to traverse it, reducing the label of the nodes in the cycle indefinitely. It is possible to test for negative cycles simply by keeping track of the number of hops in each path and halting the algorithm when this count reaches N since no simple path can contain more than $N-1$ arcs. It is not, however, possible to find the shortest path in this case; indeed no finite shortest path exists.

There are other variants of Bellman's algorithm, as well as other algorithms, for finding the shortest path from one node to all others in a variety of special situations. Excellent surveys of these are given in [Dre69] and [RDK79].

4.2.3 Floyd's Algorithm

While it is essentially just as easy to find the shortest path from one node to all others as it is to find the shortest path between two specific nodes, the problem of finding the shortest path between all pairs of nodes is N times harder. One possibility is to use Bellman's or Dijkstra's algorithms N times, starting at each source. Another possibility, particularly attractive when the network is dense, is to use Floyd's algorithm.

Floyd's algorithm [Flo62] relies on the same type of recursion described before in the introduction of Dijkstra's algorithm, but it uses the recursion in a subtly different

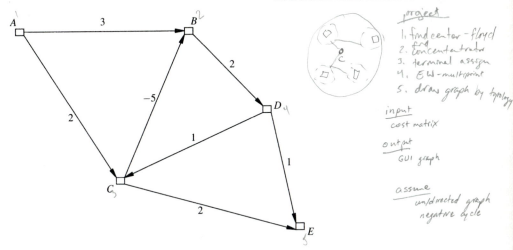

FIGURE 4.8

A graph with a negative cycle.

way. Here, $d_{ij}(k)$ is defined as the shortest path from i to j, which uses only nodes numbered k or lower as intermediate nodes. Thus, $d_{ij}(0)$ is defined as l_{ij}, the length of the link from i to j, if (i, j) exists, or as infinity otherwise. Therefore,

$$d_{ij}(k) = \min(d_{ij}(k-1), d_{ik}(k-1) + d_{kj}(k-1))\qquad(4.3)$$

That is, we consider using node k as a via point for each i, j-path. Then the entire algorithm becomes:

```
array[n][n] <- Floyd( n , dist )
    dcl dist[n][n], pred[n][n], sp_dist[n,n]

    for_each ( i , n )
       for_each ( j , n )
           sp_dist[i,j] <- dist[i,j]
           pred[i,j] <- i

    for_each ( k , n )
       for_each ( i , n )
          for_each ( j , n )
              if( sp_dist[i,j] > sp_dist[i,k] + sp_dist[k,j] )
                  sp_dist[i,j] <- sp_dist[i,k] + sp_dist[k,j]
                  pred[i,j] <- pred[k,j]

    return( pred )
```

`pred[i,j]` gives the next to last node in the path from i to j and can be used to backtrack the path from i to j. Like Bellman's algorithm, Floyd's algorithm works for negative arc lengths. If negative cycles are present, Floyd's algorithm terminates but does not guarantee shortest paths. Negative cycles can be detected by the presence of negative numbers on the main diagonal of the sp_dist array.

Example 4.9. Consider the graph in Fig. 4.8 again, with the change that the (C, B) arc has length 5 instead of -5. The initial distance and predecessor arrays are:

		to				
		A	B	C	D	E
	A	0	3	2	–	–
f	B	–	0	–	2	–
r	C	–	5	0	–	2
o	D	–	–	1	0	1
m	E	–	–	–	–	0

sp_dist

		to				
		A	B	C	D	E
	A	A	A	A	A	A
f	B	B	B	B	B	B
r	C	C	C	C	C	C
o	D	D	D	D	D	D
m	E	E	E	E	E	E

pred

Note that sp_dist has zeros on the main diagonal and infinity (denoted by '-' here) wherever no link exists. Note also, that since the graph is directed and asymmetric, sp_dist is asymmetric.

Considering A as an intermediate node has no effect on these arrays because A has no incoming arcs and hence no paths pass through A. Considering B, however, results in changes in the (A, D) and (C, D) positions in these arrays:

		to				
		A	B	C	D	E
	A	0	3	2	5	–
f	B	–	0	–	2	–
r	C	–	5	0	7	2
o	D	–	–	1	0	1
m	E	–	–	–	–	0

sp_dist

		to				
		A	B	C	D	E
	A	A	A	A	B	A
f	B	B	B	B	B	B
r	C	C	C	C	B	C
o	D	D	D	D	D	D
m	E	E	E	E	E	E

pred

Further, considering C, D, and E as intermediate nodes results in the following arrays:

		to				
		A	B	C	D	E
	A	0	3	2	5	4
f	B	–	0	3	2	3
r	C	–	5	0	7	2
o	D	–	6	1	0	1
m	E	–	–	–	–	0

sp_dist

		to				
		A	B	C	D	E
	A	A	A	A	B	c
f	B	B	B	D	B	D
r	C	C	C	C	B	C
o	D	D	C	D	D	D
m	E	E	E	E	E	E

pred

4.2.4 Incremental Shortest Path Algorithms

Sometimes, in the context of design and analysis of networks, it is necessary to find the shortest paths between all pairs of nodes (or some pairs) after a modification has been made to the length of an arc. This includes having to add or delete an arc (in which case the length of the arc can be said to have gone from infinite to finite or vice versa.) Thus we assume that the shortest path between all pairs of nodes is known and that the problem is to determine what changes, if any, result from the change in one arc's length. The following algorithm is due to Murchland [DF79], and considers the cases of increasing and decreasing the length of the arc separately. These algorithms work on directed graphs and can handle negative arc lengths. As always, however, they cannot deal with negative cycles.

4.2.4.1 DECREASED ARC LENGTH. Suppose the length of arc (i, j) is decreased. Since shortest paths nest (i.e., if the shortest path from i to j includes k, then the i, j path must be the i, k path concatenated with the k, j path), if (i, j) is not the shortest path from i to j after it has been shortened (in which case it could not have been the shortest path before its length was changed either) then it is not part of any shortest path and the change can be ignored. Similarly, if (i, j) is to be part of the path from k to m then it must also be part of the paths from k to j and i to m. In fact, the new shortest path from k to m must be precisely the concatenation of the old shortest path from k to i, the link (i, j), and the old shortest path from j to m. This is illustrated in Fig. 4.9.

Therefore, it is only necessary to scan nodes i and j to find appropriate sets K and M containing nodes k and m as shown in Fig. 4.9 and then to consider pairs of nodes, one from each set. Then include i in set K and j in set M, and then test if

$$d_{km} > d_{ki} + l_{ij} + d_{jm}$$

and if so, update d_{km} and the predecessors. This gives rise to the following algorithm. The algorithm returns `sp_dist` and `pred`, the updated shortest path and predecessor arrays. It takes as input the current link-length array, `dist`, the endpoints (i and j), of the newly shortened link, and the new (shortened) `length`. Φ is the empty list.

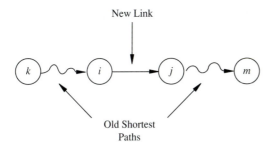

New Link

Old Shortest
Paths

FIGURE 4.9
Incremental shortest path when (i, j) is shortened.

```
(array[n,n] , array[n,n]) <-
    sp_decrease( n , i , j , length , *dist , sp_dist , pred )

    dcl dist[n,n], pred[n,n], sp_dist[n,n] ,
        setk[set] , setm[set]

    dist[i,j] <- length
    if( length >= sp_dist[i,j] )
        return( sp_dist , pred )

    setk <- Φ
    setm <- Φ
    for_each ( k , n )
        if( sp_dist[k,j] > sp_dist[k,i] + length )
            append( k , setk )
    for_each ( m , n )
        if( sp_dist[i,m] > length + sp_dist[j,m] )
            append( m , setm )

    for_each( k , setk )
        for_each( m , setm )
            if( sp_dist[k,m] >
                sp_dist[k,i] + length + sp_dist[j,m] )
                sp_dist[k,m] <-
                    sp_dist[k,i] + length + sp_dist[j,m]
                if( j = m )
                    pred[k,m] <- i
                else
                    pred[k,m] <- pred[j,m]
    return( sp_dist , pred )
```

It can be seen that the worst case complexity of the above procedure is $O(n^2)$ since it involves executing two nested loops of at worst $O(n)$. In practice, it is rare that both sets are $O(n)$ and so the actual complexity is usually considerably lower.

4.2.4.2 INCREASED ARC LENGTH.

We now consider the case where the link, (i, j) is lengthened or removed from the graph (in which case, its length can be thought of as infinite). If (i, j) was not part of the shortest path from k to m before its length was increased, it is certainly not part of the path afterwards. So it is only necessary to check pairs (k, m), whose shortest path lengths satisfy:

$$d_{km} = d_{ki} + l_{ij} + d_{jm}$$

Note that if l_{ij} is not part of the shortest path from i to j, then no paths change. This gives rise to the following algorithm. In this case, pairs is a set of node pairs which need to be checked. Thus, the elements of pairs are node pairs. This algorithm has the same inputs and outputs as the preceding one.

The following algorithm is essentially the same as Floyd's algorithm except that it only works with selected pairs possibly using the altered link before it was lengthened.

```
(array[n,n] , array[n,n]) <-
   sp_increase( n , i , j , *dist , length , sp_dist , pred )

   dcl dist[n,n], pred[n,n], pairs[set]

   // old dist
   dist[i,j] <- length
   if( dist[i,j] > sp_dist[i,j] )
      return( sp_dist , pred )   // do nothing if longer edge

   // find affected paths  n->k
   pairs <- Φ
   for_each ( k , n )
      for_each ( m , n )
         if( sp_dist[k,m] =
               sp_dist[k,i] + dist[i,j] + sp_dist[j,m] )
            append( (k,m) , pairs )
            sp_dist[k,m] <- dist[k,m]   // revert back edge

   // update new edge
   dist[i,j] <- length

   for_each ( a , n )
      for_each( (k,m) , pairs )
         if( sp_dist[k,m] > sp_dist[k,a] + sp_dist[a,m] )
            sp_dist[k,m] <- sp_dist[k,a] + sp_dist[a,m]
            pred[k,m] <- pred[a,m]

   return( sp_dist , pred )
```

The complexity of this procedure is $O(np)$, where p is the number of node pairs in the set `pairs`. In the worst case, it is possible for `pairs` to contain $n^2/2$ node pairs and for the complexity to be $O(n^3)$, like Floyd's algorithm. In practice, however, p tends to be much smaller in most cases.

Example 4.10. Consider the network shown in Fig. 4.10. The edges shown are bidirectional. The lengths of the shortest paths between all pairs of nodes are given in Table 4.2.

Now add an arc, (B, E), with $l_{BE} = 1$. Since

$$d_{BE} > l_{BE}$$

we proceed. Further, it is apparent that

$$d_{BC} > l_{BE} + d_{EC}$$

but that

$$d_{Bx} \le l_{BE} + d_{Ex}$$

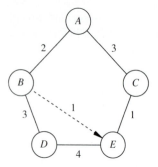

FIGURE 4.10

TABLE 4.2
Shortest paths

	A	B	C	D	E
A	0	2	3	5	4
B	2	0	5	3	6
C	3	5	0	5	1
D	5	3	5	0	4
E	4	6	1	4	0

for all other nodes, x. Therefore,

$$set_m = C, E$$

Similarly,

$$set_k = A, B$$

Now investigate all pairs

$$(k, m), \text{ where } k \in set_k \text{ and } m \in set_m$$

(i.e., the pairs (A, C), (A, E), (B, C) and (B, E)). We find that all but (A, C) are improvements and so we update the shortest paths and predecessors. The shortest path matrix is now as shown in Table 4.3, below.

Notice that this matrix is no longer symmetric because an arc (unidirectional) has been added to the network.

Now suppose that we set $l_{BE} = 5$. Examining the shortest path matrix, it is apparent that before the change to l_{BE} that

$$d_{BE} = l_{BE}$$

TABLE 4.3
Shortest paths with (B, E) added

	A	B	C	D	E
A	0	2	3	5	3
B	2	0	2	3	1
C	3	5	0	5	1
D	5	3	5	0	4
E	4	6	1	4	0

TABLE 4.4
Shortest paths with $l_{BE} = 5$

	A	B	C	D	E
A	0	2	3	5	4
B	2	0	5	3	5
C	3	5	0	5	1
D	5	3	5	0	4
E	4	6	1	4	0

We now check all node pairs, (k, m), and find that

$$d_{km} = d_{ki} + l_{ij} + d_{jm}$$

only for the pairs (A, E), (B, C) and (B, E). We therefore set

```
pairs <- { (A,E) , (B,C) , (B,E) }
```

and then loop on all intermediate nodes, checking for the shortest paths for these pairs only. As this is done, notice that the shortest (A, E) path becomes 4 (through C, as before) and the shortest (B, C) path becomes 5 (through A, as before). The shortest (B, E) path, however, remains the arc (B, E). This leaves us with Table 4.4.

4.3 SINGLE COMMODITY NETWORK FLOWS

Given a network topology and a single requirement from source s to destination d, it is desired to find a feasible flow pattern, that is, a set of link flows that get the requirement from s to d without any link's flow exceeding its capacity. The topology is given as a set of links, l_{ij}, with associated capacities, c_{ij}. Since most real networks are sparse, it is possible to store the topology as an explicit list and take advantage of the sparsity. Alternatively, it is possible to store the capacities as a matrix and set c_{ij} to zero when l_{ij} does not exist.

The problem then is to find one or more paths from s to d and to send flow over these paths satisfying the requirement. The sum of the flows is to be equal to the requirement magnitude and the sum of the flows on each link is not to exceed the link's capacity.

There are several forms of this problem. In the first, as stated above, the attempt is to find flows satisfying a given requirement. Alternatively, it is possible to try to maximize the amount of s, d-flow, subject to the capacity constraint. Finally, with a knowledge of the a cost-per-unit flow for each link, we can attempt to find a flow satisfying a given requirement at minimum cost. The solutions to all these problems are closely related and will be discussed further. Additionally, the solution to this problem is the basis for the solution to a more complex problem, called the multicommodity flow problem, where there are many requirements between different sources and destinations. This is one of the central problems in network design and is discussed in a later chapter.

Formally, we wish to find flows, f_{ij}, satisfying:

$$\sum_j f_{ij} - \sum_j f_{ji} = r_{ij} \qquad \text{for } i = s$$

(1)
$$\sum_j f_{ij} - \sum_j f_{ji} = -r_{ij} \qquad \text{for } i = d$$

$$\sum_j f_{ij} - \sum_j f_{ji} = 0 \qquad \text{otherwise}$$

$$f_{ij} \le c_{ij} \tag{4.4}$$

$$f_{ij} \ge 0; \qquad \forall i, j \tag{4.5}$$

Note that this is a directed link model (i.e., there are separate capacities c_{ij} and c_{ji}). It is also possible to deal with undirected networks by replacing each undirected link, l_{ij}, with two directed links with separate capacities. As will be seen, there is only flow in one direction in any given link at any time during the solution of this problem.

4.3.1 The Ford-Fulkerson Algorithm

The best known algorithm for the solution of the single-commodity, network flow problem is due to Ford and Fulkerson [FF62]. The algorithm identifies s, d-paths and sends as much flow as possible over each path without violating the capacity constraint. Indeed, one is tempted to simply identify paths and fill them with flow.

Consider, for example, the network shown in Fig. 4.11. Suppose all links have capacities of 1. It is possible to send one unit of flow on the $SABD$ path and one unit of flow on the $SEFD$ path. Since the sum of the capacities of the links leaving S is 2, and each unit of SD-flow must use a unit of this capacity, it is fairly clear that no more flow than this can be sent. In fact, since each unit of SD-flow must utilize at least one unit of capacity in any SD-cut (set of links whose removal separates S from D), the maximum SD-flow can be no greater than the capacity of any cut (total capacity of all links in the cut). Therefore,

Lemma 4.1. (Ford and Fulkerson) The maximum SD-flow can be no greater than the capacity of the minimum SD-cut.

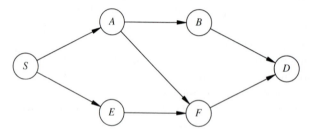

FIGURE 4.11

In fact, the maximum SD-flow is exactly equal to the capacity of the minimum SD-cut. This is the celebrated Max Flow–Min Cut Theorem of Ford and Fulkerson. The proof is constructive; their algorithm finds the flow.

Constraint (**1**) above is called the conservation of flow constraint. It says that the flow into a node must equal the flow out of the node unless the node is the source or the destination of the requirement, in which case the flow out (or in) must equal the source destination flow. Any SD-cut partitions the nodes into two sets, X and Y, with S in X and D in Y. If constraints (**1**) are summed for all nodes in X, we find that the total flow from X to Y, minus the flow from Y to X is equal to the SD-flow. Note that the sum of the left hand sides is the flow in links, with one end in X and the other end in Y, minus the sum of the flow in links, with one end in Y and the other in X. Links with both ends in X contribute nothing to this sum since they appear in the sum with both positive and negative signs. Links with neither end in X do not appear in the sum at all. S contributes the requirement to the right hand side; all the other nodes contribute nothing.

Thus, for the theorem to be satisfied,

1. The total flow across the minimum capacity cut must equal its capacity, that is, all links in the cut must be saturated (flow equal to capacity).

2. The flow backwards across this cut must be zero.

In fact, all minimum capacity cuts must be saturated and they are at the end of the algorithm. The algorithm proceeds by identifying paths with spare capacity and sending flow over these paths. When it can not find any path with spare capacity, a saturated cut is identified and the algorithm terminates. All other minimum capacity cuts are also saturated, but the algorithm does not identify these.

In its simplest form, the Ford-Fulkerson algorithm follows. n is the number of nodes. m is the number of links. Each node has a label

(maxflow, pred, sign)

that gives the maximum flow possible on the current path so far, the predecessor of the node on the current path, and the direction of flow through the link. The symbolic value U stands for undefined; the actual value of U should be distinguishable from any valid value.

The algorithm returns the flow in each link. The total s, d-flow can be found by summing the flow out of s (or into d). The algorithm identifies s, d-paths using a modification of Bellman's algorithm. The modification permits the use of links, (i, j) in the forward direction (from i to j) if f_{ij}, the flow from i to j is less than its capacity, c_{ij}. It also permits the use of links in the reverse direction (i.e., the (i, j) link is used to move flow from j to i), but this is only permitted if there is already flow from i to j. In this case, flow is actually removed from the (i, j) link.

The maximum amount of flow in the (i, j) direction is $c_{ij} - f_{ij}$. The maximum amount of flow in the (j, i) direction is f_{ij}. mxf, in the label of each node, identifies the maximum amount of flow that can be sent on a path.

Inside the `while` loop below, we start at s and try to label d. If there is success, we backtrack from d to s, following `pred` back from d. Increment the flow in each forward link and decrement the flow in each reverse link in the path. If there is no label for d, the algorithm terminates. This identifies the maximum flow; the links (i, j) with i labeled and j unlabeled form the saturated cut.

```
number <- FFflow( n , s , d , cap , *flow )
     dcl cap[n][n] , flow[n][n] , pred[n] ,
         sign[n] , mxf[n] , scan_queue[queue]

    void <- Scan( i )
         for each( j , n )
             if( pred[j] = U )
                if( flow[i,j] < cap[i,j] )
                    mxflow <- min( mxf[i] , cap[i,j] - flow[i,j] )
                    mxf[j] , pred[j] , sign[j] <- mxflow , i , +
                  else if( flow[j,i] > 0 )
                    mxflow <- min( mxf[i] , flow[j,i] )
                    mxf[j] , pred[j] , sign[j] <- mxflow , i , -

    void <- Backtrack( )
         n <- d
         tot_flow <- tot_flow + mxf[d]
         while( n != s )
            p <- pred[n]
            if( sign[n] = + )
              flow[p,n] <- flow[p,n] + mxf[d]
            else
              flow[n,p] <- flow[n,p] + mxf[d]

    tot_flow <- 0
    flow <- 0
    flag <- TRUE

    while( flag )
       pred <- U
       Initialize_queue( scan_queue)
       Push( scan_queue , s )
       mxf[s] <- INFINITY
       while( !( Empty(scan_queue) & (pred[d] = U) )
             i <- Pop( scan_queue )
             Scan( i )
       if( pred[d] != U )
         Backtrack( )
       flag <- ( pred[d] != U )
    return( tot_flow )
```

The `Scan` function is $O(n)$. A more efficient form of this algorithm, with `Scan` of $O(d)$ (the degree of the node) creates an adjacency list for each node. In `Scan(i)`, replace

```
for j = 1 to n
```

by

```
for each ( j , adj_set[i] )
```

When the algorithm halts, a cut is implicitly defined. The nodes with labels not equal to U are in set X and the remaining nodes are in set Y, where X and Y are defined as before. By the way the algorithm labels, it guarantees that all the arcs in the (X, Y)-cut are saturated, and all arcs in the (Y, X)-cut have zero flow. This can be seen by noting that the algorithm halts when it cannot continue the labeling. Any unsaturated arc in an S, D-cut or any arc in a D, S-cut, with non-zero flow could be used to continue the labeling. The fact that we cannot continue means that no such arcs exist. Thus, the SD-flow is equal to the capacity of the (X, Y)-cut and the max-flow min-cut theorem is proven constructively.

> **Example 4.11.** It is important to consider using arcs in the reverse direction. If this is not done, there is no guarantee that the flow is maximum. Consider the network in Fig. 4.11. Suppose the first path is the SAFD path. One unit of flow is sent over this path. Then another path is sought. S cannot label A because the SA arc is saturated. S labels E and E labels F. F cannot label D as the FD arc is saturated. Note that there is no arc from F to A; the FA arc is directed from A to F. It is important that the algorithm be able to use the AF arc in the reverse direction, removing a unit of flow from it. This allows F to label A. A then labels B and B labels D. Thus, a second path has been found, $SEFABD$, with the FA arc used in the reverse direction. The net effect of sending flow on the two paths is to send one unit of flow from S to E to F to D, and another unit of flow from S to A to B to D. The original unit of flow on the AF link is canceled in the second path and the net flow is zero on this arc. The two paths found by the algorithm have merged into two new paths.

As shown, for a network with N nodes and E edges, one pass of this algorithm to find a single path has complexity $O(N^2)$, since each node is scanned at most once (nodes are not relabeled), and the scan is $O(N)$. With the modification using an adjacency list, each arc is examined at most once from each end and the pass has complexity $O(E)$. The effort in setting up the adjacency lists is $O(E)$; a single pass through the arcs, inserting nodes in the adjacency lists, is sufficient. Thus, for sparse networks, the effort is justified.

The complexity of the entire algorithm can be seen to be the product of the complexity of finding a single path, times the number of paths found. If the capacities of the arcs are integers, each path adds a minimum of one unit of flow to the network. The number of paths is therefore bounded by the final flow, F. Thus the overall complexity is $O(EF)$.

> **Example 4.12.** In general F can be very large. Consider, for example, the network shown in Fig. 4.12. All the arcs except (A, B) have capacity K, a very large number. (A, B) has capacity 1. Suppose the first path found is $SABD$. Because arc (A, B) has

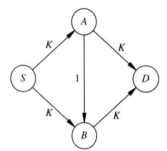

FIGURE 4.12

capacity 1, only one unit of flow can be sent over the path. Next, suppose the $SBAD$ path is found. Since only one unit of flow can be removed from (A, B), again only one unit of flow is sent over this path. The algorithm proceeds finding $2K$ paths, $SABD$ and $SBAD$, over and over again, each with one unit of flow, and the worst case complexity is attained.

It turns out that such problems cannot arise if the algorithm finding the paths finds minimum hop paths. The breadth first search above does this. In this case, it can be shown (c.f., [Law76], [PS82], [GM84]) that the maximum number of paths is $O(E)$. In fact, [Din70] presents an algorithm which is worst case $O(N^3)$. The maximum flow problem has been much studied and there are many algorithms and refinements. These references discuss many of these.

In practice, a good implementation of the previous algorithm will do well in most cases to solve either the maximum flow problem or the problem of finding a flow of a specific value. We now turn to the problem of finding minimum cost flows.

4.3.2   Minimum Cost Flows

Suppose there is a given cost per unit flow, c_{ij}, associated with each link. It is desired then to find a minimum cost s, d-flow of a given value, where the cost of a flow is defined as the sum of the flow over all links times the cost per unit flow for that link. Similarly, we may wish to find a minimum cost flow of maximum magnitude. For example, we wish to minimize cost, subject to the restriction that we produce a flow of maximum magnitude.

The simplest way to find a minimum cost flow is to modify the Ford-Fulkerson algorithm to find shortest paths, instead of just min-hop paths, where cost per unit flow is used as the link lengths. Bellman's algorithm or any of the other shortest path algorithms can be adapted to this purpose. All that is required is to keep track of the flow in each link and, as in the original Ford-Fulkerson algorithm, to only use unsaturated links in the forward direction, and only use links in the reverse direction if they have positive flow.

Alternatively, one can think of implementing the algorithm by modifying the Ford-Fulkerson algorithm. The label now contains a fourth entry, p, the length of the path. This value is updated the same way it is in Bellman's algorithm. For example, a node with path length p gives its neighbor a path length of q, where q is the sum of p and the length of the link joining the two nodes.

Example 4.13. In Fig. 4.13, each link is labeled (cost/flow, capacity). The links are bi-directional. For example, it costs \$4 to send a unit of flow between A and B in either direction. Use the Ford-Fulkerson algorithm already given, modified to keep track of path lengths, and allowing a node to be relabeled if its path length can be improved to solve the problem. Thus, each node has a label

(pathlength, maxflow, pred, sign) S has the label (0, INFINITY, PHI, PHI)

indicating that there is a cost (path length) of 0 from the source, no limit on the amount of flow, and no predecessor. All other nodes begin without labels or, equivalently, with labels of

(INFINITY, INFINITY, PHI, PHI)

A label with infinite pathlength is equivalent to no label because any relabeling with a finite pathlength replaces such a label.

S is placed on the scan list and is scanned first. S gives C the label $(2,4,S,+)$, and C is placed on the scan list. Since S is at distance 0 from itself and has no restriction on the amount of flow it can pass, the path length is simply the length of the link from S to C, and the maxflow is its capacity. S gives A the label $(2,3,S,+)$, and A is placed on the scan list. The order in which C and A are labeled is arbitrary. It is a function of how the adjacency list at S was set up.

C is then scanned. It attempts to label S, but fails since S already has a label with path length 0, while C would give it a label with path length 4. C does succeed, however, in giving label $(8,3,C,+)$ to E. The path length of 8 is the sum of 2 (the path length in the current label on C) and 6 (the length of the link from C to E). The maxflow of 3 is the min of 4 (the maxflow in the label on C) and 3 (the capacity of the link from C to E minus its current flow, 0). E is placed in the scan list. Similarly, C gives B the label $(11,4,C,+)$ and B is placed in the scan list.

A is then scanned. It can relabel B with a smaller path length and gives B the label $(6,2,A,+)$. Note that B is relabeled with the smaller path length, even though it also

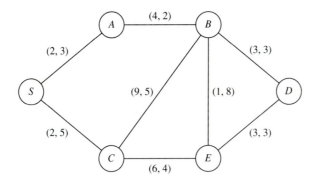

FIGURE 4.13
Minimum cost flow.

results in a smaller maxflow in the label. This can diminish the amount of flow in this path, but it will not diminish the total flow eventually sent to D; it can simply require more paths to deliver that flow. Although B is relabeled, it is not placed in the scan list because it is already there.

E is then scanned, giving D the label $(11,3,C,+)$. D is the destination and need not be placed in the scan list. Although D has been labeled, it is important to continue until the scan list is empty because there may yet be a better path. E fails to relabel B since B already has a label with path length 6, and E could only give it a 9. B is scanned and relabels D with $(9,2,B,+)$.

The scan list is now empty. Backtracking the path from D, there are nodes B (the predecessor in the label on D), A (the predecessor in the label on B), and S. Add two units of flow (the maxflow in the label on D) to links (S, A), (A, B), and (B, D). All three of these links now have positive flows and are eligible to be used in the reverse direction. Link (A, B) is saturated in the forward direction and eligible for use only in the reverse direction.

The second pass of the algorithm finds the path S-C-E-D with a length of 11 and a flow of 3. The third pass finds the path S-C-E-B-D with a length of 12 and a flow of 1. The reader is encouraged to work through the details of these two passes.

On the fourth pass all nodes except D are labeled, but D cannot be labeled and the algorithm halts. This corresponds to the cut with capacity 6 between the rest of the nodes and node D. Thus, there is again a maximum flow equal to the capacity of a minimum cut. This produces a total cost of

$$(9 \times 2 + 11 \times 3 + 12 \times 1) = 63$$

If we only wanted to send three units of flow it could be done at a cost of

$$(9 \times 2 + 11 \times 1) = 29$$

using just the first and second paths. Therefore, this algorithm can be used to solve both the minimum cost, maximum flow problem, or the problem of finding the minimum cost flow of a given magnitude. In the latter case, it is possible to stop the algorithm when the flow reaches the desired value. In the former case, as before, the algorithm is run until no more paths can be found. The fact that only shortest paths are accepted at each step does not alter the fact that the algorithm saturates a cut and finds a maximum flow, since eventually any path with spare capacity becomes the shortest path.

This extension to the basic Ford-Fulkerson algorithm is straightforward. Its only drawback is that we have lost our assurance on the computational complexity. There is no longer a depth first search, and it is now possible to have to find $O(L)$ paths for a flow of magnitude L. In practice, minimum length paths also tend to be min-hop and there is rarely significant degradation in running time. Nevertheless, in theory this can happen. This has led to the development of more complex algorithms with lower worst case complexities. Many of these algorithms, called dual algorithms, start by using the original Ford-Fulkerson algorithm to find a maximum flow (or a flow of a given value) and then seek to reroute flow along a cycle of negative length, moving flow from a higher cost path to a lower cost path. The interested reader is referred to [PS82] [GM84] for details.

EXERCISES

4.1. We define a complete binary tree as follows. It has a distinguished node, called the root, with two neighbors. Each of these two nodes in turn has two additional neighbors. Each of these four new nodes in turn has two additional neighbors; etc. Eventually, a level in the tree is reached where the nodes have no additional neighbors. These nodes are called leaves. The number of nodes in a path from the root to a leaf is called the number of levels in the tree.

(a) How many nodes are there in a complete binary tree with L levels?

(b) How many possible orders can the nodes be visited in by a depth first search?

(c) How many possible orders can the nodes be visited in by a breadth first search?

(d) How many possible orderings of the nodes exist without any restriction on how the nodes are visited?

(e) Under what circumstances might Prim's and Kruskal's algorithms select different trees?

4.2. Modify `BfsTree` so that it works on undirected trees.

4.3. Modify `DfsTree` so that it works on undirected trees.

4.4. Modify `BfsTree` so that it works on general undirected graphs as `Dfs` does.

4.5. Suppose we implement Kruskal's algorithm, keeping track of components by pointing from one node in a component to another (as was described in the text), but we do not collapse chains. What will the complexity of the algorithm be in the worst case?

4.6. Carry out Floyd's algorithm on the graph in Fig. 4.8.

(a) Where are there negative numbers on the main diagonal?

(b) What do these numbers (in these particular places) signify?

4.7. What is the difficulty with finding the longest path in a graph in general?

4.8. Under what circumstances is it possible to find the longest path in a graph?

4.9. Roughly how many paths are there between two nodes in a complete graph with n nodes?

4.10. Give an efficient algorithm for finding the shortest paths between all pairs of nodes in a tree. What is the complexity of your algorithm?

4.11. Extend the maximum flow algorithm just created to the case where there are many sources and a common destination, by creating a common source with arcs leading into all the other sources.

BIBLIOGRAPHY

[Bel58] Bellman, R.: "On a routing problem," *Quart. Appl. Math.*, 16:87–90, 1958.

[DF79] Dionne, R. and M. Florian: "Exact and approximate algorithms for optimal network design," *Networks*, 9:39–59, 1979.

[Dij59] Dijkstra, E. W.: "A note on two problems in connexion with graphs," *Numerische Mathematik*, 1:269–271, 1959.

[Din70] Dinic, E. A.: "Algorithm for solution of a problem of maximum flow in a network with power estimation." *Soviet Math. Dokl.*, 11, 1970.

[Dre69] Dreyfus, S. E.: "An appraisal of some shortest-path algorithms." *Journal of the OR Society of Amer.*, 17:395–412, 1969.

[FF62] Ford, L. R., and D. R. Fulkerson: *Flows in Networks*, Princeton Univ. Press, Princeton, NJ, 1962.

[Flo62] Floyd, R. W.: "Algorithm 97: Shortest path," *Comm. ACM*, 5:345, 1962.

[GM84] Gondran, M., and M. Minoux: *Graphs and Algorithms*, John Wiley, New York, 1984.

[Ker81] Kershenbaum, A.: "A note on finding shortest path trees," *Networks*, 11:399–400, 1981.

[KS72] Kershenbaum, A., and R. Van Slyke: "Computing minimal spanning trees efficiently," In *Proc. Assoc. for Computing Machinery Annu. Conf.*, pages 518–527, Aug. 1972.

[KT81] Karp, R. M., and R. E. Tarjan: "Linear expected time algorithms for connectivity problems," *J. Algorithms*, 1:374–393, 1981.

[Law76] Lawler, E.: *Combinatorial Optimization: Networks and Matroids*, Holt, Rinehart & Winston, New York, 1976.

[Moo59] Moore, E. F.: "The shortest path through a maze," in *Proc. Internat. Symp. Switching Th., 1957, Part II, Harvard Univ. Press*, volume 4, pages 285–292, 1959.

[Pri57] Prim, R. C.: "Shortest connection networks and some generalizations," *BSTJ*, 36:1389–1401, 1957.

[PS82] Papadimitriou, C., and K. Steglitz: *Combinatorial Optimization: Algorithms and Complexity*, Prentice-Hall, Englewood Cliffs, NJ, 1982.

[RDK79] Karney, F., F. Glover and D. Klingman: "A computational analysis of alternative algorithms and labeling techniques for finding shortest path trees." *Networks*, 9:215–248, 1979.

CHAPTER
5

CENTRALIZED NETWORK DESIGN

5.1 PROBLEM DEFINITION

As was defined in Chap. 1, a centralized network (see Fig. 5.1) is a network where all communication is to and from a single site. In such networks, the other sites usually have relatively simple equipment not capable of making routing decisions. Also, with all traffic going to a single site, there is relatively little motivation (other than reliability) for including lines between other sites. This leads to a tree topology where there is only one possible path to the center (and hence, between any pair of nodes).

Sometimes terminals are connected directly to the central site. Sometimes multi-point lines are used, where groups of terminals share a tree to the center. The topology may include remote concentrators or multiplexors, which consolidate low speed lines into higher speed lines, taking advantage of the economy of scale of cost versus capacity. These devices are generically refered to as concentrators. These concentrators may themselves be cascaded, forming a tree topology, or even a mesh, connecting to the center.

In this chapter, the focus is on the basic problems that arise in choosing a topology for centralized networks that are trees. The issues relating to mesh networks are discussed in a later chapter, although some of the problems discussed in this

179

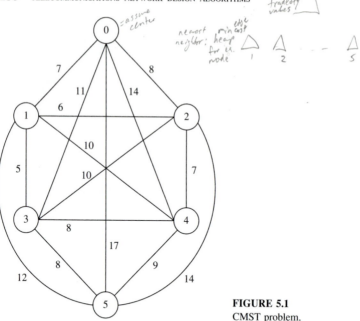

FIGURE 5.1
CMST problem.

chapter have relevance in the design of mesh networks as well. Therefore, the focus is on the following problems:

✓**1.** Multipoint line topology—selection of the links connecting terminals to concentrators or directly to the center.

✓**2.** Terminal assignment—the association of terminals with specific concentrators.

✓**3.** Concentrator location—deciding where to place concentrators, and whether or not to use them at all.

These problems are presented in the context of multipoint line networks because they were originally developed to solve such problems. These algorithms also have relevance in the design of local area networks (LANs). The basic topological structure of a LAN is a ring where there is only one (or two) paths for each packet. Thus, routing within a single ring is not an issue. When rings are interconnected by bridges, the topology at any given time is a tree (although the physical topology can be more highly connected for the sake of reliability). The partitioning of the terminals onto the rings, to avoid overloading the rings, is similar to that of partitioning terminals into multipoint lines. The bin packing algorithms discussed in this chapter have relevance in the solution of LAN layout problems. Thus, many of the problems which arise in the design of centralized networks also arise in the design of LANs and, more importantly, many of the topological problems which arise in the design of LANs can be approached using the techniques originally developed for the design of centralized networks.

5.1.1 Multipoint Line Layout Heuristics

This section focuses on heuristic algorithms used to solve the problem of determining the topology of the multipoint lines connecting the terminals to the center. In studying this problem, we begin by making some simplifying assumptions, some of which are relaxed later.

We are given N sites, which are identified by the subscripts 1 through N, and a central site denoted by the subscript 0. The traffic from site i to the center is denoted t_{i0} and, likewise, t_{0i} denotes the traffic from the center to site i. The sites can be individual terminals or locations with many terminals.

The level of detail chosen to represent the problem is an important design decision in its own right. In some cases, it is desirable to use more than one number to model the traffic. For example, it may be desirable to keep track of traffic during different times of the day, traffic with different priorities, traffic with different message lengths, traffic with different security requirements, or different applications with special needs. Such detailed information allows a more exact analysis of a given design and may help to produce designs meeting precise performance goals. It also may greatly complicate the design process. But the focus here is on general design issues, and hence the simple traffic model given previously is used.

In fact, it is usually the case that the traffic in one direction (usually from the center to the terminals) dominates. In this case, it is necessary to only consider a single traffic number for each site. We refer to this as the weight of the site and denote it by w_i.

We are also given c_{ij}, the cost of connecting terminals i and j (including the costs, c_{i0} and c_{0j}, of connecting to and from the center). We assume that costs are symmetric, as indeed they usually are and we model the cost of the network as the sum of the costs of the individual lines plus the cost of node equipment. Doing this, we ignore bulk discounts and tariffs which cost out lines as a group. Notice that even with this simple model, it is possible to account for costs such as modem costs, node port costs, and even (to some extent) node capacity costs, by adding a constant to the c_{ij} to account for line termination charges.

The focus here is on doing designs with a single line speed. Hence, there is only a single c_{ij} for each i and j. The problem can be extended to multiple speeds by sequentially considering each line speed and resolving the entire problem. By its nature, a multipoint line (actually a tree) can contain only a single speed line; all terminals share this line and different parts of it cannot operate at different speeds. (The general extension to multiple speeds is considered along with optimization of mesh networks in Chap. 7.)

However, it is possible to consider using different speeds in different multipoint lines. In practice, however, there is rarely any motivation to do so. The principal reason for using multipoint lines in the first place is that the traffic from individual terminals is not sufficient to make cost effective the use of higher speed channels. Once traffic levels rise to the point that they can fill channels to cost effective levels (e.g., 30% to 50%), there is little motivation for using multipoint lines. Point to point lines, possibly at different speeds, then come into play.

Example 5.1. Suppose there are 100 terminals with weights of between 50 char/sec and 10000 char/sec. We consider using 9600 bps (1200 char/sec) lines and 56 Kbps lines. Suppose that, based on delay and stability issues, we are willing to tolerate at most a 40 percent line utilization on the 9600 bps lines, and a 70 percent utilization on the 56 Kbps lines. Terminals with weights above 3600 char/sec occupy 56 Kbps point-to-point lines. Terminals with weights above 480 char/sec occupy 9600 bps point-to-point lines. Terminals with weights below 480 char/sec are candidates for multipoint 9600 bps lines. There is little motivation for considering the use of 56 Kbps multipoint lines since little can be saved and the network becomes much more complex. However, because the cost of a 9600 bps line is not much greater than (typically within 10% of) the cost of 4800 or 2400 bps lines, there is motivation for using multipoint 9600 bps lines.

Thus we assume that all the terminals with higher weights are placed on point-to-point lines of an appropriate speed and then consider the problem of selecting the topology of the multipoint lines for the remaining terminals. The constraint on these lines is on the aggregate weights of the terminals. Thus, it is desirable to find trees of minimum total length (cost) subject to the constraint that the aggregate weight of the terminals on any multipoint line not exceed a given weight, W. Formally, the problem is:

Find $z = \min \sum_{i,j} c_{ij} x_{ij}$
subject to:

$$\sum_{i,j \in T_k} w_i x_{ij} \leq W \quad \forall k \tag{5.1}$$

weight constraint

$$\sum_{i,j \in S} x_{ij} < |S| \quad \forall S \tag{5.2}$$

no cycles

$$\sum_{i,j} x_{ij} = N \tag{5.3}$$

N connections for N+1 nodes ∴ totally connected tree

$$x_{ij} \in \{0, 1\} \quad 0 \leq i < j \leq N \tag{5.4}$$

x_{ij} is 0 if (i, j) is not included in the topology. Therefore, z is the cost of the all links in the network. T_k is the kth multipoint line (subtree), defined as the set of links connecting a set of nodes to the center. k takes values from 1 to N and T_k can be empty for some k. Constraint 5.1 guarantees that the total weight of the terminals on each multipoint line does not exceed the limit. Constraint 5.2 guarantees that the links chosen for the topology do not include any cycles. Constraint 5.3 guarantees that enough links will be selected to connect the network (of $N + 1$ nodes.) The way the problem has been defined above, it is only necessary to consider links in one direction.

Note that there are $O(2^N)$ constraints in 5.2 and therefore, while this problem may appear at first glance to be solvable exactly as a linear programming problem, it is not. In fact, the problem is *NP*-complete [GJ79] and algorithms exist yielding exact solutions only for problems of modest size [Gav85] [KB83]. The next discussion focuses on heuristics.

5.1.2   Constrained MST Algorithms

The problem described above is known as the Constrained Minimal Spanning Tree (CMST) problem; that is, a minimal spanning tree (MST) problem with an additional constraint on the size of the subtrees rooted at the center. It is reasonable then to consider applying MST algorithms, which are fast, simple, and effective, to the solution of this problem. For example, Kruskal's algorithm, discussed in Chap. 4, can be adapted to this purpose by adding a constraint on subtree size. Thus there is the following generic greedy algorithm, which in fact forms the basis for most CMST algorithms currently used in practice.

```
list <- GreedyCMST( n , costs , weights , Wmax )
     dcl costs[n,n] , weights[n] , solution[list],

     InitializeLinks( n , costs , *m , *LinkStruc )
     l_heap <- HeapSet( m , LinkStruc )
     InitializeComponents( *CompStruc )

     Naccepted <- 0
     solution <- Φ

     while( (Naccepted < n) & (HeapEmpty(l_heap) = FALSE) )
          edge <- HeapPop( *l_heap )
          comp1 <- FindComponent( ends[edge,1] , *CompStruc )
          comp2 <- FindComponent( ends[edge,2] , *CompStruc )
          if( (comp1 != comp2) &
               TestComp( comp1 , comp2 , CompStruc , Wmax ) )
             Merge( comp1 , comp2 , *CompStruc )
             solution <- Append( edge , solution )
             Naccepted <- Naccepted + 1
     return( solution )
```

This implementation is modeled on the implementation of Kruskal's algorithm given in Chap. 4. It begins by setting up a link structure that contains information on the current set of links, which are candidates for inclusion in the solution. In its simplest form, this could be vectors that give the endpoints and lengths (costs) of all links. This is exactly the information used by Kruskal's algorithm. There are other appropriate structures below. A heap is then set up on the current candidate links so it is possible to select the next best candidate efficiently.

Next, a component structure is set up which allows us to keep track of which component (in this case a multipoint line) each node is in. The FindComponent procedure, discussed in the presentation of Kruskal's algorithm, which maintains a pointer towards the root of each component, can be used for this purpose. Thus, the vector next can be part of CompStruc.

This structure also contains the data necessary to test whether the merging of two components violates any constraints. Thus far, the only constraint is the weight of the component. Thus, it is necessary to keep track of the total weight of each component. Initially, each component contains a single node and the weight of the component is the weight of the node. As components are merged, their weights are added.

Links are popped from the heap and tested. If a link connects nodes in different components and these components can be merged without violating any constraints, then the link is accepted into the solution and the components are merged. If not, the link is discarded.

In merging components, we must deal carefully with the case where one of the endpoints is the center. From the graph theoretic point of view, a component is a connected part of the graph. From the perspective of keeping track of the multipoint lines and their properties, however, it is necessary to keep the multipoint lines separate. The procedure, `Merge`, shown below deals with this by keeping a boolean variable, `ConnCenter`, as part of the component structure. Initially, this is set to `FALSE` for all components. When a component (multipoint line) is connected to the center, the value is changed to `TRUE`. The `TestComp` procedure also references this vector to avoid merging two multipoint lines which are already both connected to the center.

The algorithm terminates when all links have been tested, or when n links have been accepted, forming a set of multipoint lines connecting all nodes to the center. The first case is an error, indicating that something is wrong with the input data or the implementation of the algorithm; that is, the weight of a single node exceeds `Wmax`.

```
void <- Merge( comp1 , comp2 , *CompStruc )
    dcl CompStruc structure [ CompWt , next , ConnCenter ]

    if( comp1 = CENTER )
        swap( *comp1 , *comp2 )
    if( comp2 = CENTER )
        ConnCenter[comp1] <- TRUE
    else
        CompWt[comp1] <- CompWt[comp1] + CompWt[comp2]
        next[comp2] <- comp1
    ConnCenter[comp1]
                    <- ConnCenter[comp1] | ConnCenter[comp2]

boolean <- TestComp( comp1 , comp2 , CompStruc , Wmax )
    dcl CompStruc structure [ CompWt , next , ConnCenter ]

    if( (CompWt[comp1] + CompWt[comp2] <= Wmax) &
        (NOT( ConnCenter[comp1] & ConnCenter[comp2] )) )
        return(TRUE)
    else
        return(FALSE)
```

The previous algorithm produces a capacitated spanning tree with subtrees (multipoint lines) meeting the capacity constraint of `Wmax`. In fact, if the constraint is sufficiently loose, it produces an MST (i.e., it becomes Kruskal's algorithm). However, unlike Kruskal's algorithm, it cannot guarantee optimal results in all cases. ✓

Example 5.2. Consider, for example, the 6-node problem shown in Fig. 5.1. Suppose that the weight of each node is 1 and that `Wmax` is 3. The costs of the links are shown in the figure. The greedy algorithm selects links (1,3), (1,2), and (0,1), in that order. These links are all part of the MST, and the components formed by them do not violate the capacity constraint. At this point, however, the (2,4) and (3,5) links (which would complete the MST) must be rejected as they create subtrees, which violate the capacity constraint of 3. The solution is finally completed with links (4,5) and (4,0). The total cost of the network is 41. On the other hand, observe that lower cost feasible trees exist. For example, (0,1), (1,3), (0,2), (2,4), (4,5), which has a cost of 36. Thus, it is apparent that the greedy algorithm need not yield an optimal solution. In the following section, we explore approaches to improving the quality of the solutions obtained.

5.1.3 The Esau-Williams Algorithm

One of the ways that the generic greedy algorithm given above fails is that it can leave nodes which are far from the center stranded because all their neighbors are absorbed into subtrees that fill up. Therefore, in the last example, nodes 3 and 4 are left with no neighbors, and the relatively expensive link (0,3) is left to be included in the solution.

One way of dealing with this problem is to pay more attention to nodes that are far from the center, giving preference to links incident upon them. The Esau-Williams algorithm [EW66], which is widely used in practice, accomplishes this by forming a tradeoff function, t_{ij}, associated with each link (i, j). t_{ij} is defined by

$$t_{ij} = c_{ij} - c_{c_i}$$

where c_{c_i} is the cost of connecting the component containing node i to the center. Initially this is simply the cost of connecting node i directly to the center. As i becomes part of a component containing other nodes, however, this changes. Specifically,

$$c_{c_i} = \min_{k \in c_i} c_{kC} \tag{5.5}$$

where c_{kC} is the cost of connecting k directly to the center.

This tradeoff function can be thought of literally as a tradeoff between connecting each component directly to the center and interconnecting two components. Thus, it is possible to think of this process as starting with each node connected directly to the center, then greedily exchanging links to the center for links of lower cost, thereby directly connecting nodes to one another. The process stops when no cost effective exchanges can be found. This is, in effect, what the Esau-Williams algorithm does.

Note that t_{ij} is negative for any j closer to i, than i is to the center. t_{ij} is 0 for j equal to the center. It is always feasible to connect i to the center (if it is not already part of a component which is connected to the center). Thus, the algorithm terminates, connecting all remaining nodes to the center, when it reaches the point where the smallest t_{ij} is 0.

As the algorithm progresses, c_{c_i} may decrease and t_{ij} will then increase. The notion of "greed" then becomes somewhat more complex. It is always desirable to pick the link with the smallest t_{ij}, but as the algorithm proceeds, the t_{ij} are changing. Thus, it is not possible to simply sort the t_{ij}, once at the beginning of the algorithm, and then go through them in order. Instead, it is necessary to update them and repeatedly find the smallest one. A heap, which is a partial ordering, is a convenient data structure for repeatedly finding the smallest among a set of changing elements.

It is possible to maintain a heap of the t_{ij}, and update it as the t_{ij} change. This saves us from repeatedly having to search the entire set of t_{ij} in order to find the smallest one. Every time a value changes, however, it is necessary to update the heap. This requires $O(\log N)$ effort per update. It is also necessary to update the heap every time a value is popped out, again requiring $O(\log n)$ effort. Thus, the overall complexity of the algorithm becomes $O(n^2 \log n)$. This is good, but it is possible to do better in practice.

Observe that all the t_{ij} for a given i are formed using the current value of c_{c_i}. Thus, if c_{ij} is less than c_{ik}, then t_{ij} must also be less than t_{ik}. So, for any i it is only necessary to consider the t_{ij} for j, the nearest unexamined feasible neighbor of i. We will refer to the value of the current best (smallest) tradeoff for node i as t_i. Therefore, we need only consider a total of n values of t_i, instead of n^2 values of t_{ij}. Furthermore, when components merge and c_{c_i} changes for one component, it is only necessary to update one t_i for each i in that component.

Also observe that, as was mentioned above, the t_{ij} (and, hence the t_i) can only increase. This means it is not necessary to update any t_i until it reaches the top of the heap. This is because if the old value of t_i is not at the top of the heap (smallest), the new value (which must be the same or larger) could not possibly belong there. We do, however, need to check the value at the top of the heap to make sure it is still current (i.e., that it is equal to the length of the link, minus the current value of the cost from the component to the center). If the value is not current, update it and then update the heap. By deferring the updating, the algorithm's complexity is further reduced.

In the worst case, if the link lengths are pathologically chosen, the complexity of the algorithm is still $O(n^2 \log n)$. In practice, however, if implemented as above, the complexity is usually lower and rarely is it necessary to examine all the links.

Finally, realize that the smallest t_i at any given time, is always for a link joining a component to its nearest neighbor. We know from the discussion in Chap. 4 that this is a property of MST algorithms. Thus, in the absence of constraints, the Esau-Williams algorithm yields an MST. This is reassuring. It is reasonable to expect that when the constraints are loose that it yields a tree very close to an MST. This has been observed in practice but there are no guarantees. To date no reasonably tight bounds on the performance have been found.

An implementation of the Esau-Williams algorithm is given below. It is done in terms of heaps, as previously described. A function, set_heap, sets up a heap in its first argument, given a vector of values. The function top returns the index of the element currently at the top of the heap. The function heap_replace, replaces a value in the vector of heap values, and then adjusts the heap so that the value is put in the proper position. We use cost[i,*] to denote the ith row of the array cost. nHeap is an array of heaps of neighbors, one for each node (except the center).

`FindComponent` is the function described in Chap. 4, which finds the component containing a node and maintains the data structure. In this case, `compPtr`, does this efficiently.

Example 5.3. Consider the network shown in Fig. 5.1. Suppose again that `Wmax` is 3, and that all the node weights are 1. The Esau-Williams algorithm is applied in a straightforward way, simply computing tradeoffs as needed.

Begin by computing the tradeoffs for the nearest neighbors of each node:

$$t_{13} = 5 - 7 = -2$$

$$t_{21} = 8 - 6 = -2$$

$W = 3$

$$t_{31} = 5 - 11 = -6$$

$$t_{42} = 7 - 14 = -7$$

$$t_{53} = 8 - 17 = -9 \quad \leftarrow \min$$

Consider (5,3). It connects two nodes which are in different components; based on the weights of these components, it is feasible to merge them. Thus, we accept the link and merge the components. A new tradeoff is computed for node 5, since its previous nearest neighbor has been considered and its cost to the center has changed (since it is now part of a component including node 3, whose cost to the center is 11). So,

$$t_{54} = 9 - 11 = -2$$

Next, consider (4,2) and add it to the solution. Only the tradeoff for node 4 is affected. Therefore,

$$t_{43} = 8 - 8 = 0$$

Note that it is not necessary to check the feasibility of the link at this time, but simply to compute the new tradeoff.

Next, consider (3,1) and accept it too. (We've been very lucky so far.) Now recompute the tradeoff for node 3 and find

$$t_{34} = 8 - 7 = 1$$

We now consider (5,4) and find that accepting it would create a component of weight 5, and so we reject it. Proceeding, we examine and reject (1,3), and (2,1), both with tradeoffs equal to -2.

Next, examine (5,4) with $t_{54} = -2$, and discover that t_{54} is out of date. So, recomputing it,

$$t_{54} = 2$$

and it is pushed back into the tradeoff heap.

Examine and then reject (1,2) and (2,4), with tradeoffs of -1. Then examine and accept (2,0) and (1,0), completing the tree. Note that the algorithm always terminates with such links, directly into the center with tradeoffs of 0. Such links are always feasible (assuming the input data is reasonable and individual nodes can feasibly occupy a line alone). Thus, we need never consider any links with positive tradeoff.

Thus, the complete solution has a cost of 35, a noticeable improvement over simply using Kruskal's algorithm. It is in fact the optimal solution to this problem, which is known only by exhaustively trying all possible solutions. The Esau-Williams algorithm usually finds better solutions than Kruskal's, but it cannot guarantee an optimal solution.

```
void <- EsauWilliams( nn , center , cost , weight ,
                          weight_limit , ends )
    dcl cost[ nn , nn ] , weight[ nn ] , ends[nn,2]
        tHeap[ heap ], nHeap[ n , heap ]
        compPtr[nn] , connCtr[nn] , cDCtr[nn] , compweight[nn]
        tradeoff[nn]

/*                                                            */
/* nn            - # of nodes                                 */
/* center        - Index of the center                        */
/* cost[n,n]     - cost of connecting nodes                   */
/* weight[n]     - Weight (traffic) of a node                 */
/* weight_limit  - max. total weight for nodes in a           */
/*                    component                               */
/* nHeap[n]      - Heaps of neighbors of each node            */
/* tHeap[c]      - Heap of tradeoff values                    */
/*                     ( c[i][j] - c[i][0] )                  */
/* compPtr[c]    - Pointer to another node                    */
/*                    in the same component                   */
/* connCtr[c]    - 1 if the component is connected            */
/*                    to the center                           */
/* cDCtr[c]      - Cost of connecting component               */
/*                    to the center                           */
/* compWeight[c] - Weight (traffic) of the nodes              */
/*                    in component                            */
/* tradeoff[c]   - Tradeoff values                            */
/* ends[n,2]     - Endpoints of the links in                  */
/*                    the solution                            */
/* pred[n]       - (return) Predecessor of                    */
/*                    each node in CMST                       */

/*  Initialize the neighbor heaps and component               */
/*    structure for each( node )
      compWeight[node] <- weights[node]
      compPtr[node] <- node
      cDCtr[node] <- cost[node][center]
      connCtr[node] <- FALSE
      set_heap( nHeap[node] , nn , cost[node,*] )
    connCtr[center]<- TRUE;
    compWeight[center] <- 0;

/*  Initialize the tradeoff heap  */

    tradeoff[center] = INFINITY;
    for each ( node )
      if( node \cneq center )
          j <- top( nHeap[node] )
          tradeoff[node] <- cost[node][j] - cDCtr[node]
    set_heap( tHeap , nn , tradeoff )
```

```
/*  Select links for inclusion in the solution  */

nlinks <- 0    // # of links selected so far
while( nlinks < nn-1 )    // Continue until a tree is found
   n1 <- top( tHeap )        /get i th tix
   n2 <- top( nHeap[n1] )    /get closest neighbor of ti
   c1 <- FindComponent( n1 , compPtr )
   c2 <- FindComponent( n2 , compPtr )
   if( tradeoff[n1] != (cost[n1][n2] - cDCtr[c1]) )
      tradeoff[n1] <- cost[n1,n2] - cDCtr[c1]
      heap_replace( tHeap , n1 , tradeoff[n1] );
      continue    // Return to the beginning of the while loop
   if( TestComponent( c1 , c2 , weight_limit , connCtr ,
                      compWeight ) )
      Merge( c1 , c2 , center , compPtr , connCtr ,
             compWeight , cDCtr )
      ends[nlinks] <- n1 , n2
      nlinks <- nlinks + 1
   heap_replace( nHeap[n1] , n2 , INFINITY )
   j <- top( nHeap[n1] )
   tradeoff[n1] <- cost[n1][j] - cDCtr[c1]
   heap_replace( tHeap , n1 , tradeoff[n1] )
                               LW
/* Test the feasibility of merging two components */   if duplicate, cycle, reverse, if remotely stranded
                                            ↑
boolean <- TestComponent(
            c1 , c2 , weight_limit , connCtr , compWeight )
  dcl connCtr[nn] , compWeight[nn]

  if( ( c1 = c2 ) || ( connCtr[c1] && connCtr[c2] ) ||
      ( compWeight[c1] + compWeight[c2] > weight_limit )
     return( FALSE )
  return( TRUE )

// Merge two components
void <- Merge(
     c1 , c2 , center , compPtr , connCtr , compWeight , cDCtr )
  dcl compPtr[nn] , connctr[nn] , compWeight[nn] , cDCtr[nn]
  if( c1 = center ) swap( *c1 , *c2 )
  if( c2 = center )
    connCtr[c1] <- TRUE
    cDCtr[c1] <- 0
  else
    compPtr[c2] <- c1
    compWeight[c1] <- compWeight[c1] + compWeight[c2]
    connCtr[c1] <- connCtr[c1] | connCtr[c2]
    cDCtr[c1] <- min( cDCtr[c1] , cDCtr[c2] )
```

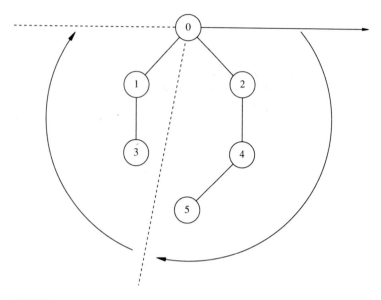

FIGURE 5.2

Sharma's algorithm.

5.1.4 Sharma's Algorithm

We note that the nodes in any multipoint line are connected in an MST. Thus, the problem can be thought of as one of partitioning the nodes into the sets which occupy each multipoint line. Given this partition, it is then possible to complete the solution by forming the MSTs, which can easily be done using an ordinary MST algorithm (e.g., Kruskal's).

[SEB70] proposed that this partition is formed by dividing the region containing nodes into sectors, sweeping out angularly and including as many nodes as possible in each sector without violating the weight constraint. Thus, in Fig. 5.2, which shows the same problem as Fig. 5.1, it is possible to sweep out clockwise from the center (node 0), and include nodes 2, 4, and 5 in the first multipoint line. Then start another multipoint line and include in it the remaining nodes, 3 and 1. The resulting solution is two MSTs, one on nodes 0, 2, 4, and 5 and the other on nodes 0, 1, and 3, with a total cost of 36. This is again a noticeable improvement over Kruskal's and almost as good as Esau-Williams.

If we had swept in a counter-clockwise direction, we would have formed the partition {1,3,5}, {2,4} and would have formed and produced the same solution as the Esau-Williams algorithm. It is also possible to start the sweep at any angle. Sharma [SEB70] suggested iterating the algorithm starting at each node. This increases the complexity of the algorithm but, in general, leads to better solutions.

The complexity of Sharma's algorithm, without this iteration, is essentially $O(N^2/k)$, where N is the number of nodes and k is the average number of nodes per multipoint line. This comes from forming N/k MSTs, each containing k nodes. The effort in finding an MST on k nodes is, as seen in Chap. 4, $O(k^2)$. With the iteration, the algorithm is $O(N^{3/2}k)$. The quality of the solutions obtained with the iteration is comparable to those obtained with Esau-Williams. The Esau-Williams algorithm, however, is usually significantly faster if it is implemented carefully. Sharma's algorithm is easier to implement. It requires just an MST algorithm and a sectoring algorithm.

The sectoring can be done by generating polar coordinates for the nodes and then sorting them by angle. Polar coordinates are easily generated. Begin by translating the x and y coordinates of each node to make the center, say node 0, be at coordinates $(0,0)$. Thus,

$$x_i \leftarrow x_i - x_0 \qquad y_i \leftarrow y_i - y_0 \qquad \leftarrow \text{shift cartesian system wrt center node as } (0,0)$$

This produces the angle a_i, for each node,

$$a_i \leftarrow (\tan)^{-1}(y_i/x_i) \qquad \leftarrow \text{convert to polar } \& \text{ to "sweep" by sorting as}$$

In doing this, be careful not to divide by 0 (if x_i happens to be 0) and to choose the correct quadrant. If x_i is 0 and y_i is positive, a_i is $90°$. If x_i is 0 and y_i is negative, a_i is $270°$. If $\tan^{-1}$ returns a value between 0 and 180, set a_i to $180 + a_i$ if y_i is negative. Note that Sharma's algorithm requires x and y coordinates in addition to the cost information all other algorithms require. This additional requirement is ordinarily not a problem since coordinates are usually obtainable (e.g., telephone company V&H coordinates or latitudes and longitudes).

5.1.5   The Unified Algorithm

The notion of a component weight, w_i, can be generalized [KC74] to encompass a family of widely used algorithms, including Esau-Williams' (EW), Kruskal's, and others. A Unified Algorithm can then be implemented as outlined below.

The implementation relies on the following data structures and functions, which are generalizations of those used in the implementation of the EW algorithm given above.

ComPStruct—The component structure, contains information about each component, (e.g., its weight, size, member nodes, distance to the center, etc).

NeighborStruct—The neighbor structure, contains information about the neighbors of each node and facilitates finding the next neighbor to be considered for each node.

V, VFunction—Vfunction associates a value, V, with each component. This value is used in forming the tradeoff governing the order in which the links are considered. In the EW algorithm, VFunction simply gives the distance from the component to the center.

InitializeComponentStruct,   InitializeNeighborStruct— These functions set up the structures which keep track of the components and neighbors (i.e., initialize heaps and predecessor structures). These are implemented as inline code, in the implementation of the EW algorithm.

FindBestNeighbor, FindNextNeighbor—These functions find the next best neighbor to be considered for a given node. FindNextNeighbor also modifies the neighbor structure, eliminating the neighbor selected from further consideration. In the EW algorithm, we simply popped a heap. In general, it is also possible to test feasibility at this point. Also, structures other than a heap can be used to maintain the neighbor structure. It is even possible to identify the set of neighbors "on the fly."

FindComponent—This function identifies the component containing a node, as in the implementations of Kruskal's algorithm and the EW algorithm above.

ResetV—This function motifies the V values for merged components. In the EW algorithm, the distance is set to the center. This allows the merged component to be the minimum of the values for the two components merged.

```
void <- UnifiedAlgorithm( nn , center , cost , weight ,
                          weight_limit , ends )
   dcl cost[nn,nn] , weight[nn] , ends[nn,2]
       tHeap[ heap ], tradeoff[nn] , V[nn]
       CompStruct[ nn , structure ]
       NeighborStruct[ nn , structure ]

 // Initialization

 for each( node )
    V[node] <- VFunction( node , CompStruct )
 InitializeComponentStruct( CompStruct )
 InitializeNeighborStruct( NeighborStruct )
 for each ( node )
     bestJ <- FindBestNeighbor( node , NeighborStruct )
     tradeoff[node] <- cost[node][bestJ] - V[node]
 set_heap( tHeap , nn , tradeoff )

 // Select links for inclusion in the solution

 nlinks <- 0    // # of links selected so far
 while( nlinks < nn-1 )    // Continue until a tree is found
    n1 <- top( tHeap )
    n2 <- FindBestNeighbor( n1 , NeighborStruct )
    c1 <- FindComponent( n1 , CompStruct )
    c2 <- FindComponent( n2 , CompStruct )
    if( tradeoff[n1] != (cost[n1][n2] - V[c1]) )
        tradeoff[n1] <- cost[n1,n2] - V[c1]
        heap_replace( tHeap , n1 , tradeoff[n1] );
        continue
    if( TestComp( c1 , c2 , weight_limit , CompStructure ) )
        Merge( c1 , c2 , center , CompStructure )
        ends[nlinks] = n1 , n2
    j <- FindNextNeighbor( n1 , NeighborStruct )
    ResetV( V , c1 , c2 )

    tradeoff[n1] = cost[n1][j] - V[c1];
    heap_replace( tHeap , n1 , tradeoff[n1] );
```

By using different VFunctions, it is possible to implement a variety of algorithms:

1. *Kruskal's algorithm*: V is set to 0 for all components; the tradeoff is simply the length of the links.
2. *EW algorithm*: V is set to the distance from the component to the center and is updated as components are merged.
3. *Vogel's approximation method* (VAM) [RV58]: Let C_{i1} be the cost of connecting component i to its nearest (feasible) component. Let C_{i2} be the cost of connecting component i to its second nearest (feasible) component. Then compute V_i as

$$V_i = C_{i2} - C_{i1}$$

This difference represents the increment in network cost if it is necessary to connect component i to its second choice rather than its first. If component i is not immediately connected to its nearest component, there is the risk of that component filling up; and then component i would have to connect elsewhere. By giving preference to components that would suffer the most by not connecting to their nearest neighbors, they are considered sooner, making it less likely that they would "lose" their first choice.

This function can be implemented in a manner similar to that used to implement the EW algorithm. Specifically, the neighbors of each node are placed in a heap where they can be quickly accessed. Note that the two nearest neighbors must be in two of the three top heap positions. The only serious question is whether to base the tradeoff on feasible neighbors or any neighbors. (This is actually a question in implementing the EW algorithm as well.)

It is usually more convenient to base the tradeoff on all neighbors. In this case, however, unlike in the EW algorithm, the tradeoff function does not consistently increase. As neighbors become infeasible, C_{i1} consistently increases as will C_{i2}, but the difference can increase or decrease. The strictest implementation of the VAM algorithm would thus be to reevaluate all the tradeoffs after each merge, significantly increasing the complexity of the algorithm. In particular, the position of many values in the heap would change when a component (which might be their first or their second choice) fills up.

In practice, it is usually sufficient to base the tradeoff on all neighbors. This keeps the implementation simple and competitively fast (when compared with the EW algorithm). The quality of the solutions obtained is also comparable, although the EW algorithm tends to do a few percent better on average when VAM is implemented this way. When VAM is implemented with the tradeoff based on feasible neighbors, it is considerably slower, but the results obtained are then more competitive with EW.

4. *Martin's algorithm* (Reverse Prim's): Martin [Mar72] suggests an algorithm where the node farthest from the center (most expensive) is used as the root of a component. When Prim's algorithm is run, starting at this node, it only accepts nodes which do not violate the weight constraint. When the component attaches itself to

the center (because no node closer to the component than the center is feasible), another component is started with the remaining node farthest from the center. The algorithm proceeds until all nodes are connected.

This algorithm has the advantage that it fills up multipoint lines, unlike most of the others mentioned thus far. This is sometimes an advantage. But it has the disadvantage that it sometimes strands nodes far from the center between two clusters which grow in from opposite directions. In practice, this algorithm tends to produce results which are slightly worse than EW. It is, however, one of the simplest algorithms to carry out by hand.

To implement it as part of the Unified Algorithm, initially set the V for all components to a large negative number. When a new root is selected, set its V to 0. This guarantees that the tradeoff function favors the nearest neighbor of this node. As new nodes join the component, they become part of the component, with weight 0, and links to their neighbors receive consideration. Thus, the V updating process is orderly and this algorithm is competitive with the others in terms of speed.

5. *Prim's algorithm*: By simply setting the weight on the center to 0, and all other weights to a large negative number, it is possible to implement Prim's algorithm as part of the Unified Algorithm. As a node is brought into the tree, its weight is updated to 0. Prim's algorithm, by its nature tends to strand nodes far from the center and so it usually produces somewhat more expensive networks. It is, nevertheless, sometimes convenient to have an implementation of Prim's algorithm available in this way.

6. *Weight based algorithm*: It is possible to set the V for each component based on its weight (e.g., to a multiple of the weight of the nodes in the component). As components are merged, the V increases. This tends to favor large nodes, which are more easily stranded. It also tends to fill up the components because as a component grows its V increases, improving its tradeoff value. This algorithm, while competitive with the others in terms of speed, tends to produce slightly worse results because it ignores distance from the center.

7. *Hybrid algorithms*: By taking a linear combination of the V functions above, it is possible to implement hybrids of these algorithms. By multiplying the V for any single algorithm it is also possible to create new algorithms. So, for example, if taking cost to the center works well in the EW algorithm, why not set V to twice the cost from the center? Chou and Kershenbaum [KC74] experimented with this idea and found that the EW algorithm worked as well as any hybrid in most cases. In specific instances, however, different algorithms (sometimes even Kruksal's) worked best. Thus, it can be worthwhile to run the Unified Algorithm with several different V functions and then keep the best result obtained. The improvement obtainable is typically only a few percent, however, so this is appropriate only in the final stages of the design process.

The nodes on a multipoint line are usually polled by the center. The size of tables used in polling often impose a constraint on the number of terminals on a line. For the sake of reliability, or due to common carrier restrictions on the quality of the signal, there can be constraints on the number of nodes in cascade, or on the degree of nodes in the multipoint line. There can also be prohibitions against certain combinations of nodes being on a given line; for example, for the sake of reliability, it may be desirable not to have all the nodes of a particular type (e.g., printers) on the same line. For full duplex lines, there is a separate constraint on the total traffic in each direction.

All of these constraints can be taken into account by the Unified Algorithm (or in fact by any of the algorithms mentioned thus far, except Sharma's). The Unified Algorithm can also take additional constraints into account in the V function (that is, the V function can reflect these constraints *and* help call attention to critical components early enough to avoid having them stranded). Multiple constraints can also be accommodated; when merging two components, check as many constraints as is necessary. In some cases, the presence of additional constraints necessitates keeping additional information in the cluster structure (e.g., the number of nodes in the component as well as the total traffic).

The only real restriction placed on the constraints themselves is that they be monotone, in the sense that adding a node to a component makes things worse. This is Property 1 discussed when the Greedy Algorithm was presented in Chap. 4 (i.e., every subset of a feasible set is feasible). Thus, it is expected that at worst it is possible to connect all nodes directly to the center (if not, there is no feasible solution), and that once it is infeasible to merge two components it remains infeasible to do so after additional nodes join them. This is a reasonable assumption since all the constraints mentioned above are monotone in this sense.

There is, however, one important type of constraint not handled by these algorithms. The center (which can be a remote concentrator to which a group of nodes have been assigned) can be degree limited. For example, it can be a device with a port constraint. In this case, it is not feasible to simply connect all nodes directly to the center. It becomes important to minimize the number of multipoint lines. To the extent that this is an active constraint, the Unified Algorithm can help by adjusting the V function to favor full lines (e.g., by shifting towards Martin's or weight based algorithms). It is also possible to consider other approaches to the problem, some of which are discussed in later sections.

5.1.6   Extensions to Other Topologies

The previous algorithms can be naturally extended to other local access topologies. Specifically, by starting with a topology other than a tree it is possible to form a tradeoff function and perform exchanges among links to produce low cost topologies of other useful types.

The most obvious problem with a tree is that a failure of a single link disconnects part of the network. If this is a serious concern, a loop topology is often considered (see Fig. 5.3) as an alternative. Loop topologies are almost as simple as tree topologies. The terminals all share a single line and do not carry out any routing. All traffic ordinarily travels in one direction around the loop, say clockwise. If, however, a link breaks, the terminals may have the capability of recognizing this and of temporarily using the remaining portion of the loop in the other direction. This can require a manual switchover and some rearrangement (usually at the software level) at the central site. Also, loop networks are usually not much more expensive than trees.

A natural way of approaching this problem, which is a generalization of the EW algorithm, is called the Clarke-Wrighte algorithm [CW64]. The idea is illustrated in Fig. 5.3. We start (conceptually) with each node on a separate loop (two direct connections) to the center. As the algorithm proceeds, loops are merged, subject to the restriction that the set of terminals on the merged loop not violate any constraints (e.g., number of terminals, total traffic). Thus, the tradeoff associated with link (i, j) is

$$t_{ij} = c_{ij} - c_{i0} - c_{j0}$$

In this case, t_{ij} is only defined for nodes i and j which are still connected to the center (i.e., in the end position on a loop). Thus, the algorithm limits itself to considering nodes in the end positions of the loops. It proceeds until no profitable feasible links remain.

The implementation of the algorithm and its complexity are similar to the EW algorithm. The restriction to only considering nodes at the end of a loop can be used to an advantage in accelerating the algorithm by removing nodes in the interior of loops from further consideration.

The algorithm accomplishes two things. It partitions nodes into loops, and it connects the nodes on a loop in a low cost manner. Given the partition into loops, it is often possible to improve the solution by running a Traveling Salesperson (TSP) algorithm, which finds a minimum length tour on a given set of nodes. The TSP problem is itself difficult, but many heuristics exist for its solution, as well as optimal methods [Law76] that are effective on the small problems that typically arise in this

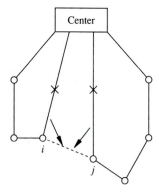

FIGURE 5.3
Loop local access network.

context. It is possible to run a TSP algorithm on the nodes in a particular loop and improve the solution quality.

Even using loops, terminals can become disconnected from the rest of the network if the center to which they are connected fails. One way of avoiding this situation is to form dual-homed loops, as is illustrated in Fig. 5.4. In this case, instead of attaching the loop to a single node, attach each of its ends to two different centers. Similar to before, begin with each node directly connected to two centers, perhaps the two nearest ones. Again, attempt to merge two loops containing nodes i and j at their ends, subject to whatever constraints are present. A tradeoff is formed as before,

$$t_{ij} = c_{ij} - c_{ia} - c_{jb}$$

where a and b are centers to which the loops containing i and j are attached. In addition to whatever restrictions exist on the nodes in the merged component, there is now also a constraint that the loop must remain connected to two centers. Thus, when the (i, a) and (j, b) edges are deleted, the other ends of the loops containing nodes i and j must be connected to different centers. This further restricts the allowable tradeoffs and must be checked by the algorithm. Note that this algorithm can move nodes from one center to another, changing the load on the centers. Limitations on the capacity of the centers can impose further restrictions on allowable exchanges.

5.1.7 Bin Packing Algorithms

In the case where there are groups of terminals in the same city, it may be beneficial to approach the line layout problem as a bin packing problem. There is a good deal of literature on the bin packing problem [GJ79], which has application in many other areas of optimization in addition to multipoint line layout.

Informally, there are objects of various sizes and bins of a fixed size, larger than the objects. The problem is to pack the objects into the smallest possible number of bins. In this case, if there are many terminals in the same city, enough to fill several multipoint lines, the problem is one of packing the terminals onto the smallest possible number of lines. The cost of interconnecting the terminals in the city is a constant; the issue is to minimize the number of lines.

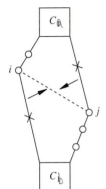

FIGURE 5.4
Dual-homed loops.

The CMST heuristics discussed earlier in this section do not do a particularly good job of solving this problem. They focus on link cost rather than on the number of multipoint lines created. In the case where many terminals are in the same city, many costs are the same, and such algorithms have little information on which to base link selection decisions. While the bin packing problem is known to be NP-complete [GJ79] there are a number of effective heuristics for its solution.

One of the simplest and most effective algorithms is called First Fit Decreasing. It is a greedy algorithm which orders objects (terminals), largest first, and then places each terminal into the first bin with sufficient capacity. Other alternatives are Best Fit Decreasing and Worst Fit Decreasing, where again objects are sorted largest first, but in this case the feasible bin with the least remaining slack (Best Fit) or most remaining slack (Worst Fit) is chosen. All of these algorithms work well, but the First Fit is somewhat faster than the others.

The implementation shown below includes both the Best Fit and First Fit algorithms. In both cases, we begin by sorting the objects in decreasing order of size. The sort produces a permutation of the object indices, returned in `permu`, but does not disturb the actual order of the object sizes. We use `permu` to consider the objects in decreasing order of size.

The parameter `alg` controls which algorithm is run. In each case (FIRST or BEST) a function is called to find the appropriate bin (first or best) for each object. The First Fit is faster because it breaks out of the selection loop as soon as it finds a fit, whereas the best fit must examine all bins in all cases. Both algorithms open up a new bin if the object being considered does not fit in any existing bin. The algorithm returns the number of bins required and the bin associated with each object.

The algorithms are comparable in terms of the number of bins they pick. Both have worst case complexities of $O(N \log N + NB)$, where N is the number of objects and B is the number of bins.

Example 5.4. Suppose there are 8 terminals with traffic equal to:

$$7\ 2\ 6\ 5\ 2\ 6\ 9\ 3$$

and multipoint lines are capable of handling 20 units of traffic. Begin by sorting the terminals in decreasing order of traffic:

$$9\ 7\ 6\ 6\ 5\ 3\ 2\ 2$$

Applying either First Fit or Best Fit, we begin by placing the 9 and 7 on Line #1. The 6 does not fit, so we open up a new line, Line #2, and place the first 6 there. The second 6 and the 5, likewise, do not fit on Line #1, and are also placed on Line #2. At this point, Line #1 has 16 units of traffic and Line #2 has 17.

Now the algorithms differ. The 3 fits on either line. First Fit would place it on Line #1 (the first feasible line) while Best Fit would place it on Line #2 (where it fits more tightly.)

TABLE 5.1

Comparison of bin packing algorithms

First Fit Solution		Best Fit Solution	
Line	Terminals	Line	Terminals
1	9 7 3	1	9 7 2 2
2	6 6 5 2	2	6 6 5 3
3	2		

Proceeding with First Fit, the first 2 would be placed on Line #2. At this point, both lines have 19 units of traffic and the final 2 would have to be placed on a third line. In comparison, with Best Fit, after placing the 3 on Line #2 and using up all its spare capacity, Line #1 still has 4 units of slack which allow it to hold both 2s.

Thus, in this case, Best Fit did a better job. In general, however, there is usually little or no difference between these algorithms and First Fit can do better in some cases. It is important, however, that the terminals be sorted in decreasing order as the quality of the solution degrades significantly if they are not. Thus, the important part of these algorithms is the "decreasing" part, not what kind of fit is used.

```
void <- BinPack( nn , alg , bin_size , size , *nbin , bin )
      dcl size[nn] , bin[nb] , permu[nn] , values[nn]
          slack[nb]

/*  nn   - the number of objects                    */
/*  size - the size of each object                  */
/*  bin_size - the capacity of a bin                */
/*  nbin - # of bins (output)                       */
/*  bin - the bin each object is put into (output)  */


/*  Sort into decreasing order   */

sort( nn , size , permu , D )

/*  Pack into bins  */

nbin <- 0
for_each ( i )  /*  For each object to be packed  */
    j <- permu[i]
    if ( alg = BEST )  /* Best Fit */
        bin[j] <-
            find_best_bin( nbin , bin_size , size[j] , slack )
    else if ( alg = FIRST )
        bin[j] <-
            find_first_bin( nbin , bin_size , size[j] , slack )
```

```
/*    First Fit: Find the first bin with enough slack    */

int <- find_first_bin( bin_size , *size , slack , *nbin )
     dcl slack[nb]

   best_i <- nbin;
   for_each ( b , nbin )
      if( size <= slack[b] )
          best_i <- i    /*  Found a bin  */
          break

   if( best_i = nbin ) {    /*  Make a new bin  */
      slack[nbin] <- bin_size
      nbin <- nbin + 1
   slack[best_i] <- slack[best_i] - size
   return( best_i )

/*    Best Fit: Find the feasible bin with the smallest slack    */

int <- find_best_bin( bin_size , size , *slack , *nbin )
     dcl slack[nb]

   best_slack <- INFINITY
   best_i <- nbin
   for_each ( b ,   nbin )
      if( (size <= slack[i]) && (slack[i] < best_slack) )
          best_slack <- slack[i]    /*  Found a bin  */
          best_i <- i

   if( best_i = *nbin )    /*  Make a new bin  */
      nbin <- nbin + 1
      slack[*nbin] <- bin_size
   slack[best_i] <- slack[best_i] - size
   return( best_i )
```

It is also possible to use bin packing in conjunction with the other multipoint, line layout heuristics in cases where there are many terminals in large cities, as well as other terminals spread out over many smaller cities. In this case, bin packing would be done to take care of the larger cities, forming nearly full multipoint lines. Each of these lines would then be treated as a large terminal and these "terminals," along with the terminals in the smaller cities, would then be input to one of the other multipoint line layout algorithms.

5.2 TERMINAL ASSIGNMENT

In the preceding section, we assumed that terminals were already assigned to centers. Sometimes, this assignment is obvious. If there is only one natural center where all traffic originates or is destined, all nodes are assigned to it. If the "centers" are actually concentrators which have sufficiently large capacities to hold as many terminals as necessary, it is possible to simply connect each terminal to the nearest (least cost connection) concentrator. The Esau-Williams algorithm adapts nicely to this situation. It starts with each terminal connected to the nearest concentrator but allows clusters to merge even if they are attached to different concentrators. This helps to reduce the cost of the multidrop lines by allowing more freedom in forming them, but can result in an imbalance in the loads on the different concentrators.

In other situations, however, there is an active capacity limitation on the concentrators and it is necessary to make the assignment more carefully. In this section, assume that there is a fixed capacity given for each concentrator and a fixed cost for connecting each terminal to each concentrator. Assume that terminals connect directly to the concentrators. If, in fact, multidrop lines are used, the algorithms discussed here can be used to assign the terminals to concentrators and the algorithms in the previous section can be used to form the multidrop lines.

Thus, we are given T terminals, i, and C concentrators, j. The cost of connecting terminal i to concentrator j is c_{ij}. The capacity of a concentrator is W. (It is actually straight forward to allow W to vary with the concentrator.) Terminal i requires w_i units of capacity at a concentrator. It is necessary to minimize the cost of connecting the terminals to the concentrators, subject to the capacity constraint on the concentrators. Formally, the problem is

$$\min z = \sum_{i,j} c_{ij} x_{ij}$$

subject to:

$$\sum_{j=1}^{C} x_{ij} = 1 \quad i = 1, 2, \ldots, T \tag{5.6}$$

$$\sum_{i=1}^{T} w_i x_{ij} \leq W \quad j = 1, 2, \ldots, C \tag{5.7}$$

$$x_{ij} \in \{0, 1\} \tag{5.8}$$

where x_{ij} is 1 if terminal i is assigned to concentrator j. The first set of constraints guarantees that each terminal is associated with a concentrator. The second set ensures that the capacity constraints are not violated.

5.2.1 Greedy Algorithms

A simple approach to this problem is to greedily assign terminals to the "nearest" feasible center. We begin by forming, for each terminal, a heap of all its possible concentrator "neighbors," in much the same way as terminal neighbors were heaped in the multidrop line algorithms described in the previous section. By popping these heaps, we then find the nearest neighbor of each terminal. Then, heap these values and pop the best one. This gives a terminal, i, and its nearest concentrator, j. If it is feasible to connect i to j, do so. Proceed in this manner, updating the heaps. Values are popped from them, until all terminals are associated with concentrators.

There are T terminals and C concentrators. In the worst case, a heap is popped TC times, requiring $O(TC \log C)$ operations. Usually, fewer pops are required. In practice, the complexity is often dominated by how the heaps are set up, which is $O(TC)$. Thus, this is a practical approach even for large problems.

The algorithm is a heuristic which has the obvious defect that it tends to strand the last terminals considered. In some cases, these last terminals might be connected to concentrators which are very far away. This is particularly likely if some regions are not properly "covered" by concentrators with sufficient capacity. In practice, if a region is not properly covered, the problem lies with the concentrator placement and sizing, not with the terminal assignment algorithm; and the problem should be corrected by revising the coverage. Smaller mismatches, however, can be dealt with directly in the terminal assignment procedure.

Specifically, it is possible to modify this greedy algorithm in a manner similar to the VAM multidrop line algorithm described in the previous section. This modification gives preference to terminals which pay a big premium for not connecting to their nearest concentrator. A tradeoff is formed,

$$t_i = c_{i1} - \alpha c_{i2}$$

where c_{i1} and c_{i2} are the costs of connecting terminal i to its nearest and second nearest neighbors, respectively. α is a parameter between 0 and 1 reflecting how much preference to give to "critical" terminals.

As with the VAM algorithm, we have the choice of forming the tradeoff based on all neighbors or feasible neighbors. The disadvantage of keeping track of each node's nearest and second nearest feasible neighbors is that these values potentially must be updated every time an assignment is made since a neighbor can become infeasible for a given node as its remaining capacity decreases. This checking and updating can slow down the algorithm significantly.

On the other hand, the updating is more important here than it was in the VAM algorithm. If the constraint is tight, the quality of the solution can degrade significantly. In fact, there is no guarantee of a feasible solution, even when one exists. There are several possibilities:

1. The total concentrator capacity, is less than the total terminal weight; that is,

$$\sum_{i=1}^{NT} w_i > WC$$

In this case, it is clear that no feasible solution exists and concentrator capacity must be added.

2. The total concentrator capacity equals or exceeds the total terminal weight, but there is no feasible assignment of terminals to concentrators. For example, there may be 20 terminals, each with weight 3, and 6 concentrators, each with capacity 10. The total weight is equal to 60, which is equal to the total capacity, but there is no way of assigning 10 units of weight to each concentrator. The problem of determining whether a feasible solution exists is itself difficult when the number of terminals is large and the terminals have different weights. In practice, it is best to avoid situations where the constraint is so tight that there are doubts as to the existence of a feasible solution. This is because it is usually less expensive to increase the total concentrator capacity than it is to force many connections between terminals and distant concentrators that have remaining capacity available.

3. The total concentrator capacity exceeds the total terminal weight, and feasible solutions exist, but the algorithm fails to find one. This is a situation we wish to avoid, especially if the feasible solution is a reasonable one. Therefore it is best to work harder updating the tradeoffs.

An implementation of the greedy terminal assignment algorithm just described follows. The algorithm takes as input the number of terminals, nt, the number of concentrators, nc, the cost of connecting terminal i to concentrator j, cost, the weights of the terminals, Wter, the limit on the weight of terminals associated with a concentrator, Wlimit, and a parameter, alpha, which controls how the tradeoff, for each terminal is formed. It returns a vector, Cassoc, giving the concentrator associated with each terminal. It also returns a boolean value indicating whether or not it succeeded in associating every terminal with a concentrator. If it fails, it returns FALSE, and Cassoc is −1 for unassociated terminals. The functions UpdateTradeoff, TradeF, Test and Assign are internal to the main procedure and share its variables. However, for the sake of clarity they are shown outside the body of the main procedure. (We continue to do this from now on, for the sake of brevity.)

It is best to maintain the lists of neighbor concentrators for each terminal as sorted vectors rather than heaps. This is more convenient because access is necessary to both the nearest (in terms of cost) and second, nearest concentrators. Each row of the array Neigh gives a list of concentrators for each terminal, nearest concentrator first. There are also two vectors, First and Second, giving the position in Neigh of the current nearest and second nearest, feasible concentrators. Throughout the algorithm we keep track of feasibility and update these vectors. Thus, the nearest feasible concentrator for terminal t is Neigh[t,First[t]].

The current value of the tradeoff associated with each terminal is held in the vector Vtrade. These values form a heap, Theap, which gives ready access to the smallest value. Generally, since values of Vtrade can increase or decrease as the algorithm proceeds, Theap must be updated whenever a value in Vtrade changes. This is particularly important when a terminal "goes critical." If the problem is tightly constrained, it is possible that a terminal with a large weight may have only one remaining feasible concentrator. It is then critical that this terminal be associated with that concentrator. Its tradeoff is set to $-\infty$ in this case, and it is important that it bubble to the top of the heap immediately.

The current weight of the terminals associated with each concentrator is maintained in the vector Wconc. Lists of terminals, considering each concentrator as first and second neighbors, are maintained in lists Clist. When a terminal is associated with a concentrator, it is removed from these lists as it no longer needs a tradeoff value. Also, a terminal is removed from the list for any concentrator that is no longer feasible for it (i.e., does not have sufficient weight left to accept it). The function `Delete` is used to remove terminals from lists; the implementation of `Delete` is dependent on the implementation of the lists.

The algorithm proceeds, greedily associating terminals with concentrators. The larger the parameter `alpha` is, the more attention the algorithm pays to terminals that would suffer if their current nearest concentrator fills up, denying them access to it. The smaller `alpha` is, the more the algorithm simply pays attention to connecting terminals to nearby concentrators. In practice, values of `alpha` near 1 seem to work well.

After associating a terminal with a concentrator, the algorithm invokes `UpdateTradeoff`. This procedure checks that the tradeoffs for all terminals are still valid, updates the ones that are not, and also updates the tradeoff heap. This involves checking that the first and second nearest neighbors (concentrators) of each terminal are still feasible; that is, that the neighbors still have enough slack weight to hold the terminals. This is done by the `Test` function. Since only the slack of a single concentrator has decreased (i.e., the one associated with a new terminal), only terminals using it in the definition of their tradeoff must be checked. This is why, for each concentrator, we maintain `Clist`, the list of terminals using it in their tradeoff.

As a terminal is associated with a concentrator, it no longer needs a tradeoff since its decision has been made. It is removed from all the `Clist`s and its tradeoff is removed from the heap. A terminal which is found to be no longer feasible for attachment to a specific concentrator is removed from the `Clist` for that concentrator and a new neighbor is sought, by examining the next concentrators in the sorted list of neighbors for that terminal. If no other feasible neighbor is found, there are two possibilities. The first is that the terminal has gone from two feasible neighbors to one. In this case, the terminal has "gone critical." It is essential that it be associated with its only surviving neighbor. In this case, its tradeoff is set to $-\infty$. This brings it to the top of the heap (except, perhaps for other critical terminals).

The second possibility is that the terminal has just lost its last feasible neighbor. This can happen if many terminals go critical at the same time, as a concentrator fills up, denying them all a neighbor. Then, as some of the critical terminals are processed, others lose their remaining neighbor. In this case, the algorithm halts, declaring failure. It is possible that a feasible solution exists, but this algorithm will not find it. Once it associates a terminal with a concentrator, it never takes back the decision. This is the nature of greedy algorithms. In practice, if this situation arises, it is usually an indication that the problem is too tightly constrained and that more concentrator capacity should be added. It is not known whether or not a feasible solution exists. In the next section we explore algorithms that take back their initial decisions and continue searching for better solutions.

```
boolean <-
    GreedyTA( nt , nc , cost , Wter ,  Wlimit , alpha , *Cassoc )
        dcl cost[nt,nc] , Wter[nt] , Cassoc[nt]
            Neigh[nt,nc], First[nt], Second[nt], Vtrade[nt]
            Wconc[nc], Clist[nc, list], Theap[heap]
```

```
/* Initialize the concentrator structures */
   Wconc <- 0
   Clist <- Φ   /* empty list */

/* Initialize the terminal structures */
   Cassoc <- -1
   First <- 0
   Second <- 1
   for_each( t )
      Neigh[t,*] <- sort( cost[t,*] )
      c1 <- Neigh[t,0]
      c2 <- Neigh[t,1]
      append( t , Clist[t,c1] )
      append( t , Clist[t,c2] )
      Vtrade[t] <- TradeF( cost[t,c1] , cost[t,c2] , alpha )

/* Initialize the tradeoff heap */
    set_heap( tHeap , nt , tradeoff );

/* Associate terminals with concentrators */
nassigned <- 0;
while( nassigned < nt )
   ter <- pop( tHeap )
   if( ter = -1 ) return FALSE   /* Failure */
   conc <- Neigh[ter,First[ter]]
   Assign( ter , conc , pred , Wter , Wconc )
   nassigned <- nassigned + 1
   i <- UpdateTradeoff( conc )
   if( i = FALSE ) return FALSE   /* Failure */
return TRUE   /* Success */

/* Update tradeoffs based on new assignment to a concentrator */
boolean <- UpdateTradeoff( conc )
   for_each ( ter , Clist[conc] )
      if( Cassoc[ter] >= 0 )
         delete( ter , Clist[conc] )
      else if( Test( Wter[ter] , Wconc[conc] , Wlimit ) = FALSE )
         delete( ter , Clist[conc] )
         pos2 <- Second[ter] + 1
         flag <- FALSE
         while( (flag=FALSE) & (pos2 <= nc) )
            conc2 = Neigh[ter][pos2];
            flag <- Test( Wter[ter], Wconc[conc2], Wlimit )
         if( Neigh[ter,First[ter]] = conc )
            First[ter] <- Second[ter]
         Second[ter] <- pos2
         if( First[ter] > nc ) return FALSE
         if( Second[ter] > nc )
            Vtrade[ter] <- -INFINITY
         else
```

```
            c1 <- Neigh[ter,First[ter]]
            c2 <- Neigh[ter,Second[ter]]
            Vtrade[t] <- TradeF( cost[t,c1], cost[t,c2], alpha )
            append( ter , Clist[c2] )
        heap_replace( tHeap , ter , Vtrade[ter] );
    return TRUE


/* Tradeoff function: C(nearest) - Alpha * C(second_nearest) */
int <- TradeF( cost1 , cost2 , alpha )
  return( cost1 - alpha * cost2 )


/* Test the feasibility of merging two components */
boolean <- Test( Wter , Wconc , Wlimit )
  if( Wter + Wconc > Wlimit )
      return FALSE
  else return TRUE


/* Associate a terminal with a concentrator */
void <- Assign( ter , conc , Cassoc , Wter , Wconc )
      dcl Cassoc[nt], Wter[nt] , Wconc[nc]
  Cassoc[ter] <- conc
  Wconc[conc] <- Wconc[conc] + Wter[ter]
```

5.2.2 Exchange Algorithms

The greedy algorithm given in the preceding section works reasonably well in practice. It usually finds good feasible solutions and does not run too long. Recognize, however, that checking the tradeoffs after each terminal assignment slows it down and complicates the implementation. A simple greedy algorithm that forms the tradeoffs only once runs faster and is easier to implement. Also, it is disturbing that the algorithm gives up when it can't find a feasible solution. Closely related to this is the problem (which actually arises in tightly constrained problems) that the last terminals considered have neighbors, but these feasible neighbors are very far away. This leads to a solution with some obviously misassigned terminals.

Therefore, it is best to use another class of algorithms, either directly or as a follow up to a greedy algorithm. This leads to the class of algorithms known as semi-greedy algorithms or, more formally, as **alternating chain algorithms**.

The simplest form of exchange in this problem, illustrated in Fig. 5.5, is to exchange the assignments of a pair of terminals. Thus, if terminal i is assigned to concentrator j, and terminal k is assigned to concentrator m, if

$$c_{ij} + c_{km} > c_{im} + c_{jk}$$

we can switch the assignments and lower the cost of the network.

Similarly, if assignments are permitted that violate the capacity constraints, it is possible to look for exchanges which help make the solution feasible. Thus, the exchange above can be made because w_i (the weight of terminal i) is greater than w_k and concentrator m has slack to accommodate w_i, while j does not.

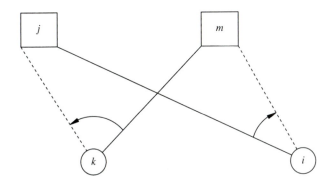

FIGURE 5.5
Terminal exchange.

If the greedy algorithm terminated with some terminals unassigned, it is possible to temporarily assign them to concentrators without slack and then exchange them to concentrators which do have slack. While this still cannot guarantee a feasible solution, it will find one more often than a simple greedy algorithm does. Also, these exchanges will compensate for many of the most serious errors a greedy algorithm can make in fitting in the last terminals.

The one problem with this approach is that it can get out of hand computationally. The greedy algorithms above are essentially $O(TC)$ in practice. A single exchange is $O(T^2)$, that is, it is possible to exchange any terminal with any other. There are a number of ways of dealing with this.

1. The problem may be small enough, or the number of exchanges small enough, that there is minimal complexity. It is possible then to seek out exchanges in a straightforward way, considering all pairs of terminals.

2. If the goal is feasibility, it is possible to identify "giver" concentrators (which are overloaded) and "taker" concentrators (which have slack). Then, it is possible to limit the exchanges to those which move weight in the right direction.

3. If the goal is to reduce cost, it is possible to identify terminals which are assigned to concentrators at a cost much greater than the minimum, or assigned to their kth best choice for k much larger than 1. We also identify terminals that can easily move to another concentrator without significantly increasing their connection cost. Exchanges can then be limited to such pairs.

This is all somewhat ad hoc, but can work in practice, especially if the problem is not too tightly constrained. However, it is only part of the solution. As was mentioned earlier, often the problem is that a whole region is congested, that is, the distribution of terminal weight does not match the distribution of concentrator capacity. Sometimes, the best thing to do is to change the distribution of concentrator capacity. Other times, this may not be possible. And, sometimes the distribution of capacity is actually correct, but the greedy algorithm is not sensitive to it.

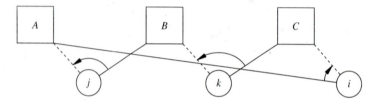

FIGURE 5.6

A more complete exchange.

> **Example 5.5.** What may be needed is to make more complex exchanges involving more
> than two terminals. This is illustrated in Fig. 5.6. Terminal i is misassigned. It is desirable
> to assign it to concentrator C, but there is no slack. If terminal k were moved, there would
> be room for i. Terminal k could be moved to concentrator B at very little cost, but there
> is not enough slack at B. Terminal j could move to concentrator A, either because there
> is slack or because removing i would create slack. No single exchange between two
> terminals makes sense, but a three way exchange does.

Similarly, it might be profitable to perform exchanges between four or more terminals. There are $O(T^k)$ possible exchanges among k terminals, however. We clearly
cannot afford to examine all possibilities. This leads to a more orderly approach. In
fact, as shown in the following discussion, in some cases, a provably optimal solution
can be obtained for this problem.

The algorithm described next is based on the following observations:

1. Ideally, every terminal would be assigned to the nearest concentrator. The only
   reason for not doing so is that the capacity constraint prevents this.
2. If a terminal is already assigned to a nearby concentrator, the only reason for
   moving it to one that is farther away is to make room for another terminal that
   would have to detour even farther. Thus, it is only necessary to move terminals
   on concentrators that are full, and then only to make room for another terminal.
3. Given an optimal partial solution with k terminals assigned (i.e., these k terminals
   are assigned to concentrators at a minimum possible total cost), an optimal partial
   solution with $k + 1$ terminals can be found by finding the least expensive way of
   adding the $k + 1$st terminal to the k terminal solution. Note that this may involve
   reassigning some of the first k terminals.

Stated this way, its is clear that with sufficient effort it is possible to find an
optimal solution, since an entirely new solution can be found by reassigning as many
terminals as desired. Clearly, any solution, including the optimal one, is obtainable
this way. The trick is to find the best one without looking at all T^C possibilities.

The step from a solution with k terminals to one with $k + 1$ is called an **augmentation** (augmenting the solution by one more terminal) and the sequence of re-
assignments moving from one solution to the other is called an **augmenting path**.
Augmenting path algorithms are the natural generalizations of greedy algorithms in
terms of their complexity and effectiveness.

It was observed in Chap. 4 that greedy algorithms guarantee optimal solutions to problems defined on a matroid. Recall that in such problems the objective is to find an optimal, independent set (e.g., maximum weight), and that independent sets have the following two properties:

1. Any subset of an independent set is independent.
2. Given two independent sets, S_1 and S_2, with S_2 containing more elements than S_1, there is at least one element in S_2 which can be added to S_1 without destroying independence.

It can be shown ([Law76]) that augmenting path algorithms can be used to find optimal solutions to problems defined on the intersection of two matroids. Such problems are characterized by solutions which simultaneously satisfy two notions of independence, each of which has the properties of a matroid.

Suppose there is a matrix containing positive numbers. The objective is to select elements of the matrix, maximizing the total of the numbers picked. There is a restriction, however, to pick at most one number from each row. This is a problem defined on a matroid (called a partition matroid). A solution corresponds to a set of numbers, at most one from each row. Any subset of a solution is also a solution. Given a solution, S_2, containing more elements than S_1, S_2 must contain an element in a row not included in S_1. This element can be added to S_1 forming a new feasible solution. This satisfies the conditions for a matroid. A greedy algorithm, picking the largest element in each row, is guaranteed to find an optimal solution.

Similarly, if the objective is to pick numbers maximizing the total, subject to the constraint that at most one from each column is picked, we have another matroid defined on the same set of elements. Again, a greedy algorithm, picking the largest in each column, would find the optimal solution.

Now suppose the constraint is to pick the largest total, subject to constraints that at most one element from each row and at most one element from each column are picked. This is an intersection of two matroids. A greedy algorithm, picking the largest in a row or column, is no longer necessarily optimal. For example, given the matrix

$$
\begin{array}{ccc}
3 & 8 & 9 \\
6 & 5 & 8 \\
8 & 1 & 2
\end{array}
$$

if we greedily pick the 9, the best we can do is 22, while the optimal solution, 3 8s, has a value of 24. An augmenting path algorithm, however, can find the optimal solution to this problem in a reasonable amount of time.

This problem is known as the Assignment Problem and there is a great deal of literature on it ([GN72]). Think of the rows as people and the columns as jobs. A solution then corresponds to assigning people to jobs.

The terminal assignment problem has the same structure and can be approached in the same way. In this case, the matrix corresponds to the cost of connecting terminals to concentrators. The constraints are that no more than W terminals are picked in each column and that there is one picked in each row. Note that the problem is restricted and now assumes that the constraint is on the number of terminals, not on the sum of their weights.

This problem is still defined on the intersection of two matroids. It should be possible to believe that both matroid properties still hold when the constraint is that no more than W elements are picked from each column, rather than just one. Also, while the original problem is phrased in terms of maximizing, it is possible to minimize by greedily selecting the smallest elements rather than the largest ones. Thus, an augmenting path algorithm finds the optimal solution to this problem. This is how the algorithm proceeds.

Begin by assuming that all terminals have the same weight, which without loss of generality, is assumed to be 1. A concentrator then has a capacity of W terminals. In this case, whether or not a feasible solution exists is quite clear. If there are T terminals and C concentrators, a solution exists if and only if

$$T \leq WC$$

Furthermore, it is easy to find a feasible solution if one exists. Any assignment which does not violate the concentrator constraint is feasible. Thus, in this case it is possible to test immediately if a feasible solution exists and halt the algorithm if one does not.

Otherwise, the process continues by trying to associate each terminal with its nearest concentrator. If there is success without assigning more than W terminals to any concentrator, the optimal solution has clearly been found, and we are done. If not, some, say k, terminals have been associated with their nearest concentrator and others remain unassigned. This is a partial solution which is optimal for the assigned terminals.

Now the objective is to find an optimal augmentation. This is done by forming an auxiliary graph, illustrated in Fig. 5.7a, defined in the following way:

1. There are two special nodes, S and F, which are the start and end, respectively, of all augmenting paths. The optimal (least cost) augmentation is the shortest path from S to F.

2. The remaining nodes in the graph are labelled xY, corresponding to the connection of terminal x to concentrator Y. Squares are used to represent nodes, xY, where x is currently assigned to Y and circles are used represent nodes where x is not currently assigned to Y. In Fig. 5.7a, b and e are currently assigned to G. Assume in this example that a concentrator has capacity for two terminals and thus that G and H are full while I is not.

3. There is an arc, (S, xY), for all concentrators, Y, from node S to each terminal, x (which is unassigned). This corresponds to the assignment of terminal x to concentrator Y. The length of this arc is $c(xY)$, the cost of connecting x to Y. (Note: Because x and Y are subscripted in the following discussion, it is notationally convenient to refer to these costs as $c(xY)$ rather than c_{xY}.) The arc (S, fG) is an example of such an arc.

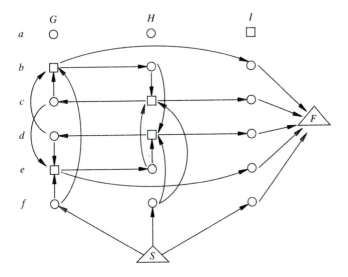

FIGURE 5.7*a*
Auxiliary graphs (original).

4. There is an arc, $(x_1 Y, x_2 Y)$, for all x_1 not currently assigned to Y, all x_2 currently assigned to Y, and all Y currently full. These arcs correspond to reassigning terminal x_2 to another concentrator to make room for terminal x_1. The length of such arcs is $-c(x_2 Y)$, since, by making this decision, the cost of the assignment is reduced by $c(x_2 Y)$. The arc (fG, eG) is an example of such an arc.

5. There is an arc, (xY, xY_2), for all x currently assigned to a full concentrator Y, and for all other concentrators Y_2. These arcs correspond to reassigning terminal x to concentrator Y_2. The length of such arcs is $c(xY_2)$. Y_2 may be full or not full. Arcs (eG, eH) and (eG, eI) are examples of such arcs.

6. There is an arc, (xY_2, F), for all x currently assigned to a full concentrator Y and for all Y_2 not full. These arcs correspond to ending the augmentation after reassigning x to Y_2. The cost of such arcs is 0. Arc (eI, F) is an example of such an arc.

Notice that there are no arcs to or from terminals currently assigned to concentrators which are not full. In Fig. 5.7 there are no arcs to or from terminal a. There is no reason to disturb the assignment of such terminals as they are already well-assigned and need not be moved to make room for other terminals. (As the concentrator fills up, it may enter into an augmenting path later in the algorithm.)

Every path from S to F corresponds to a valid augmenting path, resulting in a new solution, with one more terminal assigned and no concentrators overloaded. For example in Fig. 5.7, the path

$$(S, fG), (fG, bG), (bG, bH), (bH, cH), (cH, cI), (cI, F)$$

corresponds to assigning f to G (which is full), moving b from G (to make room for f) to H (which is also full), and moving c from H to I. Similarly, the path

$$(S, fI), (fI, F)$$

corresponds to simply assigning f to I, which is not full.

After each augmentation, the auxiliary graph has to be reconstructed. Terminal assignments change and some concentrator fills up. Thus, some previously uninvolved terminals (formerly associated with a concentrator which was not full) may enter into the process. For example, if a terminal (say b) moved to concentrator I, I would be full and aI would now have incoming and outgoing arcs.

Notice also that once a concentrator fills up, it remains full. An augmentation only moves a terminal from a full concentrator to make room for another terminal bcause it costs money to move a terminal already associated with the nearest concentrator.

Not only does each SF-path correspond to an augmentation, but each possible augmentation also has a path corresponding to it. This can be seen by observing that there are arcs present to reassign any terminal currently assigned to a full concentrator to any other concentrator. Thus, the shortest SF-path corresponds to the optimal augmentation. Therefore, to find the optimal augmentation, it is only necessary to find the shortest path from S to F, which was efficiently illustrated earlier.

Notice that the auxiliary graph contains cycles as well as negative arcs. If it contains any negative cycles, the shortest path algorithms do not work. Fortunately, there is proof that there are no negative cycles in the auxiliary graph. A cycle corresponds to exchanging a group of two or more terminals with one another. For example, the cycle

$$(dG, eG), (eG, eH), (eH, dH)(dH, dG)$$

corresponds to exchanging the assignments of d and e. If this cycle had negative length, it would mean that the current assignments of d and e were not optimal in the current partial solution. It is best to start with each terminal assigned to its nearest concentrator (which is clearly optimal) and proceed with minimum cost augmentations. The length of the augmenting path is the incremental cost of the augmentation. Thus, at each step in the algorithm, there is an assignment (of the terminals currently assigned) of minimum total cost. It is therefore not possible to reduce the cost simply by exchanging terminal assignments. Thus, there are no negative cycles in the auxiliary graph.

The auxiliary graph has $O(TC)$ nodes and $O(WTC)$ edges (round nodes associated with full concentrators have in-degree 1 and out-degree $W - 1$). Thus, an augmentation is at least $O(WTC)$ and possibly as bad as $O(T^2C^2)$. However, it is possible to do better by working with a smaller auxiliary graph. Therefore, it is necessary only to examine a smaller number of candidate paths. In particular, it is only necessary to consider the best transfer from one full concentrator to another. For example, consider the two partial augmenting paths

$$(S, fG), (fG, eG), (eG, eH), (eH, dH)$$

and
$$(S, fG), (fG, bG), (bG, bH), (bH, dH)$$

Both paths correspond to assigning f to G, moving a terminal from G to H, and moving d from H (to another concentrator, as yet undetermined). These two paths are equivalent in the sense that they can both be extended in the same way. In other words, once dH is reached, the only concern is that we have assigned f to G, moved some terminal from G to H, to make room for f, and will move d someplace else to make room for that terminal. In fact, once the decision is made to move a terminal from H, to make room for the incoming terminal, the choice is open as to which one to pick. The terminal cheapest to move is the one chosen.

Using this observation, it is possible to set up a compressed auxiliary graph of the type shown in Fig. 5.7b, which is formed as follows:

1. Again, there are two distinguished nodes, S and F, which are the start and finish of all augmenting paths.
2. There is one node (square) corresponding to each fully-loaded concentrator. There are no nodes corresponding to concentrators with spare capacity or corresponding to terminals.
3. There is an arc, (S, Y), corresponding to some as yet unassociated terminal, t, with fully loaded concentrator Y. t is selected as the (unassociated) terminal which can connect to Y at the lowest cost. Thus,

$$c(tY) = \min_{x \text{ unassociated}} c(xY)$$

The (S, G) arc in the compressed auxiliary graph "represents" all the (S, xG) arcs in the original auxiliary graph. In forming the compressed graph, it is necessary

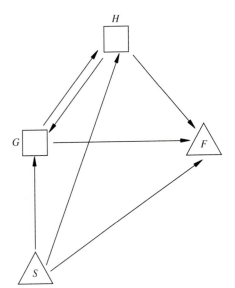

FIGURE 5.7b
Compressed auxiliary graph.

to also record the value of t which should be used in the augmentation, should arc (S, G) appear in the shortest path. Thus, arc (S, G) thus has both a length, equal to $c(tG)$, and a terminal, $TER(S, G)$, associated with it.

4. There is an arc, (Y_1, Y_2), that corresponds to moving some terminal, t, from fully loaded concentrator Y_1 to fully loaded concentrator Y_2. t must be a terminal currently associated with Y_1, and should be the terminal least expensive to move to Y_2. Thus,

$$c(tY_2) - c(tY_1) = \min_{x \text{ associated with } Y_1} c(xY_2) - c(xY_1)$$

Note that the length of arc (Y_1, Y_2) is not in general the same as that of (Y_2, Y_1).

5. There is an arc, (Y, F), that corresponds to moving some terminal, t, from fully loaded concentrator, Y, to a concentrator, Y_2, with spare capacity. Again, choose the terminal which is least expensive to move. Thus,

$$c(tY_2) - c(tY) = \min_{x \text{ associated with } Y} c(xY_2) - c(xY)$$

Now we must record the identity of Y_2 as well as that of t. Thus, (Y, F) has $CONC(Y, F)$ set to Y_2 as well as $TER(Y, F)$ being set to t.

6. There is an arc, (S, F), that corresponds to associating some previously unassigned terminal, t, with some concentrator, Y, with spare capacity. Again, the Y with minimum cost is picked. The identities of t and Y are recorded in $TER(S, F)$ and $CONC(S, F)$, respectively.

A single pass through the cost matrix can be used to set up the entire compressed graph. For each terminal, t, and each concentrator, c, examine their status and post information to the appropriate arc in the graph. The implementation that follows shows all this. After the graph is set up, find the shortest path from S to F. Then backtrack this path, using the information in $TER(x, Y)$ and $CONC(x, y)$ to do the augmentation.

After an augmentation, the graph must be updated. This can be done incrementally, since only part of the graph has changed. For the sake of simplicity, the objective is to just rebuild the graph from scratch. It is also possible to retain some of the node labels from the shortest path algorithm and to carry out further iterations on it incrementally. Again, for the sake of simplicity, this has been decided against. For very large problems, or if this algorithm were to be used in the inner loop of some larger optimization procedure, it is possible to consider these further refinements.

The complexity of the algorithm that will be implemented can now be measured. The compressed auxiliary graph has $m + 2$ nodes, where m is the number of concentrators currently full. m cannot be more than C, the total number of concentrators. Thus, the shortest path algorithm can be carried out in $O(C^2)$. As mentioned earlier, setting up the graph requires $O(TC)$ operations. Since T is usually larger than C, this latter operation dominates. The backtracking and augmentation is only $O(C)$. The first phase of the algorithm attempts to associate terminals with their nearest concentrators. This is $O(TC)$ and leaves k terminals unassociated. There should be one augmentation per unassociated terminal.

Thus, the overall complexity of this algorithm is $O(kTC)$, at worst $O(T^2C)$. In practice, if the problem is not tightly constrained, k is much smaller than T. Of course, in this case the greedy algorithm also works well. In summary, it is appropriate to use the greedy algorithm in early phases of the design process and to use this procedure in later phases when a more precise result is required.

Another issue is that, as stated, this procedure only works when the constraint is on the number of terminals or when the terminals all have the same weight. These are common cases. But often, the constraint is on the number of ports. In other cases, multipoint lines, all of which have about the same amount of traffic on them, have already been formed. In cases where the terminals have different weights, this procedure can be generalized to check that the augmentation is feasible. This complicates the implementation, both setting up the graph and doing the shortest path labelling,√ so an optimal solution is no longer guaranteed. In this case, the greedy algorithm, possibly with local exchanges, is most appropriate.

```
//         Optimal Terminal Assignment:
//
//         Inputs: nt - # of terminals
//                 nc - # of concentrators
//                 cost - n x n matrix giving cost from i to j
//                 Wlimit - maximum weight for nodes at a conc.
//
//         Output: Cassoc - assigned conc. for each node

void <- Terasg( nt , nc , Wlimit , cost , Cassoc )
     dcl cost[nt,nc] , Cassoc[nt] , Wconc[nc] ,
         Ters[n,n], dist[n,n] , pred[n] , conc[n]

   // Initialize

   Cassoc <- NULL
   Wconc <- 0
   nassigned <- 0

   // Try to connect each terminal to its nearest concentrator

   for_each( t )
      c = IndexMin( cost[t,*] )
      if( Wconc[c] < Wlimit )
         Cassoc[t] = c
         Wconc[c] <- Wconc[c] + 1
         nassigned <- nassigned + 1

   // Assign remaining terminals

   while ( nassigned < nt )
      SetupGraph( nt , nc , cost , Cassoc , Wlimit , Wconc ,
                  *nn , *conc , *dist , *Ters )
```

```
      Dijkstra( nn , S , dist , pred )
      Assign( pred , Ters , Cassoc , conc , Wconc )
      nassigned <- nassigned + 1

/* Set up the auxiliary graph for shortest path computation */

void <- SetupGraph( nt , nc , cost , Cassoc , Wlimit , Wconc ,
                    *nn , *conc , *dist , *Ters )
      dcl cost[nt,nc] , Cassoc[nt] , Wconc[nc] , conc[nn] ,
          dist[nn,nn] , ters[nn,nn] , concPos[nc]

  // nn is the number of nodes in the graph being set up

  nn <- 2    /*  Node 0 is start. Node 1 is finish  */

  // Find the full concentrators; make them nodes
  // concPos[c] is the node number associated with conc. c
  // conc[n] is the conc. associated with node n
  // conc[F] is the conc. with spare which gets a new terminal

  for_each ( c )
     concPos[c] <- F
     if( Wconc[c] == Wlimit )
        conc[*nn] <- c
        concPos[c] <- nn
        nn <- nn + 1

  // dist is the matrix of arc lengths in the auxiliary graph

  dist <- $- \infty $
  for_each ( n )
     dist[n,n] <- 0

  // Set up the graph

  for_each ( t )
     for_each ( c )
        if( Cassoc[t] == -1 )   /* Unassociated terminal */
           b <- concPos[c]
           if( dist[S][b] > cost[t][c] )   /* New best t */
              dist[S][b] <- cost[t][c]
              Ters[S][b] <- t
              if( b == F ) conc[F] <- c
        else                    /* Associated terminal */
           casc <- Cassoc[t]
           a <- concPos[casc]
           if( a == F ) continue
           b <- concPos[c]
           if( dist[a][b] > cost[t][c] - cost[t][casc] )
```

```
                    dist[a][b] = cost[t][c] - cost[t][casc]
                    Ters[a][b] <- t
                    if( b == F ) conc[F] <- c

/*  Dijkstra's Algorithm ( No frills ; complete graph )  */

void <- Dijkstra( nn , root , dist , pred )
      dcl  dist[nn,nn] , pred[nn] , scanned[nn] , label[nn]

   for_each ( n )
      label[n] <- dist[root,n]
      pred[n] <- root
      scanned[n] <- FALSE
   scanned[root] <- TRUE
   nscanned <- 1

   while ( nscanned < nn )

      // Find the next node to scan.

      bestL <- $\infty$
      for_each ( n )
         if( (scanned[n]=FALSE) $\wedge$ (label[n] < bestL) )
            bestL <- label[n]
            Iscan <- i

      // Scan node Iscan.

      for_each ( n )
         if( label[i] > label[Iscan] + dist[Iscan][i] )
            label[i] <- label[Iscan] + dist[Iscan][i]
            pred[i] <- Iscan
      scanned[Iscan] <- TRUE
      nscanned <- nscanned + 1

/*  Augment the solution  */

void <- Assign( pred , ters , Cassoc , conc , Wconc )
      dcl pred[nn] , ters[nn,nn] , conc[nn] ,
          Cassoc[nt] , Wconc[nc]

   n <- F
   while ( n != S )
     p <- pred[n]
     t <- ters[p,n]
     Cassoc[t] <- conc[n]
     n <- p

   Wconc[conc[F]] <- Wconc[conc[F]] + 1
```

TABLE 5.2
**A 6-terminal 3-concentrator
assignment problem**

		CONC		
		G	H	I
T	a	6	3	8
E	b	2	9	4
R	c	3	1	4
M	d	2	5	9
	e	1	6	3
	f	2	7	9

Example 5.6. Consider the problem defined by Table 5.2. The cost of connecting each terminal to each concentrator is given in the table. Each terminal has a weight of 1, and a concentrator has capacity for 2 terminals.

The algorithm begins by trying to assign each terminal to its nearest concentrator. The following assignments appear:

$$(a, H), (b, G), (c, H), (d, G)$$

The nearest concentrators for terminals e and f are full, so they remain unassigned.

Next, set up the compressed auxiliary graph shown in Fig. 5.8a. In the figure, each arc is labelled by its length and the terminal corresponding to that length. Thus, the $3, d$ next to link (G, H) corresponds to moving terminal d from G to H at a cost of 3. The

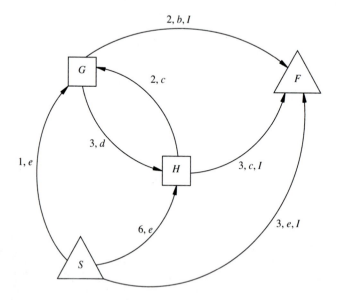

FIGURE 5.8a
Alternating path example.

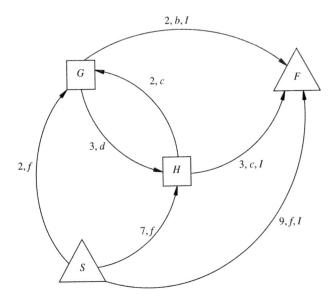

FIGURE 5.8*b*

arcs to the final node, F, are labelled by their length, terminal, and the concentrator, with slack receiving a new terminal. Thus, the (H, F) arc is labelled $3, c, I$ corresponding to moving terminal c from H to I at a cost of 3.

The shortest path from S to F is the single arc (S, F), which makes a path of length 3. This corresponds to simply assigning e to I. In this case, it is best to assign e to its second nearest concentrator rather than moving any terminal from G to make room for it.

Next, the compressed auxiliary graph is shown in Fig. 5.8*b*. Since none of the previously assigned terminals moved, most of this graph is the same as in Fig. 5.8*a*. The only difference is in the arcs from S. Now the shortest path is

$$(S, G), (G, F)$$

which corresponds to assigning f to G and moving b from G to I. The cost of this is 4 (2 to assign f to G and 2 to move b from G to I).

Thus, the total cost of this assignment, which is optimal, is 15. This corresponds to the cost of assigning the terminals a, b, c and d to their nearest concentrators (8) plus the sum of the lengths of the two augmenting paths (7).

5.3 CONCENTRATOR LOCATION

The final problem to consider in the area of centralized network design is that of locating concentrators. This problem is also an important part of the local access network design problem in distributed networks. Indeed, concentrator location is a version of the facility location problem, which also includes placement of backbone nodes in distributed networks. The focus here is on the basic concentrator location problem and we leave consideration of the more general problems for a later section.

The reason for using concentrators (or multiplexors) is to benefit from the economy of scale provided by higher speed lines, replacing separate low speed lines connecting terminals to the central site with higher speed lines between concentrators and the center. Thus, there are relatively short, low speed lines between terminals and concentrators and longer, higher speed lines between the concentrators and the center (see Figure 5.9).

In the basic concentrator location problem, there is a set of terminal locations, i, and a set of potential concentrator locations, j, along with the cost, c_{ij}, of connecting terminal i to concentrator j for all i and j. We are also given d_j, the cost of placing a concentrator at location j.

Assume initially that there are no active capacity constraints on the concentrators. This is a good assumption when concentrators are cost effective, that is, when a concentrator saves enough money (by providing terminals with a closer entry point into the network) without filling itself up with terminal traffic. In other cases, it is possible to initially ignore the capacity constraint and then run a terminal assignment algorithm to redistribute terminal load.

Also assume, for the sake of simplicity, that terminals connect directly to the concentrators. In reality, they can connect via multipoint lines. It is possible to deal with this more general problem in a number of ways. The simplest is to first lay out the multipoint lines, connecting terminals to the center, and then treat each multipoint line as a "terminal," potentially to be connected to a concentrator or the center. In this case, c_{ij} is the minimum cost of connecting a terminal on line i to concentrator j.

A more sophisticated approach would be to iterate a solution, starting with terminals formed from multipoint lines to the center and picking concentrators based on this initial clustering of terminals to multipoint lines. Then, with an initial set of concentrators picked, it is possible to lay out a new set of multipoint lines, allowing these multipoint lines to home to the concentrator sites already selected. Then, again, treat these new multipoint lines as terminals and repeat the concentrator location step. There are also algorithms [BT72] [WT73] proposed to deal with this problem directly, simultaneously performing concentrator location and multipoint line selection.

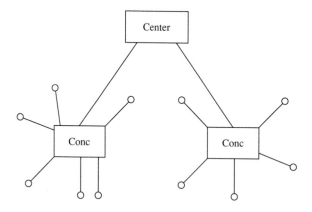

FIGURE 5.9
Centralized network with concentrators.

In the simple model considered here, there is only one type of concentrator. Without a capacity constraint, there is no motivation for considering different sizes of concentrators. Simply use the least expensive one. The cost includes both the concentrator hardware and the line connecting the concentrator to the center. This model can also include the cost of concentrator ports and modems by adding a fixed charge to the c_{ij}. Indeed, it is possible to include different costs for different terminals (perhaps based on the terminal type), if that is appropriate.

Thus, the formal statement of the basic concentrator location problem is:

$$\min z = \sum_{i,j} c_{ij} x_{ij} + \sum_{j} d_j y_j$$

subject to:

$$\sum_{j} x_{ij} = 1 \quad \forall i \tag{5.9}$$

$$\sum_{i} x_{ij} \leq N y_j \quad \forall j \tag{5.10}$$

$$x_{ij}, y_j \in \{0, 1\} \quad \forall i \quad \forall j \tag{5.11}$$

where N is the total number of terminals. Equations (5.9) ensure that each terminal is associated with a concentrator. Equations (5.10) ensure that all concentrators with associated terminals are included in the solution, that is, that y_j must be 1 if any x_{ij} is 1.

Equations (5.10) could be generalized to handle capacity limits on the concentrators by associating a weight, w_i, with each terminal and a capacity, K_j, with each concentrator. Equations (5.10) then become:

$$\sum_{i} w_i x_{ij} \leq K_j y_j \quad \forall j \tag{5.12}$$

5.3.1 Center of Mass (COM) Algorithm

One of the simplest and computationally efficient methods of locating concentrators is to simply identify natural clusters of traffic. This approach is attractive when there are not even candidate sites. The following algorithm can be used to identify such sites.

Suppose there are coordinates, (x_i, y_i) and traffic-based weights (e.g., the total traffic to and from each terminal), w_i, for each terminal, i. Begin with each terminal in a cluster by itself. Then seek to merge clusters which are close to one another, replacing two clusters, i an j, with a new cluster, k, at the center of mass of i and j. Thus,

$$x_k = \frac{w_i x_i + w_j x_j}{w_i + w_j} \tag{5.13}$$

$$y_k = \frac{w_i y_i + w_j y_j}{w_i + w_j} \tag{5.14}$$

Clusters i and j are removed form further consideration and cluster k is added to the clusters considered for merging.

Clustering is restricted by the following rules:

1. There is a desired weight, W, for a cluster. Do not form clusters with weight greater than W. There may also be a minimum acceptable weight for a cluster.

2. There is a distance limit, D. Do not merge clusters whose distance is greater than D. The distance between two clusters is defined to be the distance between their centers of mass.

3. There is a desired number of clusters, N. Stop merging when there are N clusters.

In practice, these constraints are contradictory and all cannot be enforced simultaneously. Thus, there may be a choice between leaving a number of clusters larger than N and forming clusters of weight larger than W. Similarly, there may be a decision between retaining a cluster with weight below the minimum and merging it with another cluster outside the distance limit. Thus, tradeoffs must be formed among the constraints.

There is also the issue of which clusters to merge. A natural tendency is to merge the two clusters which are closest together. This requires repeatedly finding the nearest pair of clusters as clusters are merged and new clusters are created. The heap of heaps structure discussed in the implementation of the Unified Algorithm facilitates this, leading to the following algorithm.

The array cost gives the cost of connecting each pair of terminals. In the algorithm shown, Euclidean distance is used to represent cost. In practice, a more realistic cost function, possibly a piecewise linear function, based on a tariff, might be used. It is not possible to use an actual tariff, since this algorithm forms clusters at the center of mass, of groups of terminals, which in general do not fall at real locations. Thus, some form of approximation is necessary.

Each terminal has x and y coordinates and a weight based on total traffic to and from the terminal. The parameters nClus, Wlimit, and Cfac control the algorithm. The algorithm, as implemented does not form a cluster with total weight above Wlimit. It tries to consolidate the terminals into nClus clusters, and stops if it succeeds in doing so. Cfac controls the maximum cost (distance) between clusters to be merged. The algorithm computes, Cmax, the maximum cost (distance) between terminals. It then sets Cthresh, the threshold on cost allowed between two merging clusters, as Cmax times Cfac.

The algorithm outputs a vector, Cassoc, the cluster associated with each terminal. Terminals, t, with Cassoc[t] equal to t are cluster heads, but have no real significance. In practice, if this clustering is to be kept, another algorithm would be run to pick the actual centers of each cluster. Here, the only interest is the clustering itself.

Note that cost is passed by value not by reference. The algorithm destroys half of the cost matrix during processing. Specifically, to avoid processing both terminal i as a neighbor of terminal j and terminal j as a neighbor of i, the cost[i,j] is infinite for j, greater than or equal to i. Thus, if the language used for the actual implementation does not allow call by reference for arrays (and most languages do not), a local copy of the cost matrix must be made.

The algorithm sets up a heap of neighbors of each node. This facilitates finding the next nearest neighbor of each node during processing. It then sets up a heap, called sHeap, of the current nearest neighbor of each node. The values in this heap, held in the vector sel_val, are the cost to the nearest neighbors of each node. The top of this heap is the closest pair of nodes. As the algorithm proceeds, neighbors are considered and popped off the individual nodes' heaps. sHeap is updated to reflect this.

As nodes (then clusters) merge forming clusters, one node ceases to exist, and the other becomes the head of the merged cluster. cNum[t] helps to find the cluster number associated with node t. It is the number of another node in the same cluster. It is possible to find the actual cluster number by following the chain of cNums back to the cluster head, h, recognizable by the fact that cNum[h] equals h. This is the same cluster representation used in implementing Kruskal's algorithm. And Tarjan's method of collapsing these chains can be used to shorten them. That was not done here because it would have complicated the implementation somewhat and would not have reduced the overall complexity, which is dominated by other factors.

As the algorithm proceeds, nodes that no longer exist pop to the top of the neighbor heaps and sHeap. As they are recognized (by the fact that cNum[c] is less than 0), they are removed from further consideration.

```
void <- COM( nt , cost , weight , x , y , nClus , Wlimit , Cfac ,
             *Cassoc )
      dcl cost[nt,nt] , weight[nt] , x[nt] , y[nt] , Cassoc[nt]
          nHeap[nt,heap] , sHeap[heap] , selVal[nt] , cWt[nt]


/* Initialization */

Cmax <- 0  // Find the maximum cost
for_each( t , nt )
   for_each( t2 , nt )
      if ( t2 \cgeq t ) // Consider only lower numbered neighbors
         cost[t][t2] <- INFINITY
      else
         Cmax <- max( Cmax , cost[t][t2] )
Cthresh <- Cmax * Cfac  // Set the cost threshold

for_each( t , nt )  // Set up the neighbor heaps
   set_heap( nHeap[t] , nt , cost[t] )
   sel_val[t] <- top( nHeap[t] )  // Cost to nearest neighbor
   cNum[t] <- t  // Cluster number
set_heap( sHeap , nt , sel_val )  // Heap of nearest costs
```

```
cWt <- weight   // Cluster weight
nc <- nt   // Current number of clusters

/* Form clusters */
while( nc > nClus )
   c <- pop( sHeap )   /* Node with best nearest neighbor */
   if ( c == -1 )      /* Heap is empty; no more merges  */
      break                            do not update ≈ lazy update
   if ( cNum[c] < 0 )                  /* Cluster no longer exists */
      heap_replace( sHeap , c , INFINITY )
                                       /* Remove from consideration */
      continue
   c2 <- pop( nHeap[c] )   // Nearest neighbor
   if ( TestMerge( c , c2 ) )   // Test the validity of merging
      Merge( c , c2 )   // Merge cluster c2 into cluster c
      nc <- nc - 1
   heap_replace( sHeap , c , top(nHeap[c]) )
                                              update to next nearest neighbor

// Set the associated cluster for each terminal
for_each( t , nt )
   c <- t
   while( cNum[c] != c )
      c <- cNum[c]
      Cassoc[t] <- c
// Test the feasibility of merging clusters c1 and c2

boolean <- TestMerge( c , c2 )

 if( ( cNum[c2] >= 0 )  &          // clus c2 still exists
     ( cWt[c] + cWt[c2] < Wlimit ) &    // weight within limit
     ( cost[c][c2] < Cthresh )      )  // cost within limit
   return ( TRUE )
 else
   return ( FALSE )

// Merge clusters c1 and c2
void <- Merge( c1 , c2 )

cNum[c2] <- c1
x[c1] <- (x[c1]*cWt[c1] + x[c2]*cWt[c2]) / (cWt[c1] + cWt[c2])
y[c1] <- (y[c1]*cWt[c1] + y[c2]*cWt[c2]) / (cWt[c1] + cWt[c2])
cWt[c1] <- cWt[c1] + cWt[c2]
for_each( t , nt )
   if ( (c != c1) && (cNum[c] != c) )
      cost[c1][c] <- Eucl( x[c1] , x[c] , y[c1] , y[c] )
   else
      cost[c1][c] <- INFINITY
set_heap( nHeap[c1] , nt , cost[c1] )
```

Each merge between the current closest pair of nodes is tested and if it is found feasible (i.e., both nodes exist and the merged cluster does not violate the weight constraint or the distance constraint) a new cluster is formed at the center of mass of the merging clusters. The newly formed cluster carries its own number, and the other node takes that number, identifying it as no longer an existing independent cluster.

The process terminates when the requisite number of clusters is formed or when no further merges are feasible. This latter case is indicated by the fact that sHeap is empty and returns −1 when popped. Note that Pop returns an index (a node number here) while Top returns a value.

As mentioned above, this algorithm can be followed by post processors which consider tradeoffs among the constraints. For example, if the number of clusters is larger than nClus, and some clusters have weight below a weight minimum, it is possible to relax the cost (distance) or maximum weight constraints somewhat and allow further merging. Also, it may be desirable to identify better cluster heads by finding the node closest to the center of mass of each cluster. This is easily done by enumeration.

The complexity of this algorithm is dominated by popping the heaps. In the worst case, if there are N terminals, it might pop $O(T^2)$ times for a complexity of $O(T^2 \log T)$. In practice, however, expect a much smaller number of pops, and the actual complexity dominated by computing the costs and setting up the heaps, which is $O(T^2)$. It is possible to reduce the complexity further by using more sophisticated neighbor finding techniques ([KP86]) and not considering all possible neighbors.

Example 5.7. Tables 5.3 and 5.4 and Fig. 5.10 show an example with 10 terminals. For simplicity, all terminals have a weight of 1; the algorithm does not require this. The weight limit is 4 and the desired number of clusters is 2. The cost factor is 0.5. The cost threshold is computed, to the nearest integer, as 43; the (6,1) cost is 87.

The first merge considered is between nodes 9 and 8 at a distance of 13. This satisfies all constraints and is accepted. Similarly, the next 4 merges are accepted. The (9,1) merge is rejected because node 1 no longer exists as an independent node (it is part of the cluster headed by 2). The merge (not shown) between node 6 and its nearest neighbor is immediately rejected because node 6, itself, no longer exists as an independent node. The (2,4) merge is rejected based on weight; it would form as cluster of weight 5.

The final clusters are shown in Fig. 5.10 and in Table 5.3. In this case, the algorithm halted when the desired number of clusters (3) was reached.

5.3.2 Add Algorithm

Another approach to concentrator location is to use a greedy algorithm. Start with all terminals connected directly to the center and then evaluate the savings obtainable by adding a concentrator at each site. Then greedily select the concentrator which saves the most money. In the following discussion, the center is treated as a concentrator with infinite capacity at location 0. Thus, c_{ij} is the cost of connecting terminal i to concentrator j (for $j > 0$) and c_{i0} is the cost of connecting terminal i to the center.

TABLE 5.3
COM algorithm example

Node	X	Y	Weight
1	31	19	1
2	45	13	1
3	59	92	1
4	22	64	1
5	86	55	1
6	95	78	1
7	98	63	1
8	39	44	1
9	27	38	1
10	48	85	1

Weight limit = 4	Cost factor = .5	# of clusters = 3
	Merges considered	

Node	Neighbor	Cost	X	Y	Weight
9	8	13	33	41	2
10	3	13	53	88	2
7	5	14	92	59	2
2	1	15	38	16	2
7	6	19	93	65	3
9	1	22	–	–	–
6	-	24	–	–	–
2	9	25	35	28	4
9	-	25	–	–	–
8	-	26	–	–	–
2	4	38	–	–	–
10	4	39	42	80	3

Clusters:	(10 , 3 , 4)
	(2 , 1 , 9 , 8)
	(7 , 5 , 6)

If the concentrators have no capacity constraint, the savings, s_j, associated with concentrator j is simply

$$s_j = \left[\sum_{j \in I(j)} c_{ij} - c_i' \right] - d_j$$

where d_j is the cost of locating a concentrator at site j and c_i' is the cost of connecting the component containing i to the center. Initially, since i is alone in its component, c_i' is simply c_{i0}. $I(j)$ is the set of terminals which can save money by moving to j, that is, the cost of connecting to j is less than the cost of being connected to the site they are currently associated with.

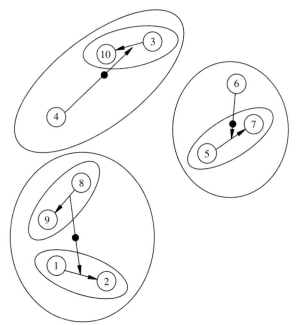

FIGURE 5.10
COM example.

Then pick the concentrator which saves the most money and choose the most cost efficient terminal to associate with it. At this point, some terminals remain associated with the center and some are associated with the newly selected concentrator. Express the cost of connecting terminal i to the concentrator with which it is currently associated as c_i'. So, c_i' is initialized to c_{i0}. The savings associated with concentrator j can now be defined at any point in the algorithm (i.e., both initially and after some concentrators have been selected) as

$$s_j = \left[\sum_{j \in I(j)} c_{ij} - c_i' \right] - d_j$$

The algorithm proceeds to greedily select concentrators until no concentrator which saves money can be found. After each concentrator is selected, the c_i' change and the concentrators must be reevaluated. Evaluating a concentrator requires that each terminal be examined. Selecting the concentrator with the best savings requires that we compare the s_j. Stated in this simple way, the effort to select the next concentrator is $O(TC)$, where T is the number of terminals and C is the number of potential concentrators. The overall complexity of the algorithm is $O(TC_2)$, since we can select at most $O(C)$ concentrators.

It is possible to improve on this complexity, however, by observing that as the algorithm proceeds and more concentrators are selected, terminals only become associated with concentrators which are closer to them. Thus, c_i' never increases. The c_{ij}, of course, do not change at all. Thus, examining the expression for s_j, it is apparent that s_j cannot increase as the algorithm proceeds. This allows us to defer reevaluation of some (actually most) of the concentrators on most iterations. Specifically, suppose concentrator c_1 has been reevaluated and the savings are found to be $s[c_1]$. Suppose we have not reevaluated concentrator c_2 and the previous savings associated with c_2 is $s[c_2]$. If $s[c_1]$ is greater than $s[c_2]$, then it is not necessary to reevaluate c_2 at this time as c_2 cannot possibly be the best concentrator to pick.

This leads to the improvement, to evaluate all concentrators once at the beginning of the algorithm and then form a heap based on the savings, with the largest value at the top of the heap. Then keep track of the iteration at which each concentrator was last reevaluated. At each iteration, pop the heap and ask if the concentrator which pops out has been reevaluated on the current iteration. If it has not, it must be reevaluated and pushed back into the heap. When a concentrator with a current evaluation pops out, select that concentrator. In the worst case, it is possible to have to reevaluate all concentrators. In practice, however, it is usually only necessary to reevaluate a few at each iteration. Thus, the worst case complexity remains $O(TC_2)$ but the complexity in practice becomes $O(TC_rC_s)$, where C_r is the average number of concentrators reevaluated per iteration, and C_s is the number of concentrators selected. In practice, C_r is usually $O(1)$ rather than $O(C)$.

This algorithm can also be extended to handle the case where concentrators are capacity constrained. Redefine the set $I(j)$ (the set of terminals that can benefit from associating with concentrator j) as the set of terminals that maximize s_j subject to the capacity constraint. This problem is a knapsack problem ([GN72]) which, while not that difficult to solve in most cases, requires more effort than is justified in the inner loop of a heuristic like this. Therefore, it is sufficient to find s_{ij}, the savings obtainable by terminal i moving to concentrator j, and then greedily filling up $I(j)$ with the terminals that have the largest s_{ij}. s_{ij} is defined as

$$s_{ij} = c_{ij} - c'_i$$

For each j, sort the terminals, i, based on the s_{ij}, largest first, and fill up $I(j)$. Stop filling when the next terminal does not fit (i.e., when w_i is greater than the remaining capacity on the concentrator), or when all terminals have been considered. The latter approach may yield a somewhat more realistic savings. The former approach may be much faster when the capacity constraint is tight. A compromise between the two approaches is to stop when the next terminal does not fit and the slack drops below a given threshold.

Strictly speaking, unless the knapsack problem is solved exactly, there is no guarantee that the s_j will not increase. Nevertheless it is possible to use the acceleration technique described above and defer the reevaluation of concentrators based on the expectation that the s_j not increase significantly.

It is possible to improve the complexity of this algorithm further by noting that a concentrator need not be reevaluated if none of the terminals in $I(j)$ have been moved. This becomes attractive if the $I(j)$ are small in comparison with the total number of terminals. This is the case if the capacity constraint is tight or if many concentrators have already been selected. To take advantage of this, however, it is necessary to keep track of which $I(j)$ each terminal belongs to. This requires additional time and space and complicates the implementation somewhat. Also, in a case where a concentrator is unnecessarily reevaluated, its value does not change. Thus, it remains at the top of the heap, and it is selected. So the greatest penalty to pay is to reevaluate one extra concentrator per iteration. Therefore this latter acceleration has not be included in the implementation that follows.

A major issue in the implementation of the Add Algorithm is whether to allow terminals to move from one concentrator to another as new concentrators are selected. If terminals are allowed to move, it is possible that a concentrator which is justified at the time it was selected (i.e., it saved money) may not still be justified at the end of the algorithm because many of the terminals previously associated with it have moved. It is possible, of course, to check for this explicitly, but it is still possible that marginal concentrators will remain.

If terminals are not allowed to move, it is possible that a concentrator selected early in the algorithm can lock in terminals that are better associated with another concentrator, which is nearer to them but has not yet been selected. This problem can be overcome by doing a terminal assignment after the completion of the Add Algorithm. However, it is possible that the latter concentrator not be selected at all because it cannot justify itself during the Add Algorithm without taking terminals already assigned to other concentrators.

If the capacity constraints are loose, it has been found in practice it is better to allow terminals to move. The concentrators selected early tend to grab too many terminals, but can still justify themselves even if some of the terminals originally associated with them move. Conversely, if the capacity constraints are tight, it is usually better to prohibit moving, or at least to discourage it. It is possible to discourage moving by penalizing a move. For example, computing the savings of moving terminal i from concentrator j to concentrator k as,

$$s_{ij} = c_{ij} - c_{ik} - P$$

where P is a penalty for moving.

Figure 5.11 shows an example of the Add Algorithm. The data is the same as in Table 5.3, except that the capacity constraint is 3 instead of 4. For simplicity (if not realism), the cost of a concentrator is simply the cost of connecting it to the center, which is assumed to be at coordinates (0,0). Costs are Euclidean distances and are given in Table 5.4.

The algorithm begins by selecting a concentrator at node 5, associating terminals 5, 6, and 7 with it. This results in a savings of 200, as follows:

```
Terminal 5: savings of 102 -  0 = 102
Terminal 6: savings of 122 - 24 =  98
Terminal 7: savings of 116 - 14 = 102
Concentrator at node 5: Cost =    102
```

This is the largest savings available from any concentrator and is larger than 0, so the concentrator at node 5 is selected.

Next to be selected is a concentrator at node 10, associating terminals 3, 4, and 10 with it, for a savings of 130. Notice that the savings are smaller. Then a

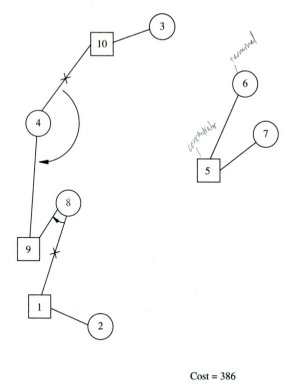

Cost = 386

FIGURE 5.11
Add algorithm.

TABLE 5.4

Cost matrix for add algorithm example

					Concentrator							
		0	1	2	3	4	5	6	7	8	9	10
T	1	36	0	15	78	45	65	87	80	26	19	68
	2	46	15	0	80	55	58	82	72	31	30	72
e	3	109	78	80	0	46	45	38	48	52	62	13
r	4	67	45	55	46	0	64	74	76	26	26	33
m	5	102	65	58	45	64	0	24	14	48	61	48
i	6	122	87	82	38	74	24	0	15	65	78	47
n	7	116	80	72	48	76	14	15	0	61	75	54
a	8	58	26	31	52	26	48	65	61	0	13	41
l	9	46	19	30	62	26	61	78	75	13	0	51
	10	97	68	72	13	33	48	47	54	41	51	0

concentrator at node 1 is selected, associating terminals 1, 2, and 8 with it, for a savings of 63.

Finally, a concentrator at node 9 is selected, associating terminals 4, 8, and 9 with it. Terminals 4 and 8 move from other concentrators. The savings is 20 and the other concentrators are still justified in this case. In other situations, in particular if the concentrators were more expensive, this might not be the case. In this instance, allowing terminals to move gave justification to the last concentrator and saved 20 units of cost.

An implementation of the Add Algorithm follows. The implementation is for the capacitated case. `nt` and `nc` are the number of terminals and concentrators, respectively. `Wlimit` is the concentrator capacity constraint. `th_move` is the penalty (threshold) for moving terminals from one concentrator to another. `Weight` is the load from each terminal. `Ccost` is the cost of placing a concentrator at each location, including the cost of connecting it to the center. `cost[t,c]` is the cost of connecting terminal `t` to concentrator `c`. Internally, the algorithm forms a heap, `sHeap`, of the expense (savings are represented as negative expense) associated with adding each concentrator.

`last_eval[c]` keeps track of the last time concentrator c was evaluated. If `last_eva[c]` is current (equal to the number of concentrators selected), then it is unnecessary to reevaluate `expense[c]` if c pops to the top of the heap.

```
void <- Add( nt , nc , Wlimit , th_move , cost ,   weight ,
             Ccost , *Cassoc )
    dcl cost[nt,nc] , weight[nt] , sHeap[heap] ,
        Ccost[nc] , Cexpense[nc] , last_eval[nc]

    // Initially associate all terminals with the center (conc 0)
    Cassoc <- 0
    // Set up a heap of expense associated with each conc.
    // ( < 0 means savings )
    nSel <- 0
    Cexpense[0] <- INFINITY
    for_each ( c , nc )        all pass locations of concentrated
        Cexpense[c] <- EvalConc( c )        <- O(c²)
        last_eval[c] <- nSel
    Set_heap( sHeap , nc , Cexpense )

    // Select concentrators
    while ( Top( sHeap ) < 0 )
        c <- Pop( sHeap )
        if ( last_eval[c] = nSel )  <- check current
            AddConc( c , Cassoc )
            nSel <- nSel + 1
        else
            Cexpense[c] <- EvalConc( c )  <- O(T)
            last_eval[c] <- nSel
            Heap_replace( sHeap , c , Cexpense[c] )  <- O(log C)
```

```
integer <- EvalConc( c )
        dcl delta[nt] , permu[nt] , Ter[nt]

    expense <- Ccost
    slack <- Wlimit
    n <- 0
    for_each ( t , nt )
      s <- cost[t][c] - cost[t][Cassoc[t]])
      if( s < 0 )
         n <- n + 1
         delta[n] <- s
         Ter[n] <- t
      }
    if ( n = 0 )   // No terminal benefitted
       return( expense )
    Sort( n , delta , permu )
    for_each ( i , n )
       p <- permu[i]
       t <- Ter[p]
       if ( delta[p] >= 0 )
          break
       else if ( ( weight[t] = slack ) &&
                 ( (Cassoc[t] = 0) | (delta[p]+th_move < 0) ) )
             expense <- expense + delta[p]
             slack <- slack - weight[t]
    return( expense )


void <- AddConc( c , *Cassoc )
    dcl Cassoc[nt] , delta[nt] , permu[nt] , Ter[nt]
    slack <- Wlimit
    n <- 0
    for_each ( t , nt )
      s <- cost[t][c] - cost[t][Cassoc[t]]
      if( s < 0 )
         n <- n + 1
         delta[n] <- s
         Ter[n] <- t
    Sort( n , delta , permu )
    for_each ( i , n )
       p <- permu[i]
       t <- Ter[p]
       if ( delta[p] >= 0 )
          break
       else if ( ( weight[t] <= slack ) \cwedge
                 ( (Cassoc[t] = 0) | (delta[p]+th_move < 0) ) )
             Cassoc[t] <- c
             slack <- slack - weight[t]
```

If the concentrator which pops to the top of the heap is current, select it and greedily associate with it the terminals that save the most money. If the concentrator evaluation is not current, reevaluate it and push it back into the heap.

The evaluation and add routines are very similar. Both form a list of terminals which can save by (re)associating with the current concentrator. Next, sort the list based on cost (delta), smallest first. Note that delta is negative and is really a savings. Sort leaves the values of delta untouched and returns a permutation vector, permu, giving the sorted order of the values of delta.

The heap routines used work with a heap of fixed size, in this case the number of concentrators. Pop moves the popped concentrator to the bottom of the heap. Heap_replace replaces a value and bubbles the item to its proper position.

The algorithm returns Cassoc, the concentrator associated with each terminal. The set of selected concentrators is found by examining the values appearing in Cassoc. Alternatively, an explicit list of selected concentrators could be returned or a boolean vector could be set.

5.3.3   Drop Algorithm

An alternative greedy approach is to start with all concentrators selected and to drop concentrators to save money. As in the Add Algorithm, evaluate how much can be saved by dropping each concentrator, then drop the one that saves the most.

The implementation of the Drop Algorithm is very similar to that of the Add Algorithm, but there are important differences as well. Again it is found that the savings decrease as the algorithm continues. Again a heap of the concentrators is formed, based on savings, and only the concentrator which pops to the top is reevaluated. Proceed until no further savings are obtainable.

The main difference is when capacity constraints are present, the evaluation of each concentrator involves solving a terminal assignment problem. Specifically, start by solving a terminal assignment problem with all concentrators present. Then it is necessary to solve another terminal assignment problem with a specific concentrator removed in order to evaluate each potential drop. Since even a heuristic solution to the terminal assignment problem is $O(TC)$ for T terminals and C concentrators, the algorithm would be $O(TC^3)$. This is computationally prohibitive except for small problems. Therefore consider the uncapacitiated case in the following discussion. The algorithm presented would work for the capacitated case as well, if the EvalDrop procedure was replaced by a terminal assignment algorithm. Since the performance of the Add and Drop algorithms are comparable in terms of the quality of the solutions they yield there is little justification for using the Drop Algorithm in the capacitated case when its complexity is so much higher.

The implementation below is for the uncapacitated case. EvalDrop computes the savings obtainable from dropping a given concentrator. This is the cost of the concentrator (which is a savings) minus the cost of moving terminals to new homes (which is an expense). As before, form a heap based on savings and greedily drop the concentrator which saves the most money.

The algorithm returns the number of concentrators selected (nSel), an explicit list of selected concentrators (Conc_list), as well as which concentrator each terminal is associated with (Cassoc). The Add Algorithm could have done this too.

Example 5.8. Figure 5.12 shows the solution of the problem given in Table 5.4, obtained by the Drop Algorithm in the unconstrained case. The solution value is 367 and two concentrators are chosen. This compares with a solution value of 374 obtained by the Add Algorithm for the same problem with no capacity constraint.

The Drop Algorithm begins with each terminal connected to its nearest neighbor. In this case, since the concentrator sites are the same as the terminal sites and Euclidean distances are used, each terminal is connected to a concentrator at the same site.

Next, dropping each concentrator is evaluated. Recall that the cost associated with each concentrator in this example is simply the cost of connecting it to the center. The greatest savings is obtained by dropping concentrator 6 (saving 122) and rehoming terminal 6 (which is the only terminal associated with concentrator 6) to its next nearest neighbor, concentrator 7 (at an increased cost of 15-0). Thus, the net savings is 107.

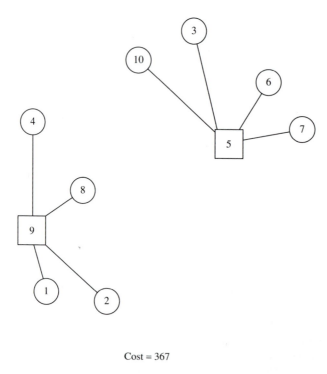

Cost = 367

FIGURE 5.12
Drop algorithm.

Next, concentrator 3 is dropped and terminal 3 moves to concentrator 10, for a net savings of 96. Then concentrator 7 is dropped and terminals 6 and 7 (currently both associated with it) rehome to concentrator 5. In general, the terminals can rehome to any of the concentrators, but in this case their next nearest neighbors were the same concentrator.

```
void <- Drop( nt , nc ,   cost , Ccost ,
                *nSel , *Conc_list , *Cassoc )
     dcl cost[nt,nc] , Ccost[nc] , Conc_list[nc] , Cassoc[nt] ,
         sHeap[heap] , Csavings[nc] , last_eval[nc]

  // Initially, all concentrators are selected

  nSel <- nc
  for_each ( c , nc )
     Conc_list[c] <- c

  // Associate each terminal with its nearest conc.

  for_each ( t , nt )
     best_cost <- INFINITY
     for_each ( c , Conc_list )
        if ( cost[t][c] < best_cost )
           best_conc <- c
           best_cost <- cost[t][c]
     Cassoc[t] <- best_conc

  // Set up a heap of savings associated with each concentrator

  Csavings[0] <- INFINITY  // 0 is the center; do not drop it
  for_each ( c , nSel )
     Csavings[c] <- EvalDrop( c )
     last_eval[c] <- nSel
  set_heap( sHeap , nc , Csavings )

  // Select concentrators

  while ( top( sHeap ) < 0 )
     c <- pop( sHeap )
     if ( last_eval[c] = nSel )
        DropConc( c )
     else
        Csavings[c] <- EvalDrop( c )
        last_eval[c] <- nSel
        heap_replace( sHeap , c , Csavings[c] )

number <- EvalDrop( conc )
```

```
// We save the cost of the conc by dropping it

expense = -Ccost

// If we drop the conc., we must pay to move the terminals
// associated with it to the next closest conc.

for_each ( t , nt )
    if ( Cassoc[t] = conc )
        new_cost <- INFINITY
        for_each ( i , nSel )
            c <- Conc_list[i]
            if ( (c != conc) && (cost[t][c] < new_cost) )
                new_cost <- cost[t][c]
        expense <- expense + new_cost - cost[t][conc]
return( expense )


void <- DropConc( int conc , *nSel , *Conc_list , *Cassoc )
    dcl Conc_list[nc] , Cassoc[nt]

// Rehome all ters associated with conc

for_each( t , nt )
    if ( Cassoc[t] = conc )
        new_cost <- INFINITY
        for_each ( i , nSel )
            c <- Conc_list[i]
            if ( (c != conc) && (cost[t][c] < new_cost) )
                new_cost <- cost[t][c]
                new_conc <- c
        Cassoc[t] <- new_conc
flag <- FALSE

// Remove conc from the list of concentrators

for_each ( i , nSel-1 )
    if ( Conc_list[i] = conc )
        flag <- TRUE
    if ( flag )
        Conc_list[i] <- Conc_list[i+1]
nSel <- nSel-1
```

5.3.4 Relaxation Algorithm

The previous algorithms are all heuristics. Thus there is no guarantee on the quality of the solutions they yield. Erlenkotter [Erl78] developed an algorithm that yields a lower bound on the quality of the solution it yields, which also produces a good selection of concentrators that can be used as a solution or as a set of candidates for input to the other heuristics previously discussed.

Erlenkotter's algorithm is based on the idea of relaxation discussed in Chapter 1. Recall that the relaxation of a problem involves temporarily removing some of the constraints or replacing the cost function with a simpler one which yields lower bounds on the real function.

Erlenkotter's algorithm will be discussed in the context of uncapacitated problems. The relaxation proposed is to modify the cost function. In particular, the original cost function is replaced by,

$$\sum_{i,j} c_{ij} x_{ij} + \sum_j d_j y_j$$

where c_{ij} and d_j are the cost of connecting terminal i to concentrator j and the cost of concentrator j, respectively, by

$$\sum_{i,j} c_{ij} x_{ij} + \sum_{i,j} p_{ij} x_{ij} = \sum_{i,j} \left(c_{ij} + p_{ij} \right) x_{ij} \tag{5.15}$$

p_{ij} is a penalty assessed for connecting terminal i to concentrator j. Part of the algorithm involves assigning values to the p_{ij} subject to the restriction that

$$\sum_i p_{ij} \le d_j$$

This restriction guarantees that the new cost function is a lower bound on the original cost function because for any feasible solution, the sum of the p_{ij} added to the cost function is no greater than the sum of the d_j removed from the cost function. The algorithm attempts to make the lower bound as tight (large) as possible in order to get a good estimate on the quality of the solution obtained.

The tightest lower bound is obtained if the p_{ij} assigned by the algorithm equal the d_j for the concentrators selected. If this occurs, the lower bound and the feasible solution have the same values. If, in addition, the optimal solution to the relaxed problem is found, the optimal solution to the original problem is also found.

The motivation for this relaxation is that it is easy to find the optimal solution to the relaxed problem. Given values for the p_{ij}, the optimal solution is obtained simply by assigning each terminal (setting $x_{ij} = 1$) to its nearest concentrator, where nearest is defined by the sum of c_{ij} and p_{ij}.

In order to make the bound as tight as possible, assign large values to the p_{ij} selected in the solution, subject to the constraint that the sum of the p_{ij} for a given j not exceed d_j. The interesting part of this is that the larger the p_{ij}, the less likely it is to be picked, since increasing p_{ij} makes it less likely that $c_{ij} + p_{ij}$ will be smallest for a given i.

We begin by describing the penalty assignment process informally and then give a formal description of the procedure. The process of setting the p_{ij} can be thought of as "selling" a concentrator to the terminals. Initially, the p_{ij} are all 0. Then go to each terminal and ask how much it is willing to pay for concentrator j. Explain that if the terminals won't pay for it, it won't be selected. A terminal that is closer to some other concentrator will pay nothing for concentrator j; it saves nothing from concentrator j being selected. A terminal, i, which is closest to concentrator j, however, is willing to pay up to $c_{ik} - c_{ij}$ where c_{ik} is the cost of connecting i to the next nearest concentrator. Therefore, set p_{ij} equal to $c_{ik} - c_{ij}$ with the assurance that the p_{ij} will not be wasted, that is, it will appear in the solution. (It is appropriate to break the tie in favor of selecting x_{ij}.)

The process of assigning p_{ij} continues so that each terminal is forced to pay something for its nearest concentrator. At this point it is still possible to increase some of the p_{ij}, as the p_{ij} do not yet equal the d_j. If p_{ij} is assigned to any j which is not the nearest (in terms of $c_{ij} + p_{ij}$) to i, however, it will be wasting the penalty since terminal i will not select it. Therefore, it is necessary to also assign some penalty to other concentrators so that $c_{ij} + p_{ij}$ is still (tied for) smallest.

> **Example 5.9.** Suppose that $c_{i1} = 3$, $c_{i2} = 5$, $c_{i3} = 1$, and $c_{i4} = 8$. It is possible to set $p_{i3} = 2$. Thus, terminal i will have to pay 3 to connect to a concentrator. If there is spare penalty available for use, we could then set p_{i1} to 2 and increase p_{i3} to 4. Terminal i would then have to pay 5 to connect to a concentrator. Finally, if there was enough spare penalty left, it would be possible to increase p_{i1}, p_{i2} and p_{i3} by 3, forcing terminal i to pay 8 to connect. If at any time during this process, the spare penalty runs out at any of the concentrators involved, it is not useful to increase the penalties on this terminal further and we would look to assign penalties to other terminals.

The more p_{ij}'s assigned in a row i (i.e., to a terminal), the more likely it is that they will be wasted; only one penalty will actually be picked. Thus, it is preferable to assign only one p_{ij} in any row, i, and always to assign as few as possible. On the other hand, it is possible to assign a total of d_j of penalties in column j (corresponding to the cost of concentrator j). Therefore, we are willing if necessary to assign several p_{ij}'s in the same row, hoping to use some of the penalty rather than waste it by not assigning it at all. This leads to the procedure of first assigning 1 p_{ij} in each row, then assigning 2 in each row and then assigning 3, etc. The following example illustrates this procedure.

> **Example 5.10.** Table 5.5 gives the cost of connecting each of 4 terminals to each of 4 concentrators. Assume the cost of each concentrator is 5. (Note that in general, the number of terminals differs from the number of concentrators and the costs of the concentrators differs from one another.)

The procedure begins by setting `Value` for each terminal to be equal to the smallest value in each row; that is, the cost of connecting each terminal to its nearest concentrator. This sum (10 here) is certainly a lower bound on the cost of a solution.

Then try to assign a single p_{ij} in each row. Set $p_{13} = 3$, making `Value` $= 5$ for terminal 2 and `spare` $= 2$ for concentrator 3. Since there is a tie for best on row 2, we would have to assign 2 p_{ij} in that row and we are not willing to do this yet. We thus go on to row 3, setting $p_{31} = 2$, `Value` $= 7$ in row 3 and `spare` $= 3$ for concentrator 1. Finally, we set $p_{43} = 1$, increasing `Value` to 3 for row 4

TABLE 5.5
Relaxation algorithm

		c	o	n	c	
		1	2	3	4	Value
t	1	5	9	2	6	$\to$ 2
e	2	3	1	4	1	$\to$ 1
r	3	5	8	9	7	$\to$ 5
m	4	9	3	2	4	$\to$ 2
spare		5	5	5	5	

sum 10 ← lower bound

TABLE 5.6
Relaxation algorithm

assign one p_{ij} per row

		c	o	n	c	
		1	2	3	4	Value
t	1	5	9	5	6	5 ← +3
e	2	3	1	4	1	1 ← skip
r	3	7	8	9	7	7 ← +2
m	4	9	3	3	4	3 ← +1
spare		3	5	1	5	

sum = 16 ← lower bound

TABLE 5.7
Relaxation algorithm

assign two p_{ij} per row

		c	o	n	c	
		1	2	3	4	Value
t	1	6	9	6	6	6 +1
e	2	3	3	4	3	3 +2
r	3	8	8	8 9	8	8 +1
m	4	9	3	3	3 4	3 skip (omit)
spare		1	3	0	2	

sum = 20 ← lower bound

and decreasing `spare` to 1 for concentrator 3. This is summarized in Table 5.6. The numbers in the interior of the table are sums of c_{ij} and p_{ij}. The lower bound has now been increased to 16.

Next, consider assigning two p_{ij} per row. To begin, increase p_{11} and p_{13} by 1, increasing `Value` to 6 in row 1 and decreasing `spare` to 2 and 0 for concentrators 1 and 3, respectively. Similarly, we increase `Value` to 3 in row 2 and to 8 in row 3. It is not possible to increase `Value` in row 4 because doing so requires `spare` from concentrator 3, and no `spare` is left. The status of the solution at this point is summarized in Table 5.7. The lower bound is now 20.

Now consider assigning three p_{ij} in a row. Row 1 requires `spare` from concentrator 3, so its `value` cannot be increased. The `value` in row 2 can be pushed

TABLE 5.8
Relaxation algorithm *assign three p' row*

		c	o	n	c	
		1	2	3	4	Value
t	1	6	9	6	6	6
e	2	4	4	4	4	4 +1
r	3	8	8	9	8	8
m	4	9	3	3	4	3 *(cmit)*
	spare	0	2	0	1	*sum=21 ← lower bound*

needed concentrators ⟹

C 1 3
T {2,3} {1,4}
...cost = 22 → add/drop to handle capacity

up to 4, exhausting the last unit of `spare` in concentrator 1. After this, no further progress can be made. The final result is given in Table 5.8. The lower bound is 21.

Concentrators 1 and 3 have "sold" themselves by using up all their `spare`. If these two concentrators are selected and each terminal is assigned to the nearer of them, terminals 2 and 3 are assigned to concentrator 1 and terminals 1 and 4 to concentrator 3, for a cost of 22. Thus, the lower bound is reasonably tight and there is assurance that the solution is within 1 of being optimal. Another solution is to only select concentrator 3 and assign all 4 terminals to it. This too has a cost of 22.

It is possible to consider all subsets of the concentrators with no remaining `spare`. In general, this would require an exponential amount of computation. In practice, if the number of concentrators without `spare` is small, this is possible. In the implementation below, the Drop Algorithm is used to select from among the set of concentrators with no `spare`.

Note that in general, while this procedure yields a good bound and often yields a provably optimal solution (e.g., feasible solution equal to the bound), it does not guarantee an optimally tight bound. Depending on the ordering of the terminals, the bound varies as concentrators run out of spare in different orders. Erlenkotter [Erl78] gives a secondary ascent procedure which adjusts the p_{ij} and tightens the bound at the expense of increasing the complexity of the procedure, but this is beyond the scope of the discussion here. Guignard and Spielberg [GS79] give an extension of this procedure to capacitated problems but this is also beyond the scope of the current discussion.

Now a look at the implementation of the algorithm. First, sort the neighbors (concentrators) of each terminal in terms of cost, nearest first. This produces, for each terminal, t, a permutation vector, `permu[t]` , containing the concentrators in sorted order, nearest first. The cost matrix itself is not reordered. Thus, `permu[t][1]` gives the index of the concentrator, c, nearest to t and `cost[t][permu[t][1]]` gives the distance (connection cost) from t to c.

Next, set the level of each concentrator. The level is the number of p_{ij} which must be increased to push up the value in row t (`row_val[t]`). Initially, `level` is set to 1 for all t and `row_val[t]` is set to the cost of connecting t to its nearest concentrator. The spare for each concentrator is initialized to its cost.

Now consider levels, l, from 2 to the number of concentrators. Any terminal below level $l-1$ has failed to reach the preceding level because doing so required spare at some concentrator which has no spare left. Thus, this terminal also cannot reach level l and is not considered further. highest_level records the highest level any terminal has reached thus far. If highest level is less than $l-1$, no terminal has reached the preceding level and there is no point in proceeding; the algorithm moves into its final phase.

If a terminal has reached the preceding level, try to push it up to the current level. Then compute delta, which is initially set to the difference in cost between the $(l-1)^{st}$ nearest concentrator and the lth nearest. If it is possible to push row_val[t] up by this value, terminal t has reached the current level. In order to do this, however, it is necessary to decrement the appropriate spares by delta. It is only allowable to push delta up by the amount of spare remaining. Thus, set delta to the minimum of its current value as well as the amount of remaining spare at each relevant concentrator (i.e., those closer than the lth nearest.) Then push row_val[t] up by delta and decrement the appropriate spares. If the next level is reached, it bumps up level[t] and highest_level.

The lower bound is the sum of the row_val's. We then compute a feasible solution. There are many ways to do this. Here, the choice is to simply call the Drop Algorithm, limiting the choices of concentrators to those with no remaining spare. This is done by setting the costs of the other concentrators to ∞. Finally, a feasible solution along with a lower bound is found by the Drop Algorithm.

Note that the center can be included as a concentrator by allotting it a column and giving it a cost of 0. This is usually appropriate since it guarantees a feasible solution and prevents row_val[t] from exceeding the cost of connecting t to the center. Thus, it deals neatly with the case where no concentrators are merited; none of the spares (except for the center) is 0 at the end of the algorithm.

```
long <- Relax( nt , nc , cost ,   Ccost ,
            *nConc , *Conc_list , *Cassoc )
    dcl cost[nt,nc] , Ccost[nc] , Conc_list[nc] , Cassoc[nt] ,
        level[nt] , row_val[nt] , spare[nc] , permu[nt,nc] ,
        local_Ccost[nc]

// Order the neighbors of each terminal

for_each ( t , nt )
    sort( nc , cost[t] , permu[t] )

// Set the initial contributions of each terminal

for_each ( t , nt )
    level[t] <- 1
    c <- permu[t][1]
    row_val[t] <- cost[t][c]

// Set the spare for each concentrator
    slack <- Ccost
```

```
// Increase the lower bound, using up the spare
highest_level <- 1
for ( l = 2 to nc )
   if ( highest_level < l-1 ) break
   for_each ( t , nt )
      if ( level[t] < l-1 ) continue
      c1 <- permu[t][l-1]
      c2 <- permu[t][l]
      delta <- cost[t][c2] - cost[t][c1]
      for_each ( j , l )
         c <- permu[t][j]
         delta <- min( slack[c] , delta )
      for_each ( j , l )
         c <- permu[t][j]
         slack[c] <- delta
      row_val[t] <- row_val[t] + delta
      if ( row_val[t] = cost[t][c2] )
         level[t] <- l
         highest_level <- l

// Set the lower bound
lower_bound <- 0
for_each ( t , nt )
   lower_bound <- lower_bound + row_val[t]

// Find a feasible solution
for_each ( c , nc )
   if ( slack[c] = 0 )
      local_Ccost[c] <- Ccost[c]
   else
      local_Ccost[c] <- INFINITY
Drop( nt , nc , cost , local_Ccost ,
       *nConc , *Conc_list , *Cassoc )
return( lower_bound )
```

(handwritten annotations: "slack[c]", "(ii) delta", "choose needed concentrators", "try to improve actual solution towards lower bound")

Example 5.11. Again, consider the example from the preceding section. Table 5.9 shows the progress of the row values, slack, and lower bound as the algorithm progresses through the levels. As can be seen, the lower bound is 367. Concentrators 1, 5, and 9 are the only ones (not including 0, which is the center) that ended with a slack of 0. If the Drop Algorithm is run, considering only these three concentrators (and the center) as candidates, concentrator 1 will be dropped and the remaining concentrators (5 and 9) will yield a feasible solution which also has a value of 367. (This happens also to be the solution that the Drop Algorithm found in the preceding section without guidance form the bounding procedure.) In this case, there is a known optimal solution. In more realistic cases, the Add or Drop Algorithms, even with the guidance of this bounding procedure, rarely find the optimal solution, but they often find good ones, within a few percent of the optimum. With this bounding procedure, it is possible to see how well it worked out.

TABLE 5.9
Slack and bound in Erlenkotter's algorithm

			Level			
Term.	0	1	2	3	4	5
2	0	15	30	30	30	30
3	0	13	38	45	46	46
4	0	26	26	33	38	38
5	0	14	24	45	48	48
6	0	15	24	38	43	43
7	0	14	15	48	48	48
8	0	13	26	26	26	26
9	0	13	19	19	19	19
10	0	13	33	41	47	48

Conclusion:

0	0	0	0	0	0	0
1	36	21	2	0	0	0
2	46	31	12	10	10	10
3	109	96	51	36	21	20
4	67	41	41	26	15	14
5	102	88	77	9	0	0
6	122	107	98	23	14	13
7	116	102	82	14	6	6
8	58	45	26	19	8	7
9	46	33	14	5	0	0
10	97	84	39	24	12	11
Bound	0	151	254	346	366	367

The complexity of Erlenkotter's algorithm is $O(TC \log C)$ to sort the neighbors of each terminal, plus $O(TL^2)$ for the adjustment of row values and slack, where L is the number of levels we go through. L can be no greater than C, so overall it has a complexity at worst $O(TC^2)$, which is competitive with the Add and Drop Algorithms. In practice, one may wish to try all three on a given problem.

EXERCISES

5.1. Consider the nodes shown in Figure 5.1. The weight (total traffic) of each node is 1. The center is at node 0. Find the best multipoint line configuration using Kruskal's (modified), Prim's, Esau-Williams, Sharma's and the VAM algorithms.

5.2. The multipoint line algorithms are given in the context of a single line constraint. This is most appropriate for half duplex lines where both incoming and outgoing traffic both contribute to the load on the line in both directions.

(a) What modifications need to be made to allow these algorithms to work for full duplex lines?

(b) Polling, where the central site requests inputs from the terminals, also contributes load to the lines. Describe how to modify these algorithms to take the overhead of polling into account.

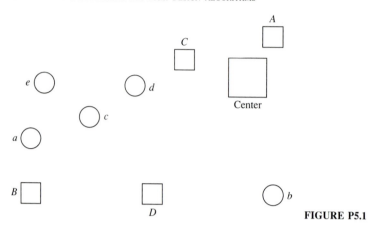

FIGURE P5.1

(c) Line errors force retransmission. Describe how to take the effect of line errors into account in these algorithms?

(d) Sometimes, because of reliability constraints or tariff restrictions the number of lines in cascade or the degree of nodes in the multipoint line are constrained. Describe how to take such constraints into account?

(e) Describe how to take a port cost at the central site into account in these algorithms?

(f) Why isn't it possible to take a degree constraint at the central site into accout in these algorithms?

5.3. Solve the problem given in the example at the end of Section 5.2 using the greedy algorithms discussed in that section, with and without exchanges.

5.4. There are 6 terminals in a city with weights 7, 9, 14, 5, 17, and 8.

(a) Use the First Fit and Best Fit algorithms to cluster these terminals onto lines with capacity 30.

(b) Another approach is to use the Worst Fit Algorithm where a terminal is placed on the line with the largest amount of spare capacity. Use the Worst Fit Algorithm to solve the above problem.

(c) Solve this problem again with a capacity of 20.

(d) What are the best possible solutions which exist for capacities of 30, 20 and 17?

5.5. Complete the execution of the Drop Algorithm and verify the solution mentioned in the example at the end of Section 5.3 and then use the Add Algorithm to find the other solution (value = 374) mentioned.

5.6. Consider Table P5.1 which gives the costs of connecting terminals to concentrators and to the center. Suppose that concentrators can hold five terminals, so there is no capacity constraint, and that concentrators cost 2000 each (including the cost of connecting them to the center). Apply the Add, Drop, and Relaxation algorithms to this problem and give the solutions obtained by each.

5.7. Consider Table P5.1 again which gives the costs of connecting terminals to concentrators and to the center. Suppose that concentrators A and D have been selcted and that a concentrator can hold three terminals. Apply the greedy and optimal assignment algorithms to find an assignment of terminals to concentrators.

TABLE P5.1

		CONCENTRATOR				
		A	B	C	D	center
T						
E	a	2965	758	2255	1884	5850
R	b	1513	2786	1627	1563	4385
M	c	2025	1223	1316	1059	4904
I	d	1491	1747	810	848	4359
N	e	2806	1368	2133	2074	5727
A						
L						

BIBLIOGRAPHY

[BT72] Bahl, L. R. and D. Tang: Optimization of concentrator locations in teleprocessing networks. In *Proc. Symp. Computer-Communication Networks and Teletraffic, Polytechnic Inst. of Brooklyn*, April 1972.

[CW64] Clarke, G. and J. Wright: "Scheduling of vehicles from a central depot to a number of delivery points," *Oper. Res.*, 12:568–581, 1964.

[Erl78] Erlenkotter, D.: "A dual-based procedure for uncapacitated facility location," *Oper. Res.*, 26:992–1009, 1978.

[EW66] Esau, L. R. and K. C. Williams: "On teleprocessing system design. a method for approximating the optimal network," *IBM System Journal*, 5:142–147, 1966.

[Gav85] Gavish, B.: "Augmented lagrangean based algorithms for centralized network design." *IEEE Trans. on Commun.*, COM–33:1247–1257, 1985.

[GJ79] Garey and Johnson: *Computers and Intraqctability: A Guide to the Theory of NP-Completeness*, W. H. Freeman, San Francisco, 1979.

[GN72] Garfinkle, R. S. and G. L. Nemhauser: *Integer Programming*, John Wiley, New York, 1972.

[GS79] Guignard, M. and K. Spielberg: "A direct dual method for the mixed plant location problem with some side constraints," *Mathematical Programming*, 17:198–228, 1979.

[KB83] Kershenbaum, A. and R. Boorstyn: "Centralized teleprocessing network design," *Networks*, 13:279–293, 1983.

[KC74] Kershenbaum, A. and W. Chou: "A unified algorithm for designing multidrop teleprocessing networks," *IEEE Trans. on Commun.*, 22:1762–1772, 1974.

[KP86] Kershenbaum, A. and S. L. Peng: "Neighbor finding algorithms for cmst calculation," *IEEE IN-FOCOM '86*, April 1986.

[Law76] Lawler, E.: *Combinatorial Optimization: Networks and Matroids*, Holt, Rinehart & Winston, New York, 1976.

[Mar72] Martin, J.: *Systems Analysis for Data Transmission*, Prentice-Hall, Englewood Cliffs, NJ, 1972.

[RV58] Reinfeld, N. V. and W. R. Vogel: *Mathematical Programming*, Prentice-Hall, Englewood Cliffs, NJ, 1958.

[SEB70] Sharma, R. and M. El-Bardai: "Suboptimal communications networks synthesis," *Proc. of the IEEE International Conf. on Comm.*, pages 19.11–19.16, 1970.

[WT73] Woo, L. S. and D. Tang: "Optimization of teleprocessing networks with concentrators," *Proc. of Nat'l Telecomm. Conf.*, pages 37C1–37C5, 1973.

CHAPTER
6

ROUTING

The following discussion focuses on routing, which is the problem of deciding how requirements should flow within a mesh network, where in general, there is more than one path available for each requirement. There is a given topology that includes the capacities of the nodes and links and the objective is to find the "best" path or paths for each requirement. Different notions of best give rise to different routing algorithms.

Routing is an important part of the ongoing maintenance of a network. As load changes, it is possible to "redesign" the network to accommmmodate the change in load simply by moving traffic from one route to another. This is by far the fastest and lowest cost method of redesign because it does not involve installing or moving any physical facilities. Indeed, if the load changes suddenly, this is usually the only alternative available. Thus, a solution to the routing problem is sometimes, by itself, the solution to an entire design problem.

It is also necessary to fully understand routing in order to compare topological alternatives in the inner loop of any design procedure. If the routing problem is not dealt with quickly and effectively, there is no hope of developing algorithms for the optimization of the mesh topologies themselves.

If the routing problem is solved, it is possible to answer the most basic question that arises in the design of networks. Specifically, we can determine, given an existing network that was designed to satisfy a set of requirements (real or projected) in the past, whether this network satisfies a current set of requirements.

Routing can be studied from the perspective of designing a network or from the perspective of managing a network in real time, adjusting to momentary changes in load. The algorithms used in both cases are similar. The major difference is that in real time there is rarely a consistent picture of the load in all parts of the network. In the time it takes to estimate the load across the network, compute optimal routes, and convey these routes to all the nodes, the load has changed. Indeed, if we include flow control procedures, which throttle the flow into the network based on perceived congestion, the problems of routing and load estimation become intimately related.

Problems of routing in real time are the focus of entire books [BG87]. The focus here is primarily on the perspective of routing within the context of design. But first, here is an overview of routing procedures when a topology and capacities are already given.

As mentioned above, the same algorithms can often be used in real time to the extent that all the nodes have the same picture of the load. It is interesting to note that the real time picture, while not necessarily consistent, is often more accurate than the requirements matrix used during the design process. The former is based on fairly recent measurements; the latter is often based on projections.

6.1 ROUTING PROCEDURES

6.1.1 Flooding

One of the simplest possible routing procedures is to have each node, with the exception of the the message destination node, send a copy of any message it receives to each of its neighbors. This is known as **flooding**. Flooding relieves a node of the burden of knowledge about the topology of the network. It does not need to know all the nodes or links within the network, which nodes and links are currently operative, which are congested, where the destination is, or even whether the destination exists. All a node needs to know is its local topology; that is, the set of nodes and links currently adjacent to it.

Thus, flooding has the advantages of simplicity, speed in selecting a route, and a minimum of overhead in exchanging topology and status information among nodes. It also has the obvious disadvantage of making very inefficient use of the nodes and links in the network. In general, multiple copies of each message will be transmitted and many messages will take circuitous paths.

Clearly, some mechanism must be present to prevent copies of a message from being forwarded forever between nodes even after it has been delivered to its destination. One way of cutting off the infinite propagation of messages is to have each message carry a hop count (the number of links it has already traversed). Each node, forwarding the message, increments the hop count and when the count reaches a predetermined limit the message is discarded. Another method is to have each node record the message number of each message it forwards and refuse to forward any message it has forwarded before. Since flooding results in multiple copies of a message reaching the destination, it is also important for the destination to keep track of the numbers of the messages it has already received, in order to discard duplicates.

Although flooding may appear trivial, it is in fact widely used. It is especially attractive in networks where the topology changes rapidly (e.g., mobile networks), or where the failure rate of links is high (e.g., military networks). Since it does not itself require global topology information, it is also useful in exchanging the topology information necessary to set up more sophisticated routes. Since the volume of such data is much smaller than the user data sent through the network, the relative inefficiency of flooding is tolerable in this case.

6.1.2 Explicit Routing

In some networks, an explicit route is stored in a table for each source and destination. Sometimes, different routes are stored for different classes of traffic; for example, interactive traffic might take one path while batch traffic might take another. Also, sometimes special routes are reserved for requirements which need secure routes or where data compression is used.

Explicit routing has the advantage of not requiring any computation of routes or the exchange of information in support of this computation. It also has the advantage of allowing the network designer a great deal of flexibility in optimizing the network because routes can be selected independently of one another. Thus, it is possible to make very good use of the network's capacity if the requirements are known in advance. In practice, explicit routes are modified periodically to reflect long term changes in network load.

Sometimes, several explicit routes are stored for each requirement. The first route in the list is used unless one of its component facilities is inoperative, in which case, the next route on the list is used. This allows the network to adapt to failures.

It is also possible to store several explicit routes for individual requirements along with the fraction of the requirement which is supposed to take each route. The requirement would then be randomly split among these paths in proportion to these fractions. In theory, this allows full optimization of the set of routes, using the algorithms to be presented in the following sections of this chapter. In practice, however, this is rarely done. The **bifurcation** (splitting) mechanism complicates the implementation of the routing procedure and the traffic is rarely known to sufficient precision to make this worthwhile. Also, in networks of even moderate size (more than a dozen nodes), almost the same results can be obtained by carefully choosing a single route for each requirement, as there is a sufficient number of requirements to allow for considerable optimization.

The major disadvantage of explicit routing is that it places a great burden on the network manager to create and maintain the lists of explicit routes. These routes must be updated whenever changes in topology are made (e.g., new nodes or links are added). In general, the lists are very large in large networks. They must also be kept consistent in all the nodes. Inconsistent route tables are the source of major operational problems in many networks.

6.1.3 Static Shortest Path Routing

In many networks, a length is assigned to each link and traffic is routed over the shortest path based on these link lengths. It is also possible to assign lengths to nodes and include these node lengths in the shortest path computation. In particular, in labeling node j from node i,

$$D_j = D_i + L_i + L_{ij}$$

where D_i is the length of the path to node i (from wherever it starts from), L_i is the length assigned to node i and L_{ij} is the length of the link from i to j.

In the simplest case, these lengths are only changed infrequently; they are set administratively (hopefully based on some knowledge of the network topology and load). Often, the operational status of the nodes and links in the network is monitored and facilities that are inoperative are assigned infinite length. This allows the network to adapt to failures and route around them.

If a length of one is assigned to each link, there is **min-hop routing**. Min-hop routing is widely used. It is simple and, since it minimizes the number of nodes in a path, tends to hold down the load on the nodes, as well as hold down delays from passing through many nodes and links. Min-hop routing tends to produce ties as there are often several alternate paths with the same number of hops. Many implementations of min-hop routing break ties randomly in order to balance the load.

It is also possible to assign different lengths to different links (or nodes) in order to encourage or discourage flow. The link length might be a function of capacity (e.g., proportional to the reciprocal of capacity) to encourage flow on high capacity links, which in general have the smallest service times and most spare. Of course, if too much traffic is diverted to a high capacity link, it becomes very congested. Conversely, if low capacity links were made sufficiently long they would attract no flow at all and there would be no point in including them in the network. Thus, there is a tradeoff.

It is also possible to assign link lengths based on link cost. While this will not minimize delay, it conserves capacity on the more expensive links. This, in turn, may allow those links to be lower capacity. This is called **least cost routing**. Least cost routing is often done during network design, to optimize the cost of the network. It makes sense then to do the same type of routing in the actual network, as in this case the same capacity which was put into the network by the design procedure is used in practice. It is sensible to maintain consistency between the network design and the actual use of the network.

It is possible to assign lengths that create ties. In this case, the ties might be broken arbitrarily (but consistently in the same direction). Except in the case of very small networks with very small requirements, little is gained by splitting of this type.

By adding a constant to the length of each link, paths with a small number of hops are favored over paths with a larger number of hops because this constant is added more times to the length of paths with a larger number of links. Thus, it is possible to get most of the benefits of min-hop routing while still retaining some flexibility of balancing the load.

Note that the class of routes obtainable using distance based routes is much smaller than the class of routes obtainable using prespecified fixed routes because the shortest paths nest inside one another forming trees. Saying this another way, when min-hop routing is used, once a message reaches a node, the path (or paths) it takes has nothing to do with its original source. In explicit routing, on the other hand, it is possible to choose a path (or paths) for each requirement individually. In shortest path routing, all requirements reaching the same node and bound for the same destination follow the same path. This can be a serious hindrance to balancing the load, especially if the number of requirements is small. In larger networks, this is less of a problem.

Shortest path routing has the important advantage that it relieves the network manager of the burden of maintaining an explicit list of routes. Once the link lengths are set, all routes are implicitly defined. If nodes or links are added to the network, only a few link lengths need to be modified. Only the link lengths need be kept consistent, a small amount of data in comparison with a full set of routes. Finally, if a link's length is made large when it fails, the network will automatically adapt to failures, routing around them whenever possible.

6.1.4 Adaptive Routing

Some networks are capable of adapting to congestion as well as to failures. To do so, the network must maintain some measure of congestion on a real time basis. Typically, some statistics on the utilization of nodes and links or delays in nodes and links are maintained. Links and nodes grow "longer" as they become more congested, thereby discouraging further traffic from using them in paths.

In addition to maintaining a measure of congestion, the network must disseminate information about congestion in order to permit the nodes to make use of it in the routing process. Unlike failure information, which needs to be updated only when a node or link fails, since congestion is an infrequent event, congestion information becomes outdated very quickly. This places a burden on the network to disseminate this information frequently. If this is not done carefully, network delay might actually increase because of the capacity given up to the overhead of transmitting information to support the routing process.

Ideally, each node would always know the level of congestion in all other nodes and links. If this is the case, it is possible to minimze delay using the procedures described in the following sections. In practice, however, nodes have only an approximate picture of congestion. Therefore, only an approximate routing can be done in practice. This is tolerable. If the picture is slightly out of date some congestion may persist until nodes learn of it.

There are more serious problems, however. If the picture is not only out of date but also inconsistent, it is possible that different nodes assign different lengths to the same link. This could lead to inconsistent routing, even looping, where traffic passes through the same node more than once. This, of course, increases congestion. Indeed, if the loop is not detected (e.g., using a hop count limit), this could lead to complete network breakdown.

Another serious problem is that adaptive routing can become unstable when the network is heavily loaded. One part of the network may become congested. The adaptive routing shunts traffic away from it, congesting another part of the network. The adaptive routing then moves the traffic back to the original route. This oscillatory behavior can persist indefinitely, actually increasing congestion rather than decreasing it.

Adaptive routing also places a computational burden on the network nodes, forcing them to maintain congestion information and disseminate it, as well as recompute link lengths and recompute shortest path lengths. It does, however, save memory since the link lengths and congestion information take up less space than explicit routes.

In practice, many networks use some form of adaptive routing. Most of the difficulties with adaptive routing result from trying to adapt too quickly. Relatively slow adpatation places a relatively small overhead on the network, in terms of both link overhead for transmission and node overhead for computation. Slow adaptation is less vulnerable to looping and oscillation. Finally, what is slow to the network may be acceptably fast to a user. If the network is operating at relatively high speed, with an end to end delay of only 100 ms., slow to the network may be one second, in this case ten end to end delays. This same one second would be a quite acceptable time, from the user's point of view, for the network to adjust to the presence of a large file beginning to transfer between two nodes.

6.1.5 Distributed Routing

It is possible to avoid many of the instabilities associated with adaptive routing by doing it on a centralized basis. In this case, all nodes forward congestion information to a single node, predesignated for this function. This node computes all link lengths and shortest paths. It then sends out the routes to all other nodes. Since all routes are computed by a single node, they are consistent. The disadvantges of this approach are that there is increased overhead and delay in getting all the congestion information to a common point. Also, the computational burden on this one node can be very high; this can be alleviated by having a separate computer, possibly a large host, do the route computation.

It is possible for the routing node to compute a full set of routes between all node pairs or, alternatively, to have it respond to requests for individual routes. In this latter case, there is the additional overhead and delay of sending a message to request a route. This approach is sensible only if **virtual circuit routing** is used. In virtual circuit routing, a route is picked for all messsages in a session, at the start of the session, and is used throughout the session. This is in contrast with **datagram routing** where each message is routed individually.

More often, however, adaptive routing is done on a distributed basis. In **distributed routing** each node makes a decision only about the next hop for the message. The node at the other end of the link then makes the decision about the next hop. This reduces overhead and delay in updating congestion information, but requires more coordination among the nodes, as described above.

One of the earliest forms of distributed routing was implemented in the ARPANET [McQ77]. In this case, each node forwards to each of its neighbors an estimate of its delay to all other nodes. These estimates are formed with respect to immediate neighbors, on the basis of direct observation of the queues on outgoing links. For example, if node A is connected directly to node B, A can estimate the delay to B directly, since it can observe the length of the queue to B, or more directly, the length of the delay itself. If C is connected to A, A will forward this estimate to C. C can now update its estimate of the delay to B. C first estimates the delay to A, by direct observation. Next, it adds A's estimate of the delay to B. Then, C compares this sum with the lowest estimate it has computed for getting to B by any other path. If this new estimate via A is smaller, this becomes C's new path to B, and C forwards this updated information to each of its neighbors. It is also possible that, because A has increased a prior estimate of the delay on its link to B (because the A-B link has become more congested) that C will change its path to B from the C-A-B path to some other path which now looks better. Such estimates continue to propagate throughout the network. Nodes update all their estimates at once and forward these estimates all at once to each neighbor. This reduces inconsistencies and also reduces the overhead in transmitting the updates by replacing many very short messages with one moderate one. It is more efficient to send the one message since it is shorter than the sum of the many smaller ones (each of which has its own header), and it involves much less nodal processing (which is basically proportional to the number of messages, not their lengths).

6.2 THE FLOW DEVIATION ALGORITHM

We now consider the problem of finding a set of routes which minimizes the average end to end delay, given a network topology and a set of requirements. Assume, for the sake of simplicity, that requirements arrive independently and that message lengths are exponentially distributed. We thus use M/M/1 queueing formulae to compute delays on the links. The focus is on link delays, ignoring the delays in the nodes. Actually, this model can be extended to include node delays, line propagation delays, and many other queueing models, but these extensions add little to the discussion that follows.

The most important assumption made here is that arrivals to each link are independent of each other. This is clearly not true. If a message arrives at node A and is routed over links (A,B) and (B,C) to get to destination C, clearly a departure from the (A,B) link is followed immediately by an arrival on the (B,C) link. If many such messages arrive at A, both links become congested at the same time. Also, while it is possible for many messages to arrive at A and queue up, the departures from the first link are spaced out by their lengths. This tends to reduce queueing for the (B,C) link. Despite these apparent anomalies, it has been demonstrated in practice that if there is a reasonable mix of traffic in the network, that is, if traffic from different sources and destinations share each link, then this independence assumption holds quite well [Kle75] [Jac57].

Formally, the problem to consider is that of finding a set of link flows that satisfy the requirements and minimize the average end to end network delay. Specifically, given requirements, r_{ij}, and links, k, with capacities c_k, the objective is to minimize

$$\bar{T} = \frac{1}{\gamma} \sum_{i,j} d_{ij} r_{ij} \tag{6.1}$$

where d_{ij} is the delay suffered by r_{ij} and γ is given by

$$\gamma = \sum_{i,j} r_{ij} \tag{6.2}$$

Equivalently, since γ is an input to this problem and is not affected by the routing, it is possible to minimize the total delay, T, which is simply the average times γ. For the sake of simplicity, this is what will be done.

d_{ij} is the sum of the delays suffered on all the links in the path used by r_{ij}. Thus,

$$d_{ij} = \sum_{p} \sum_{k \in p} a_{ijp} t_k \tag{6.3}$$

where p is a path, t_k is the delay in link k, and a_{ijp} is the fraction of r_{ij} which uses path p. Thus,

$$T = \sum_{i,j} \sum_{p} \sum_{k \in p} a_{ijp} r_{ij} t_k \tag{6.4}$$

$$= \sum_{k} f_k t_k \tag{6.5}$$

where f_k is the flow on link k, which is simply the sum over all requirements of the requirement magnitude times the fraction of the requirement which uses link k.

Note it is assumed that it is possible to split (bifurcate) the flow, that is, to divide a requirement over more than one path. In practice, not all routing procedures permit this, but many do. The ability to bifurcate flow greatly simplifies the problem.

Many approaches have been proposed for the problem of finding the routing which minimizes network delay. All are based on the intuitively appealing idea that we start with some feasible solution (i.e., one where the link flows are smaller than the link capacities), and then move requirements from more congested paths to less congested ones. The way in which flow is moved from one path to another distinguishes the algorithms from one another.

One of the earliest and most easily implemented approaches to this problem is due to [FK73] and is known as the Flow Deviation (FD) Algorithm. The FD algorithm, which is closely related to the Frank-Wolfe method [MW56], is based on the observation that if a small amount of flow is moved from a path with larger incremental delay to a path with smaller incremental delay, then the total delay (hence the average) is decreased. The incremental delay on a path is just the sum of the incremental delays on all the links in the path. Incremental delay on a path is the difference in the delay suffered by the flow on that path if a small amount of flow is added to the path.

Thus, the incremental delay on a link, k, is defined as

$$I_k = \lim_{\epsilon \leftarrow 0} \frac{f_k t_k(f_k) - (f_k - \epsilon) t_k(f_k - \epsilon)}{\epsilon} \tag{6.6}$$

where $t_k(f_k)$ is the delay on link k, given it has a flow of f_k. Note that I_k is weighted by the amount of flow in link k. Thus, I_k is the change in the contribution of link k to the total delay. And if it is possible to move flow from links with larger I_k to links with smaller I_k, T is decreased. For any reasonable delay function, I_k is well defined and will be the same for ϵ approaching 0 from above or below, that is, adding an infinitesimal amount of flow increases the delay by an equal amount as subtracting the same amount of flow decreases it.

The limit above is, of course, the derivative of delay with respect to flow, and is easily found for most queueing functions. For example, for M/M/1 queues, the delay on a link, k, with service time T_{sk} and utilization U_k is

$$t_k = \frac{T_{sk}}{1 - U_k} \tag{6.7}$$

where T_{sk} is the average length of a message, L, divided by the capacity of the link, c_k, and U_k is the flow in the link, f_k, divided by the c_k. Thus,

$$t_k = \frac{L/c_k}{1 - f_k/c_k} = \frac{L}{c_k - f_k} \tag{6.8}$$

where c_k and f_k are in bits per second. We can simplify this by dividing the numerator and denominator by L, thus converting all quantities to messages per second. Therefore,

$$t_k = \frac{1}{c_k - f_k} \tag{6.9}$$

The incremental delay, I_k, is then given by

$$I_k = \frac{c_k}{(c_k - f_k)^2} \tag{6.10}$$

This approach is not particularly dependent upon the M/M/1 model. It was possible to have used the M/D/1 model if it were more appropriate. It is possible to include a term in the delay function for propagation delay, or include node delays in the path. There is not, however, total freedom in making the model. One property of the delay function heavily relied upon is that it be convex. (Convex functions were discussed briefly in Chapter 3.) The convexity if the delay function guarantees that any local minimum found is also a global minimum, that is, if it is not possible to improve delay by moving a small amount of flow, it is not possible to improve it by moving a larger amount of flow. Another important property is that the delay function should be separable, with respect to the links (i.e., it is the sum of functions defined on the links). Both these conditions hold here.

The FD algorithm proceeds by assigning lengths to the links, based on their incremental delay, then finding the shortest path from each source to each destination, based on these lengths, and loading the links based on these paths. Thus, the flow pattern found is then superposed with the previously found flow patterns, that is, a new flow pattern is formed sending part of the flow on the old paths and part of the flow on the new path. The amount of flow deviated to the new path is chosen to minimize the total delay. This is found using a simple line search such as a binary search over the interval 0 to 1. Thus, given a flow pattern F_1 (possibly itself a superposition of other flow patterns) and a new (extremal) flow pattern, F_2, based on the current link lengths, there is a new flow pattern, F_3, as

$$F_3 = xF_1 + (1 - x)F_2 \tag{6.11}$$

where x is chosen to minimize T. The link flows are now changed and it is possible to compute new shortest paths and iterate the process.

In theory, the algorithm terminates when no further progress can be made. This happens when no flow moves. However, the only way that no flow will move is when we cannot find a path for any requirement where the incremental delay is lower than that of the path (or paths) currently used by that requirement. Thus, when the FD algorithm terminates, all flow is on minimum incremental delay paths. This is intimately related to the convexity of the delay function. The following example illustrates all this.

Example 6.1. Suppose there is a the simple network shown in Figure 6.1. There are two paths from S to D, one with capacity 2 and the other with capacity 3. r_{SD} is 1. Suppose that the delay function, $t(f, c)$, is

$$t(f, c) = (f/c)^2 \tag{6.12}$$

where f is the flow in the link and c is its capacity. This is a convex function. Now consider a flow pattern (e.g., all the flow on link 1). It is found that moving a small amount of flow improves the total delay. Specifically, the total delay with all the flow

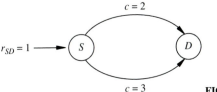

FIGURE 6.1

on link 1 is

$$T = (1)\left(\frac{1}{3}\right)^2 = \frac{1}{9} = 0.1111 \tag{6.13}$$

If a small amount of flow is moved, say 0.01, to link 2,

$$T = (0.99)\left(\frac{0.99}{3}\right)^2 + (0.01)\left(\frac{0.01}{2}\right)^2 = 0.1078 \tag{6.14}$$

there is an improvement. Continuing to move flow, it eventually is found that if 0.6 units of flow is put on link 1 and 0.4 units of flow on link 2 that

$$T = (0.6)\left(\frac{0.6}{3}\right)^2 + (0.4)\left(\frac{0.4}{2}\right)^2 = 0.04 \tag{6.15}$$

which is much better. Now attempting to move flow from either link to the other, things get worse. Specifically, if 0.01 units of flow are moved from either link to the other, T increases slightly, to roughly 0.04005.

The flow pattern with the (0.6, 0.4) split is in fact optimal. It was found incrementally by moving small amounts of flow.

The FD algorithm gives this same result. Given a link k with flow, f, a capacity, c, and the delay function $t(f, c)$ above, the incremental delay I_k, which is used as a link length in the FD algorithm, is

$$I_k = 3\left(\frac{f}{c}\right)^2 \tag{6.16}$$

Start with 1 unit of flow on link 1, the length of link 1 is 3/9 and the length of link 2 is 0. Thus, flow is moved from link 1 to link 2. If 0.01 units of flow are moved, the lengths of links 1 and 2 become 0.3267 and 0.000075, respectively. In general, it is possible to send x units of flow on link 1, and $1 - x$ units of flow on link 2 for any x between 0 and 1. The delay then is

$$T = (x)\left(\frac{x}{3}\right)^2 + (1 - x)\left(\frac{1 - x}{2}\right)^2 \tag{6.17}$$

To find the minimum of this function, find its derivative and set it to 0. And find x satisfying

$$\frac{3x^2}{9} - \frac{3(1 - x)^2}{4} = 0 \tag{6.18}$$

The solution to this equation is x equals 0.6, the same answer observed above. Note also that this equation is simply comparing the incremental lengths of links 1 and 2 (the two paths found for the requirement) and that to make these paths equal x is set.

After setting x to 0.4, the lengths of these paths are equal. Continuing to iterate the FD algorithm, the shortest path found can be either path and when we seek to deviate flow, none moves. This happens because there are only two paths and one requirement in this simple example. In general, new paths are found and they move some flow. In this case, however, everything converged absolutely to the global minimum delay flow pattern.

We now examine the implementation of the FD algorithm, shown on the following pages, more closely. The implementation relies on several basic data structures, some of which are redundant from the point of view of the information they contain, but all of which are useful for a reasonably clear and efficient implementation of the algorithm. There has been an attempt at balance here between simplicity and efficiency, taking as much of both as possible without giving up too much of either.

First, network connectivity is represented in two ways. Since the link topology does not change, it is only necessary to set up such structures once. The most basic structures are the arrays, End1 and End2, which contain the endpoints of the links. Since the links in this problem are bidirectional, each link is represented twice in these arrays, once in each direction. Thus, these arrays have nl elements, where nl is twice the actual number of links. By convention, it is possible to have pairs of links, 2k-1 and 2k, with End1[2k-1] = End2[2k] and End2[2k-1] = End1[2k]. If the network operation is full duplex, the operation of these "half" links are all independent of one another and it is possible to form these new "half" links however appropriate. If, however, the operation is half duplex, it is important to keep track of this correspondence as it is necessary to add the flow in both directions. For simplicity here, assume a full duplex operation.

From End1, an adjacency structure, Adj, is set up using the procedure SetAdj. SetAdj is not called within the FD algorithm but instead is assumed to be called elsewhere. This is appropriate since in general, such structures would be used in other places as well. The adjacency structure is represented by an array of lists, one list corresponding to the adjacency information for each node. The adjacency information for a node is a list of the links adjacent to that node.

There are two link flow arrays, Gflow and Eflow. Gflow holds the global flow (i.e., the superposition of all previous flow patterns). Eflow holds the current extremal flow, the flow pattern based on shortest paths formed from the current link lengths.

Paths are represented implicitly. The shortest path algorithm (Bellman's algorithm here) returns an array, SPpred, which contains the last link in the path between each pair of nodes. The entire path can be found by backtracking along these links, starting from the destination and working towards the source. This is exactly what is done in the LoadLinks routine below.

The FD algorithm starts with a feasible flow and then seeks to deviate flow to other paths to decrease the average delay. This feasible flow is obtained through a simple approach which is to temporarily scale up the link capacities (if necessary) so that they are all at least slightly larger than the flows assigned to them. Thus, links that are really overloaded appear simply to be congested. Continuing to reduce delay, the "congested" links are relieved of some flow and it is possible to reduce their capacity further. Eventually (assuming a feasible flow exists), all links have flows below their true capacities and the scale factor is no longer necessary. The procedure AdjustCaps scales the link capacities, producing NewCaps which is used throughout the algorithm. AdjustCaps also returns TRUE if NewCaps is greater than Caps (the true link capacity). Continue to call AdjustCaps within the main FD loop until it returns FALSE, indicating that a feasible flow has been found and no further scaling is necessary.

The FD algorithm works with flow patterns. A flow pattern is formed from a set of routes, one for each requirement. The flow pattern is the vector of link flows obtained by loading all the requirements onto these paths.

The paths are the shortest paths found using a set of link lengths. The length of each link, l, is defined as the derivative of the total delay with respect to the flow in link l. The length of a path is, of course, the sum of the lengths of the links in that path.

Thus, moving a tiny amount of flow from a longer path to a shorter path decreases the delay on the longer path and increases the delay on the shorter path. The lengths of these paths represent the amount that the average network delay changes. The net effect is to decrease the total delay and hence the average.

However, when flow is moved the shorter path becomes more congested and the longer path becomes less congested, that is, the length of the longer path decreases and the length of the shorter path increases. After moving an appropriate amount of flow (discussed next), recompute the link lengths as well as the shortest path lengths and obtain another flow pattern.

The shortest paths are obtained by calling Bellman's algorithm (discussed in Chapter 4), once for each source node. The implementation given here relies on a data structure called a queue with membership (QWM). A QWM is an ordinary queue which allows additions at the rear and deletions at the head. It functions over a fixed set of elements, in this case the nodes of the network. At any given time, each element is either in the queue or not in the queue; it cannot be in the queue in more than one place. Thus, a Push operation is ignored for elements already in the queue. However, an element can reenter the queue after it has been popped out. This is exactly the behavior necessary for implementing Bellman's algorithm. When a node is relabeled, simply Push it into the queue. If it is already there, the Push is ignored.

The procedure LoadLinks produces a flow pattern, backtracking each path and loading the links with the flow. Eflow, the flow produced in this way is called an extremal flow. This flow pattern is not optimal and in many cases it is not even feasible. It does have the property, however, that if an infinitesimal amount of flow is moved to this flow pattern the average delay decreases.

Specifically, suppose there is a flow pattern, F, which satisfies all the requirements and there is an extremal flow pattern, E, formed by sending all the requirements on the minimal incremental delay paths (i.e., paths which are shortest using the derivative of delay with respect to flow as the link lengths) found on the current iteration. If a new flow pattern, F', is formed as

$$F' = \epsilon E + (1 - \epsilon)F \tag{6.19}$$

for ϵ very small, it is a new flow pattern which still satisfies all the requirements and for which the average delay is lower. This is apparent by observing that to add an infinitesimal amount of flow, ϵ, to a link with length L, the total delay goes up by ϵL. This follows from the definition of the link lengths. Thus, adding ϵ times the requirements to all the shortest paths, means the total delay increases by ϵE times the length of the shortest paths. But, ϵF is also taken from the current flow pattern and the total delay decreases ϵE times the lengths of all the other paths (which are longer.) The net effect is to decrease the total delay and hence the average.

The argument holds for ϵ very small. As the amount of flow we deviate to the new shortest paths increases, these paths become more congested and the derivative of delay with respect to flow on the links in this these paths increases. The paths become longer. Therefore, it is necessary to find the value of x which minimizes delay when we form a new flow pattern, F' is formed, as

$$F' = xE + (1 - x)F \qquad (6.20)$$

Because of the assumptions that the delay function is a convex function of flow, it is easy to find x. The superposition of convex functions is also convex. Consider Figure 6.2. As x increases from 0, F' initially decreases. Then, it decreases more slowly, flattening out. Then, it begins to rise and continues to rise more and more quickly. This is a property of convex functions.

It is possible, therefore, to find the optimal value of x by any simple line search procedure (i.e., a procedure which searches along a single dimension for a minimum). Here, the choice is to use a binary search. There are values of T, the total delay, for x equal to 0, 0.25, 0.5, 0.75 and 1. Thus we have 5 values of T. Next identify which value of x yields the smallest value of T and halve the interval of search, keeping the interval including the minimum. In Figure 6.2, the interval from 0.25 to 0.75 would be kept. Then evaluate T at the middle of the halves of the interval (in this case, for x equal to 0.375 and 0.625). Again, there are 5 values of T, the three kept and the two new ones just found. Continue halving the interval size until the desired accuracy is obtained. Since the interval is halved each time, it is possible to get good accuracy by examining relatively few values of x (e.g., it is possible to find x at 4 places, examining 17 $(= 3 + \log_2 10000)$ values of x.

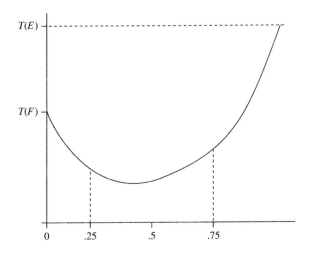

FIGURE 6.2

Then iterate this procedure, finding F_{k+1}, the flow pattern at the $(k + 1)$st iteration, from F_k and E_k, the flow pattern and extremal flow found at the kth iteration, respectively. The process terminates when no further significant progress is made in terms of the improvement in T.

The overall complexity of this algorithm is based on two factors. First, the shortest path algorithm is $O(NE)$ for each flow pattern found, where N is the number of nodes and E is the number of edges in the network. Second, the evaluation of T in the line search is $O(mE)$, where m is the number of values of x examined. In general, the shortest path computation tends to dominate because N is usually larger than m and the actual shortest path computation is more complex than the evaluation of T. Thus, the overall algorithm is of complexity $O(kNE)$, where k is the number of flow patterns examined.

Unfortunately, k can get quite large. Starting from an arbitrary flow pattern, initially good progress is made. The extremal flow pattern is significantly better than the current flow pattern and attracts a significant amount of flow, decreasing T significantly. As the algorithm continues, however, the differences among the old paths and the new ones become less significant. The objective is to evaluate the best path for each requirement individually (finding shortest paths) and then moving all requirements (an entire flow pattern) at once. This works, but in the course of approaching to the optimum, it works very slowly.

The process is not really moving in the right direction but instead is "tacking" like a sailboat heading into the wind. The result is that the last few percent of decrease in T often requires over 100 iterations. If only an approximate answer is necessary (e.g., testing if a network is reasonably loaded) the simple thing to do is to terminate the iteration early. If, on the other hand, the objective is testing the effect of adding a link to the inner loop of a design procedure (which would only affect T by a few percent), it is often necessary to resort to other methods.

One approach, known as the Extremal Flow Algorithm [CG74], is to hold on to all the individual flow patterns, F_k, and at each step to find the best superposition of them. Therefore,

$$F = \sum_k x_k F_k \tag{6.21}$$

where the sum of the x_k equals 1 and F minimizes T. This requires the replacement of the line search in the FD algorithm by a multidimensional search. There is an extensive body of literature on this problem (e.g., [FP63] [Min86]) but it is not discussed here. Instead, an alternative approach will be pursued in the following section which works with the individual requirements.

```
/////////////////////////////////////////////////////////////
//                 FLOW DEVIATION ALGORITHM               //
/////////////////////////////////////////////////////////////
void <- flow_dev( nn , nl , End1 , End2 , Cap , Req , MsgLen ,
             Adj , *Gflow , *CurrentDelay )
     dcl End1[nl] , End2[nl] , Req[nn,nn] , Adj[nn,list]
         SPdist[nn,nn] , SPpred[nn,nn] , FDlen[nl] ,
         Cap[nl] , NewCap[nl] , Gflow[nl] , Eflow[nl]
```

```
/* Initialization */


PreviousDelay <- INFINITY
TotReq <- Accumulate( Req )
Gflow <- 0


/* Find the initial routes and link flows */


SetLinkLens( nl , Gflow , Cap , MsgLen , *FDlen )
SetSP( nn , End2 , FDlen , Adj , *SPdist , *SPpred )
LoadLinks( nn , nl , Req , SPpred , Ends , *Gflow )
Aflag <- AdjustCaps( nl , Gflow , Cap , *NewCap )
CurrentDelay <-
    CalcDelay( nl , Gflow , NewCap , MsgLen , TotReq )


/* Find new routes and superpose flows


while( Aflag | (CurrentDelay < PreviousDelay*(1-EPSILON)) )
    SetLinkLens( nl , Gflow , NewCap , MsgLen , *FDlen )
    SetSP( nn , End2 , FDlen , Adj , *SPdist , *SPpred )
    LoadLinks( nn , nl , Req , SPpred , Ends , *Eflow )
    Superpose( nl, Eflow, Gflow , NewCap , TotReq , MsgLen )
    if ( Aflag )
        Aflag <- AdjustCaps( nl , Gflow , Cap , *NewCap )
    PreviousDelay <- CurrentDelay
    CurrentDelay <-
        CalcDelay( nl , Gflow , NewCap , MsgLen , TotReq )


/////////////////////////
// Set the link lengths //
/////////////////////////
void <- SetLinkLens( nl , Flow , Cap , MsgLen , *Len )
    dcl Flow[nl] , Cap[nl] , Len[nl]

  for_each ( link )
    Len[link] <- DerivDelay( Flow[link] , Cap[link] , MsgLen )


////////////////////////////////
// Find shortest path routes //
////////////////////////////////
void <- SetSP( nn , End2 , Len , Adj , *SPdist , *SPPred )
    dcl Len[nl] ,  Adj[nn,list]

  for_each ( node )
    Bellman( nn , node , End2 , Len , Adj ,
              *SPpred[node] , *SPdist[node] )
```

```
/////////////////////////////////////
// Bellman's Shortest Path Algorithm //
/////////////////////////////////////
void <- Bellman( nn , root , End2 , LinkLength , Adj ,
                *Pred , *Dist )
    dcl LinkLength[nl] , End2[nl] , Adj[nn,list] ,
        Pred[nn] , Dist[nn] , ScanQueue[QWMqueue]

  Dist <- INFINITY
  Dist[root] <- 0
  QWM_InitializeQueue( ScanQueue )
  QWM_Push( ScanQueue , root )
  while( !QWM_Empty( ScanQueue ) )
     node <- QWM_Pop( ScanQueue )
     for_each( link , Adj[node] )
        node2 <- End2[link]
        d <- Dist[node] + LinkLength[link]
        if ( Dist[node2] > d )
           Dist[node2] <- d
           Pred[node2] <- link
           QWM_Push( ScanQueue , node2 )

//////////////////////////////////////////////////
// Create a flow pattern given a predecessor matrix //
//////////////////////////////////////////////////
void <- LoadLinks( nn , nl , Req , SPpred , End1 , *Flow )
    dcl Req[nn,nn] , End1[nl] , Flow[nl]

  Flow <- 0
  for_each ( s )
     for_each ( d )
        if ( Req[s,d] > 0 )
           n <- d
           while ( n != s )
             link <- SPpred[s,n]
             p <- End1[link]
             Flow[link] += Req[s,d]
             n <- p

//////////////////////////////////////////////////
// Adjust the link capacities to ensure feasibility //
//////////////////////////////////////////////////
boolean <- AdjustCaps( nl , Flow , Cap , *NewCap , *Aflag )
    dcl Cap[nl] , Flow[nl] , NewCap[nl]

  Factor <- 1
  for_each ( link )
     Factor <- max( Factor , (1+DELTA)*Flow[link]/Cap[link]) )
  NewCap <- Factor * Cap
```

```
    if ( Factor > 1 )
        return( TRUE )
    else
        return( FALSE )


//////////////////////////////
// Calculate the average delay //
//////////////////////////////
float <- CalcDelay( nl , Flow , Cap , MsgLen , TotReq )
        dcl Flow[nl] , Cap[nl]

  float <- DelayF( link )
        return( LinkDelay( Flow[link] , Cap[link] , MsgLen )

  return ( Accumulate( DelayF , link ) / TotReq )


/////////////////////////////////////////////////////
// Superpose two flow patterns, minimizing delay //
/////////////////////////////////////////////////////
void <- superpose( nl, Eflow, *Gflow, Cap, TotReq, MsgLen )
        dcl Eflow[nl] , Gflow[nl] , Cap[nl]

  x <- FindX( nl , Gflow , Flow , Cap , TotReq , MsgLen )
  Gflow <- x * Eflow + (1-x) * Gflow


////////////////////////////////////////////
// Find the value of x minimizing delay //
// ( Binary search )                     //
////////////////////////////////////////////

float <- FindX( nl, Gflow, Eflow, Cap, TotReq, MsgLen )
        dcl Eflow[nl] , Gflow[nl] , Cap[nl]

  float <- DelayF( x )
        dcl Flow[nl]
        Flow <- x * Eflow + (1-x) * Gflow
        return( CalcDelay( nl, Flow, Cap, MsgLen, TotReq ) )

  x0 <- 0.0  ;   f0 <- DelayF( x0 )
  x4 <- 1.0  ;   f4 <- DelayF( x4 )
  x2 <- 0.5  ;   f2 <- DelayF( x2 )
  while ( x4 - x0 > EPSILON )
        x1 <- ( x0 + x2 ) / 2  ;   f1 <- DelayF( x1 )
        x3 <- ( x2 + x4 ) / 2  ;   f3 <- DelayF( x3 )
        if( (f0 <= f1) || (f1 <= f2) )
            x4 <- x2  ;   x2 <- x1
            f4 <- f2  ;   f2 <- f1
        else if( f2 <= f3 )
            x0 <- x1  ;   x4 <- x3
```

```
                f0 <- f1   ;   f4 <- f3
        else
             x0 <- x2   ;  x2 <- x3
             f0 <- f2   ;  f2 <- f3
    if ( (f0 <= f2) && (f0 <= f4) )
        return( x0 )
    else if ( f2 <= f4 )
        return( x2 )
    else
        return( x4 )

//////////////////////////////////
//     DELAY      FUNCTIONS     //
//////////////////////////////////
float <- LinkDelay( Flow , Cap , MsgLen )

    return( (MsgLen/Cap) / (1-Flow/Cap) );


float DerivDelay( Flow , Cap , MsgLen )

   f <- 1-Flow/Cap
   return( (MsgLen/Cap)/(f*f) )


float Deriv2Delay( Flow , Cap , MsgLen )

   f <- 1-Flow/Cap
   return( 2*(MsgLen/Cap)/(Cap*f*f*f) )


////////////////////////////////////////
//  Set up an adjacency structure  //
////////////////////////////////////////
void <- SetAdj( nn , nl , End1 , *Adj )
      dcl End1[nl] , Adj[nn,list]

   for_each ( link )
      node1 <- End1[link]
      Append( link , Adj[node1] )
```

Consider the network shown in Figure 6.3. There are three nodes and six directed links numbered 0 through 5. All links have capacities equal to 2. There are two requirements, a (0,1) requirement of magnitude 2 and a (1,2) requirement of magnitude 3. The objective is to find minimum delay routes using the FD algorithm.

The algorithm begins by computing link lengths based on the derivative of delay with respect to flow, with zero flows on the links. Thus, all links have lengths equal to 0.5 and the shortest paths for both requirements are direct paths. This results in the flow pattern shown in Figure 6.4a. It also results in flows that exceed the capacities of some of the links. Thus, the capacities of all the links are scaled up by slightly

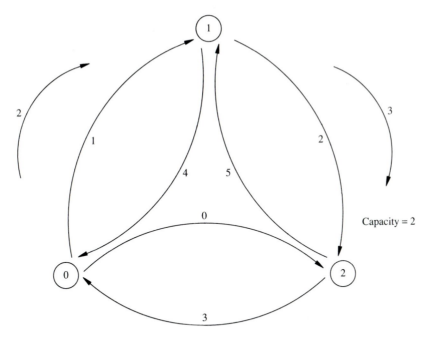

FIGURE 6.3
Min delay routing problem.

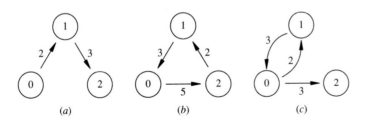

FIGURE 6.4
Flow patterns.

more than a factor of 1.5. The capacities are figured at 5 percent more than the flow. Thus, all capacities are set to 3.15.

Next, recomputing the link lengths, it is apparent that link (1,2) is very long (length = 140), and link (0,1) is long enough (length = 2.382) to cause the shortest path from 0 to 1 to be 0-2-1. The flow pattern shown in Figure 6.4*b* is the result. Superposing the two flow patterns to minimize delay (at the inflated capacity), 0.321 of the new flow pattern is taken and 0.679 of the original one. This results in the link flows shown in the Iteration 1 column of Table 6.1.

The largest flow has now been reduced to 2.037, and it is now possible to reduce the capacities to 2.139. Now link lengths are recomputed using the new link flows

TABLE 6.1
Link lengths and flows in the initial phase of the FD algorithm

Link	Iteration 0		Iteration 1		Iteration 2	
	Length	Flow	Length	Flow	Length	Flow
0	0.500	0.000	0.317	1.606	7.528	1.789
1	0.500	2.000	2.382	1.358	3.508	1.442
2	0.500	3.000	140.000	2.037	206.218	1.769
3	0.500	0.000	0.317	0.000	0.468	0.000
4	0.500	0.000	0.317	0.963	1.548	1.231
5	0.500	0.000	0.317	0.642	0.955	0.558
cap.	——		3.150		2.139	

(shown in the Iteration 1 column) and the new link capacities ($= 2.139$). Note that the link capacities have been scaled so that the most heavily loaded link has 5% spare capacity. This slack could have been smaller, if there was a willingness to risk numerical difficulties, or larger, if there was a willingness not to use the algorithm for networks whose actual utilizations exceed this slack. Note also that all links are scaled by the same factor; if this is not done the algorithm could (incorrectly) send more flow to an overloaded link because of its scaled up capacity, leaving that of another link unscaled.

The task now is to compute new link lengths (shown in the Iteration 2 column) based on these flows a new set of shortest paths, resulting in the flow pattern shown in Figure 6.4c. Superposing the flow s from Iteration 1 with this new flow pattern to minimize delay, we take 0.131 of the new pattern and 0.869 of the old, yielding the flows shown in the Iteration 2 column. These flows are feasible and it is now possible to reduce all the link capacities to 2.

The iterations of the FD algorithm are continued as above, finding new link lengths, new shortest paths, and a new flow pattern at each iteration and superposing the new pattern with the previous superposed flow pattern.

Table 6.2 shows the progress of the FD algorithm through 46 iterations. Each column shows the superposed link flows, the fraction of the new flow pattern used, and

TABLE 6.2
Link lengths and flows in the continuation of the FD algorithm

Link	It.3	It.4	It.5	It.10	It.20	It.30	It.40	It.46
0	1.751	1.788	1.740	1.742	1.701	1.677	1.665	1.661
1	1.454	1.466	1.477	1.526	1.590	1.624	1.641	1.647
2	1.795	1.756	1.783	1.732	1.709	1.699	1.694	1.692
3	0.000	0.000	0.000	0.000	0.000	0.000	0.000	0.000
4	1.205	1.244	1.217	1.268	1.291	1.301	1.306	1.308
5	0.546	0.534	0.523	0.474	0.410	0.376	0.359	0.353
frac.	0.021	0.022	0.022	0.018	0.011	0.006	0.003	0.002
delay	4.070	3.995	3.928	3.695	3.503	3.450	3.435	3.433

the average delay. It is clear that the progress of the FD algorithm after the first few iterations is very slow. Delay is reduced by roughly 0.5% on each iteration from the 10th through 20th and by 0.15% on each of the next 10 iterations. After 46 iterations, the algorithm is stopped based on a criterion of delay changing by less than 0.01% from one iteration to the next.

A closer examination reveals why convergence is so slow. The most acceptable thing to do, as evidenced by the actual change in the flows from Iteration 3 to Iteration 46, is to move flow from links 0 and 2 to link 1. Unfortunately, none of the flow patterns in Figure 6.4 allow this to happen directly, and these are the only flow patterns which arise from shortest paths. What actually happens is that we bounce back and forth between the flow patterns shown in Figures 6.4a and 6.4c. Both of these flow patterns add flow to links 0 and 2, but together they add more flow on average to link 1. Thus, there is movement toward the optimum, but only very slowly. It is, as was mentioned earlier, exactly like a boat tacking into the wind. This is illustrated in Figure 6.5.

The final link lengths and link flows are shown in Figure 6.6. Notice that the alternate path lengths for each requirement are roughly, but not exactly equal. Thus the 1-2 path has length 21.65 while the 1-0-2 path has length 21.18. The 0-1 path has length 15.99 while the 0-2-1 path has length 17.79. This latter difference is over 11%. It is apparent that even after 46 iterations, the algorithm has not really completely converged. Nevertheless, based on the small progress made in reducing the average delay on the last 16 iterations, there is confidence that the solution developed is reasonably close to the optimum. This is borne out by resetting the tolerance for convergence to 0.001%. In this case, the FD algorithm ran for 64 iterations and produced a delay of 3.431. The difference in the path lengths went down to 3.8%.

The convergence rate of the FD algorithm is affected by congestion in the network; the more congested the network is, the more slowly the algorithm converges.

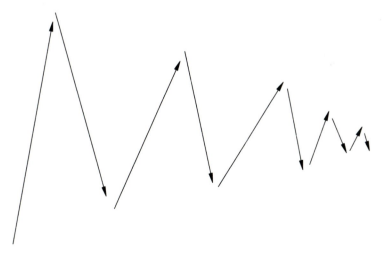

FIGURE 6.5
Movement towards the optimum.

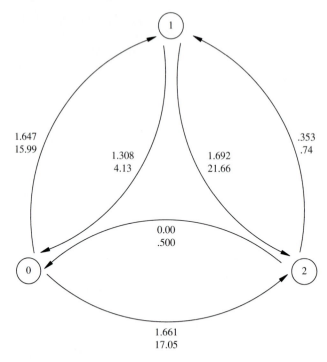

FIGURE 6.6

Flow and link lengths.

In the previous example, the critical links had utilizations of roughly 83% and the algorithm required 46 iterations to converge. The same example was run with link capacities equal to 3. This resulted in utilizations of 55% and convergence in 24 iterations. Conversely, with capacities of 1.8, the utilizations were 93% and 87 iterations were required for convergence to the same tolerance.

The previous example is for a 3-node network. What can be expected for networks of more realistic size? Figure 6.7 shows a 5 × 5 grid. We ran the FD algorithm on this network with uniform traffic of 1 unit between each pair of nodes and a delay tolerance of 0.01%. The algorithm required 69 iterations to converge, with a maximum link utilization of 60%. Thus, it is apparent that the rate of convergence is not helped by increasing the size of the problem. On the other hand, it does not appear to be hurt very much either. In particular, the number of iterations is not rising, even linearly, with the number of nodes or links.

In practice, there are a number of implementation decisions to make. The first relates to numerical factors. It is necessary to choose the tolerances for convergence for the line search and the outer iteration. The tighter these tolerances, the more slowly the algorithm will run and the more accurate the results will be. We chose 0.01% for both. While the actual values for these tolerances are a matter of choice, it is important that they be consistent. Thus, it is not sensible to choose a tighter tolerance for the

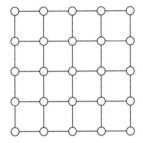

FIGURE 6.7
5 × 5 grid.

overall convergence than for the line search. It is also not sensible to set the tolerances so low that roundoff errors interfere with convergence.

Finally, there are a number of choices for what to test for convergence of the outer iteration. Here, the only thing tested was successive values of delay. Alternatively, one could test values over several iterations, values of the flows, or the relative lengths of paths for each node pair. These latter criteria are more stringent, again leading to longer running times and more accurate results.

For most purposes, simple convergence criteria and moderate tolerances are appropriate. The actual requirements are rarely known with great accuracy. However, one important exception is when using the results of the routing in the inner loop of an incremental design procedure. In this case, it may be desirable to test the effect of adding or deleting a single link. Or, for example, it may be desirable to delete the link that maximizes the change in cost (downward) over the change in delay (upward). Unfortunately, unless we run the routing algorithm with relatively tight convergence criteria, we may be badly misled. Indeed, it is possible to even see delay go down when a link ia added. The need for more accuracy in some cases leads to more sophisticated procedures, as discussed in the following section.

6.3 THE BERTSEKAS-GALLAGER ALGORITHM

An alternative approach to finding minimum delay routes was proposed by [BG87] [BG84]. Their approach (the BG algorithm) differs from the FD algorithm in two important ways. First, it moves flow for one source-destination pair at a time. Second, it computes the amount of flow to move directly rather than performing a line search. Both of these changes improve the efficiency of the algorithm significantly in most cases.

As was mentioned in the previous section, the FD algorithm often converges slowly because it is forced to move flow from all requirements at the same time. The BG algorithm deals with this by considering node pairs one at a time. Another problem with the FD algorithm is that upon approaching the optimum, the derivative of delay with respect to flow, is close to zero and the FD algorithm has difficulty identifying the right way to deviate flow. This leads to many iterations near the optimum. The BG algorithm deals with this problem by computing an approximation to the second derivative of delay with respect to flow and using this as a correction factor on the amount of flow to move. This allows us to directly compute a reasonable estimate of the amount of flow to move.

The main drawback to the BG algorithm as described in [BG87] is that it requires that the lengths of all links be adjusted and a shortest path separately found for each node pair. In comparison, the FD algorithm requires only that shortest paths be found between all node pairs simultaneously. Thus, the complexity of the BG algorithm is $O(kN^2E)$, where k is the number of iterations required for convergence. Even if k is small, which in practice it is in most cases, this is still a complexity between $O(N^3)$ and $O(N^4)$.

This problem is overcome in the implementation given below by modifying the algorithm to update link lengths and recompute shortest paths simultaneously, for all destinations for each source. This increases the number of iterations somewhat (roughly by a factor of two in practice) but leads to a substantial improvement in running time because the overall complexity is reduced by a factor of N.

The algorithm begins by computing first-derivative link lengths based on zero flow. For M/M/1 queues, the derivative of delay with respect to flow, and hence the link length, is

$$L_i \frac{C_i}{(C_i - f_i)^2} \tag{6.22}$$

where C_i is the capacity of the link and f_i is the flow on the link, both in messages per second. For f_i zero, L_i is simply

$$\frac{1}{C_i}$$

Next, as in the FD algorithm, shortest paths are computed and the requirements are loaded onto them. If necessary, the link capacities are scaled to make the flow pattern feasible. The average end to end delay is also computed, again, as in the FD algorithm.

A key difference here, however, is that the paths found are retained for each pair of nodes. The paths computed are stored in a structure updated as the algorithm proceeds. First, there is an array, `PathSet`, which is a two dimensional array of lists. `PathSet[i,j]` is the list of paths found thus far from i to j. `PathSet` is intialized to null (Φ). As a new path is found, it is added to the appropriate list. Whenever a path is found, the routine `FindPath` is called to check if the path already exists. If it does, `FindPath` returns the index of the path. If not, it returns Φ. `PathPtr` is an array of pointers to paths, which are represented as lists of links.

The array `PathFlow` holds the current amount of flow assigned to each path. Thus, at any given time during the running of the algorithm, the sum of the `PathFlow` values for all paths in `PathSet[i,j]` must equal `Req[i,j]`, the total (i, j) requirement.

As the algorithm proceeds, new link flows are computed and new shortest paths are found. For each node pair, (s,d), flow is moved from all other paths to the current shortest (s,d)-path. The amount of flow moved is determined by several factors. First, δ, the desired amount of flow to move off of path P, is computed as

$$\delta = \alpha(L_P - L_{P*})H_P^{-1} \tag{6.23}$$

where α is a stepsize, L_P is the length of path P, L_{P*} is the length of the shortest (i, j) path, and H_P is the second derivative path length.

In the original statement of the BG algorithm in [BG87], the authors suggest setting α to 1 and then decreasing it by a constant factor (e.g., 0.5) as the algorithm proceeds. Because many requirements are moved before recomputing link lengths in this implementation, the algorithm works better if α is set to a smaller value. We have chosen to set α initially to a value in the range

$$\frac{1}{N} \leq \alpha \leq \frac{L}{N^2} \tag{6.24}$$

where N is the number of nodes and L is the number of (directed) links, based on the rationale of moving roughly N requirements before recomputing link lengths, and the shortest path tree has $N - 1$ links. Thus, roughly N times as much flow is moving (which would justify moving only one Nth as much flow of each requirement) but the flow moved is spread over more links than in the original algorithm (justifying the upper bound.) The exact value of α is not critical. With α small, the algorithm tends not to oscillate as much, that is, it does not tack, moving flow back and forth among the same links. If α is too small, however, the convergence slows as the algorithm moves flow from one path to another in many small steps instead of fewer larger ones. At high utilizations, it is more important to prevent oscillation.

Assume that there are two parameters, ALPHAHI and ALPHALO, which are, respectively, the initial and final values of α. These may be constants, parameters, or may be computed during the course of running the algorithm. As mentioned before, it is recommended to set ALPHAHI to a value in the interval defined above and setting ALPHALO to a value roughly one tenth as large. Another good choice is to set ALPHALO to a much smaller value but to test for convergence by keeping track of values of delay or link flows on successive iterations with smaller values of α.

The path lengths are simply the sum of the link lengths which, in turn are given by the derivative of delay with respect to flow on each link. The second derivative path length, H_P, is actually an approximation and is given by

$$H_P = \sum_{i \in U} S_i \tag{6.25}$$

where S_i is given by

$$S_i = \frac{2C_i}{(C_i - f_i)^3} \tag{6.26}$$

and A is the set of links in the that are in P and $P*$ but not in both.

Thus, A (and hence, H_P) has to be computed separately for each pair of paths. Since this contributes significantly to the computational complexity of the algorithm, it should be done carefully. Assume the existence of several operations on lists which can be used to help compute A in the ListXOR routine. The first is ListUnion which forms the union of two (not necessarily disjoint) lists. The second in ListIntersect, which forms their intersection. The last is ListRemove, which removes all the elements of its second argument from the first argument. These operations rely on the specific representation of the lists. If the lists are simple, unordered lists, these operations have complexity of $O(E_1 E_2)$ (i.e., the product of the sizes of the lists involved). If, alternatively, the lists are ordered, these operations are $O(E_1 + E_2)$. Finally, if the lists are represented as bit vectors, these operations are $O(M/W)$, where M is the number of links in the network and W is the wordsize of the computer.

Usually, the best choice here is to use ordered lists. This complicates the `Insert` operation which is used to form the lists of the links in the paths in the `SetPath` routine, but that is done much less frequently. In particular, new paths are added, primarily in the first few iterations and then infrequently afterwards. Except in very congested networks, the actual number of paths per node pair is small.

The routine `AdjustRoutes` actually moves flow from one path to another. It begins by computing δ for each non-shortest path, P, as described above. It then checks that δ is not larger than the amount of flow currently on path P and that δ is not larger than the amount of spare capacity on the shortest path. If necessary, δ is decreased based on these considerations.

The iteration is continued, moving flows from non-shortest paths to shortest, until no more significant progress is made. In the implementation given below, progress is checked based on the value of average delay. A more stringent check could be made based on the link flows or even the path lengths (which should all converge to the same value for all paths used by any given node pair.) The strictness of the convergence test should be determined by the application. For most purposes, testing based on delay is sufficient and `DELTA`, the tolerance for convergence, might be set to 0.001. With these simple criteria, we often get convergence of the entire algorithm within 10 iterations.

Again consider the 3-node network shown in Figure 6.3. And compare the results here with those reported for the FD algorithm in the previous section. Like the FD algorithm, the implementation of the BG algorithm begins by routing all traffic on minimum incremental delay paths, that is, on shortest paths using the derivative of delay with respect to flow as link lengths. Again, starting at zero flows, this assigns the same length (0.5) to all links in this example and all flow is routed on direct paths. This results in a flow of 3 on the (1,2) link and in all link capacities being scaled up to 3.15.

Next, compute new link lengths and new shortest paths, just as in the FD algorithm. Now, however, there are several differences. First, one source at a time is dealt with. Second, we explicitly compute the amount of flow desired to move, for each requirement, from each path to the current shortest path. This proceeds in increasing order of source and destination node numbers.

The amount of flow, δ, moved from path P to shortest path $P*$ is

$$\delta = \frac{\alpha(L_P - L_{P*})}{H_P}$$

where α is the stepsize, L_P is the length of path P, L_{P*} is the length of the shortest path, and H_P is the second derivative path length, all as defined above.

Using the definitions given earlier in this section, we have

$$\alpha = \frac{L}{N^2} = 0.667$$

where L is the number of (directed) links (6), and N is the number of nodes (3).

L_P and L_{P*} are given by the sum of the lengths of the links in the path (which are the same as in column Iteration 1 in Table 6.3 since they are computed in the same way using exactly the same data). Thus, for the (0,1) requirement which originally went entirely on the direct path (link 1) and now has a new shortest path 0-2-1 (links 0 and 5),

$$L_P = 0.317 + 0.317 = 0.634$$

$$L_{P*} = 2.382$$

H_P is given by the sum of the second derivatives of delay with respect to flow for all links in either P or $P*$ but not both. In this case, the two paths have no links in common and the sum is over all links in both paths. Continuing to use the simple M/M/1 formulae described in the preceding section, and thus, the second derivative, S, for link, l, with flow F_l and capacity C_l is given by

$$S = \frac{2C_l}{(C_l - F_l)^3} \tag{6.27}$$

So, H_P is found to be 4.546. Finally, δ, the desired amount of flow to move is found to be 0.256. The flow in link 1 (the original path for the (0,1)) is reduced by 0.256 and the flows in links 0 and 5 are increased by the same amount. Based on what is already known about the final flows when this problem was run through the FD algorithm, this is quite reasonable. In particular, it was found that ultimately it was desirable to move roughly 0.34 units of flow off of link 1. Thus, even without a line search there has been a reasonable estimate of the correct amount of flow to move.

Had there been any other requirements from node 0, or any other paths, for the (0,1) requirement, similar calculations would have been carried out for them and the appropriate amount of flow would have been moved without changing either first- or second-derivative link lengths based on the change in link flows. As mentioned earlier, the original algorithm as stated by Bertsekas and Gallager did not do this, but this has been found to be an effective way of improving the speed of the algorithm.

TABLE 6.3
BG algorithm—(1,2) requirement

Link	Flow	Capacity	Length	2D Len
0	0.256	3.150	0.376	0.260
1	1.744	3.150	1.593	2.266
2	3.000	3.150	140.000	1866.663
3	0.000	3.150	0.317	0.202
4	0.000	3.150	0.317	0.202
5	0.256	3.150	0.376	0.260

Since there is nothing more to do with source 0, proceed to source 1. Table 6.3 summarizes the situation when beginning the consideration of the (1,2) requirement. Because a new source is being considered and it is necessary to find a new set of shortest paths, the link lengths (first derivative) and 2D lengths (second derivative) have been altered based on the movement of 0.256 units of flow of the (0,1) requirement.

Now compute δ for moving part of the (1,2) requirement from the original shortest path (link 2) to the new shortest path (links 3 and 1). α is still 0.667. The difference in path lengths is very large, 138.09, but H_P is much larger, 1869.13. Thus, δ is very small, 0.05. This is much smaller than desired.

The reason this happens is because link 2 has a flow which is very close to its capacity. Indeed, link 2 is really over its capacity but the capacity of all the links have been temporarily increased. The formula for the second derivative has a $(C - F)^3$ in the denominator while the formula for the first derivative only has a $(C - F)^2$. As F approaches C, $C - F$ becomes small and L_P and H_P both become large, but H_P becomes large faster. Thus, δ becomes small and only a little flow at a time moves off the very congested link. Because of this, it actually requires 29 iterations for the BG algorithm to obtain a feasible solution in this example. Possible remedies for this will be discussed.

Once a feasible solution has been found, things proceed more smoothly. Table 6.4 summarizes the next 10 iterations of the algorithm. As can be seen, things move quickly to convergence and, in fact, a slightly better result is obtained than with over 40 iterations of the FD algorithm.

All this was done with α set to 0.667. At this point, not making further progress, α is halved and continues the process. No further progress is found with α set to 0.333

TABLE 6.4
Continuation of the BG Algorithm

Link	It.29	It.30	It.31	It.32	It.33	It.34	It.35	It.36
0	1.565	1.585	1.610	1.635	1.651	1.657	1.657	1.655
1	1.545	1.557	1.572	1.589	1.608	1.625	1.638	1.647
2	1.890	1.857	1.818	1.776	1.741	1.718	1.705	1.698
3	0.000	0.000	0.000	0.000	0.000	0.000	0.000	0.000
4	1.110	1.143	1.182	1.224	1.259	1.282	1.295	1.302
5	0.455	0.443	0.428	0.411	0.392	0.375	0.362	0.353
Delay	5.157	4.394	3.900	3.624	3.500	3.454	3.438	3.433

Link	It.37	It.38
0	1.654	1.653
1	1.652	1.655
2	1.694	1.692
3	0.000	0.000
4	1.306	1.308
5	0.348	0.345
Delay	3.431	3.431

and so halve it again. This is still no further progress and halve α one last time, find no progress and halt. A total of 42 iterations have been done.

There are many possible strategies for setting α. A larger α results in larger steps, that is, more traffic being moved at each iteration. This accelerates convergence if what is done is to move the required amount of flow a little at a time instead of all at once. It results in oscillation (i.e., flow moving back and forth between several paths) if too much flow is moved. There is, unfortunately, no known way of predicting which of these situations exist.

It is possible to make the algorithm more "introspective" and vary α based on direct observation of its progress. Specifically, it can be made to observe whether flow is moving consistently in one direction or moving back and forth between paths. However, this complicates the implementation especially since different things may be happening in different parts of the network.

```
//////////////////////////////////////////////////////
// BERTSEKAS-GALLAGER MINIMUM DELAY ROUTING ALGORITHM //
//////////////////////////////////////////////////////
void <- bg( nn , nl , Req , Adj , End1 , End2 , Cap ,
            MsgLen , *CurrentDelay , *Gflow , *npath ,
            *FirstPath , *PathFlow , *PathPtr )
    dcl Req[nn MsgLen , *CurrentDelay , *Gflow , *npath ,
            *FirstPath , *PathFlow , *PathPtr )
    dcl Req[nn,nn] , Adj[nn,list] ,
        SPdist[nn,nn] , SPpred[nn,nn] , Path[list] ,
        End1[nl] , End2[nl] , Cap[nl] , Gflow[nl] ,
        NewCap[nl] , FDlen[nl] , SDlen[nl] ,
        PathSet[nn,nn] , PathFlow[np] ,
        PathPtr[np]
/************************************************************/
/*  A path is a list of link indices                       */
/*  PathPtr is a vector of pointers to paths               */
/*  PathSet is an n by n array of lists. Each list is a    */
/*     list of paths for a given pair of nodes             */
/************************************************************/

    /* Set up , load and store an initial set of paths */
    npath <- 0
    TotReq <- Accumulate( Req )
    Gflow <- 0
    PathSet <- Φ
    SetLinkLens( nl , Gflow , Cap , MsgLen , *FDlen )
    SetSP( nn , End2 , FDlen , Adj , *SPdist , *SPpred )
    for_each ( s )
        for_each ( d )
            Path <- SetPath( s , d , SPpred[s] , End1 )
            npath <- npath + 1
            StorePath( Path , npath , *PathPtr , *PathSet[s,d] )
            LoadPath( Req[s,d] , Path , Gflow )
```

```
            PathFlow[npath] <- Req[s,d]
    Aflag <- AdjustCaps( nl , Gflow , Cap , *NewCap )
    CurrentDelay <-
        CalcDelay( nl , Gflow , NewCap , MsgLen , TotReq )

    /*  Iterate, redirecting flow  */
    alpha <- ALPHA_HI
    while( alpha > ALPHA_LO )
        PreviousDelay <- INFINITY
        while( Aflag || (CurrentDelay < PreviousDelay*(1-DELTA) )
            for_each ( s )
                SetLinkLens( nl , Gflow , NewCap , MsgLen , *FDlen )
                SetSDLinkLens( nl , Gflow , NewCap , MsgLen , *SDlen )
                Bellman( nn , s , End2 , FDlen , Adj ,
                        *SPpred[s] , *SPdist[s] )
                for_each ( d )
                    if( (s = d) || (Req[s,d] = 0) ) continue
                    Path <- SetPath( s , d , SPpred[s] , End1 )
                    p <- FindPath( PathSet[s,d] , PathPtr , Path )
                    if( p = Φ )
                        npath <- npath + 1
                        StorePath( Path , npath , *PathPtr ,
                                *PathSet[s,d] )
                        PathFlow[npath] <- 0
                    AdjustRoutes( p , alpha , FDlen , SDlen ,
                        NewCap , PathSet[s,d] , *PathFlow , *Gflow )
            if ( Aflag )
                Aflag <- AdjustCaps( nl , Gflow , Cap , NewCap )
            PreviousDelay <- CurrentDelay
            CurrentDelay <-
                CalcDelay( nl , Gflow , NewCap , MsgLen , TotReq )
////////////////////////////////////////////
// Find Second Derivative Link Lengths //
////////////////////////////////////////////
void <- SetSDLinkLens( nl ,  Flow , Cap , MsgLen , Len )
    dcl Flow[nl] , Cap[nl] , Len[nl]
    for_each ( link )
        Len[link] <- SDerivDelay( Flow[link] , Cap[link] , MsgLen )

////////////////////////////////////////////////////
// Set up an sd-path given the predecessor matrix //
////////////////////////////////////////////////////
list <- SetPath( src , dst , Pred , End1 )
    dcl Pred[nn] , End1[nl] , Path[list]

    Path <- Φ
    node <- dst
    while ( node != src )
        link <- Pred[node]
```

```
       Insert( link , Path )
       node <- End1[link]
    return( ListSort(Path) )


/////////////////////////////////////
// Store a path in the path structure //
/////////////////////////////////////
void <- StorePath( Path , *npath ,  PathPtr , FirstPath )
      dcl Path[list] , dcl PathPtr[np,list]
   npath <- npath+1
   PathPtr[npath] <- Path
   Append( npath , PathList )


/////////////////////////////////////////////////////////////
// Return the index of a matching path or Φ if none exists //
/////////////////////////////////////////////////////////////
int <- FindPath( PathSet , , PathPtr , Path )
      dcl PathSet[list,list] , Path[list] , PathPtr[np,list]
    for_each ( p , PathSet )
       if ( ListEqual( Path , PathPtr[p] )
          return( p )
    return( Φ )


/////////////////////////////////////////////////////////////
// Find the XOR of two paths (i.e., links in exactly one path) //
/////////////////////////////////////////////////////////////
list <- PathXOR( Path1 , Path2 )
      dcl Path1[list] , Path2[list] [list] , Dups[list]
    return( ListRemove( ListUnion( Path1 , Path2 ) ,
                        ListIntersect( Path1 , Path2 ) ) )


/////////////////////////////////////////////////////////////
// Incrementally add or delete flow on the links in a path //
/////////////////////////////////////////////////////////////
void <- LoadPath( amount , Path , Flow )
      dcl Path[list] , Flow[nl]
    for_each( link , Path )
       Flow[link] += amount


/////////////////////////////////////////
// Move flow to the current best path //
/////////////////////////////////////////
void <- AdjustRoutes( best_path , alpha , FDlen , SDlen ,
            Cap ,  PathPtr , PathSet , *PathFlow , *Gflow )
      dcl FDlen[nl] , SDlen[nl] , Cap[nl] , PathPtr[np] ,
          PathFlow[np] , PathSet[list] , DPath[list]
    TotalMoved <- 0
    BestLen <- PathLen( PathPtr[bestp] , FDlen )
    slack <- FindSlack( PathPtr[bestp] , Cap , Gflow )
```

```
for_each ( p , PathSet )
    if ( slack <= 0 ) break
    if ( (p = bestpath) || (PathFlow[p] = 0) ) continue
    DPath <- PathXOR( PathPtr[p] , PathPtr[bestpath] )
    HpInv <- 1.0 / PathLen( Pathd , D2len )
    delta <- alpha * HpInv *
                ( PathLen( PathPtr[p] , FDlen ) - BestLen )
    delta <- min( delta , slack , PathFlow[p] )
    if ( delta > 0 )
        LoadPath( -delta , PathPtr[p] , Gflow )
        PathFlow[p] -= delta
        slack -= delta
        TotalMoved += delta
if ( TotalMoved > 0 )
    LoadPath( TotalMoved , PathPtr[bestp] , Gflow )
    PathFlow[bestp] += TotalMoved

////////////////////////////////
// Find the length of a path //
////////////////////////////////
float <- PathLen( Path , LinkLen )
    dcl Path[list] , LinkLen[nl]
  result <- 0
  for_each( link , Path )
    result += LinkLen[link]
  return( result )


////////////////////////////////////////
// Find the minimum slack on a path //
////////////////////////////////////////
float <- FindSlack( Path , Cap , Flow )
    dcl Path[list] , Cap[nl] , Flow[nl]
  result <- INFINITY
  for_each ( link , Path )
    result <- min( result , (1-DELTA2)*Cap[link]-Flow[link] )
  return( max(result,0) )
```

If α is too large, too much flow will be moved. The delay then increases and will falsely indicate that the algorithm has converged. In the implementation here, make α smaller and continue, to help compensate for the earlier error. It has been decided to halve α three times, thus exploring values of α that roughly span an order of magnitude. Other strategies exist. One can continue to halve α until no progress is made on several successive iterations. As before, it is possible to measure progress in terms of delay, link flows or path length.

One conclusion plausibly reached, based on the previous examples, is that a larger value of α may be appropriate before feasibility is reached and a smaller value afterwards. In the previous example, starting with α set to 10, feasibility was reached in 10 iterations as compared with the 29 required starting with α set to 0.667. Flows moved more smoothly, however, with the smaller value of α, especially after feasibility was reached. This points possibly to a hybrid approach, using large α before feasibility and smaller α afterwards.

Another possibility is to use the FD algorithm, which found a feasible solution in 4 iterations, initially and then switch over to the BG algorithm. Note that the BG algorithm (or the FD algorithm, for that matter) can start from any flow pattern. Note, in conclusion, that there are many choices facing the implementor of these algorithms based on the specifics at hand, but that there is considerable leeway in doing so. The algorithms work well in most cases unless the network is very close to saturation.

EXERCISES

6.1. There is a network in Figure P6.1. All link capacities are full duplex and have capacities of 56 Kbps. There is a requirement of 7 Kbps. between each pair of nodes in each direction. The objective is to minimize the maximum link utilization.

(*a*) Find the best routing you can.

(*b*) What is the minimum possible value for the maximum link utilization? Why?

(*c*) How close was the answer?

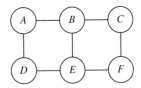

FIGURE P6.1

6.2. There are two paths, P_1 and P_2, between a given pair of nodes with capacities of 10 Kbps and 30 Kbps, respectively. It is required that a total of 10 Kbps of traffic be sent over these two paths, X over P_1 and $10000 - X$ over P_2.

(*a*) Assuming independent arrivals and exponential message lengths with an average length of 1000 bits, find the optimal value of X using binary search. Check the answer by setting the derivative of delay with respect to X to zero.

(*b*) Suppose P_2 is a satellite channel with an added propagation delay of 250 ms. What is the optimal value of X now?

6.3. There is a 4-node ring network (i.e., with links (1,2), (2,3), (3,4) and (4,1)). All links are 1000 bps and are half duplex. There are requirements of 500 bps between node pairs (1,2), (2,3), and (3,4). There is a requirement of X bps between node pair (4,1). The average packet length (exponentially distributed) is 1000 bits.

(*a*) Assuming min-hop routing, what is the average end-to-end delay if X is 900 bps?

(*b*) What is the optimal (minimum delay) routing? What is the average end-to-end delay with this routing?

(*c*) What is the largest value of X for which min-hop routing is optimal in this network?

6.4. There is a 4-node network with requirements and half duplex links shown in Table P6.1. It is desirable to find a routing for the requirements which minimizes delay.

(*a*) Use the Flow Deviation algorithm to find routes

(*b*) Use the Bertsekas-Gallager algorithm to find routes

(*c*) Give the link flows, link utilizations, link delays, requirement delay and average end-to-end delay for the network.

TABLE P6.1

Requirements:

	A	B	C	D
A	0	2027	1351	1617
B	2069	0	2895	1343
C	2404	1452	0	1989
D	1249	1167	1779	0

Selected links:

Ends		Capacity
1	2	9600
0	2	19200
1	3	9600
2	3	9600

BIBLIOGRAPHY

[BG84] Gafni, E., Bertsekas, D. and R. Gallager: "Second derivative algorithms for minimum delay distributed routing in networks," *IEEE Trans on Commun*, 32:911–919, 1984.

[BG87] Bertsekas, D. and R. Gallager: *Data Networks*, Prentice-Hall, Englewood Cliffs, NJ 1987.

[CG74] Cantor, D. and M. Gerla: "Optimal routing in a packet-switched computer network." *IEEE Trans. on Computers*, 23:1062–1068, 1974.

[FK73] Gerla, M., L Fratta. and L. Kleinrock: "The flow deviation method: An approach to store and forward communication network design." *Networks*, 3:97–133, 1973.

[FP63] Fletcher, R. and M. J. D. Powell: "A rapidly convergent descent method for minimization." *Computer Journal*, 6:163–168, 1963.

[Jac57] Jackson, J. R.: "Networks of waiting lines," *Oper. Res.*, 5:518–521, 1957.

[Kle75] Kleinrock, L.: *Queueing Systems, Volume 1: Theory*. Wiley-Interscience, New York, 1975.

[McQ77] Quillan, John M.: "Routing algorithms for computer networks: A survey." In *NTC77 Conference Record*, Vol. 2, pages 86–91, 1977.

[Min86] Minoux, M.: *Mathematical Programming: Theory and Algorithms*. John Wiley, New York, 1986.

[MW56] Frank, M. and P. Wolfe: "An algorithm for quadratic programming." *Naval Res. Logist. Quart.*, 3:149–154, 1956.

CHAPTER
7

MESH TOPOLOGY
OPTIMIZATION

The general mesh topology optimization problem is complex. It involves the selection of links, the assignment of capacities to these links, and the routing of requirements on these links. Ideally, all these are jointly optimized, leading to a minimum cost network which meets given objectives on delay and throughput. In practice, these problems are often dealt with separately and a solution iterated. The routing problem by itself has already been discussed. In this chapter the focus is on the selection of links, the assignment of capabilities to these links and how the three problems interact.

7.1 CAPACITY ASSIGNMENT

This section begins by considering the problem of assigning capacities to the links in the network given the link topology and link flows. The objective is to minimize delay subject to a constraint on the total cost. Equivalently, since it is possible to trade delay for cost, it is possible to minimize cost subject to a constraint on delay. The focus here is on the first form of this problem.

In a sense, this is a somewhat artificial problem since, usually, the routing, and hence the link flows, are chosen based on the link capacities, rather than choosing the link capacities based on the link flows. In fact, ideally, it would be preferable to optimize both simultaneously. However, the more general problem is also quite complex, and therefore, these subproblems are often solved separately. Thus, the capacity assignment problem comes up as part of the solution to the more general and more realistic problem. Also, finding the solution to this problem helps to understand the solution to the more general problem as well.

7.1.1   Continuous Capacities

To begin with assume that capacity is available in any quantity, that is, that the link capacities are continuous variables. This is unrealistic, but gives rise to a solution which, again, gives insight into the more general case. The development here follows [Kle75], [Sch77], and [Ger73].

The procedures that are described below assume that capacities can be assigned independently to links in each direction. This is a valid assumption if the links are unidirectional, as would be the case with satellite links or fiber. Most other types of links, however, are bidirectional. The assumption is still valid if link flows are symmetric because the same capacity would be assigned to the link in both directions anyhow. The assumption is also valid if the links operate in half duplex mode, because in this case, the flows on the link are added in both directions. Finally, if the requirements themselves are bidirectional (e.g., voice), the assumption is valid. In other cases, it is an approximation.

There are many forms to this problem. The simplest one is to consider the case where the constraint is on capacity rather than cost. Thus there is a given constraint, C, on the total capacity.

Formally, the objective is to minimize the average time delay, T_a, subject to a constraint that the total link capacity not exceed C. Begin by assuming that links are independent M/M/1 queues and that all message lengths are the same. Generalizations of this model are explored with this motivation to find the link capacities, c_l, which minimize T_a, defined as

$$T_a = \frac{1}{\gamma} \sum_l \frac{f_l}{c_l - f_l} \tag{7.1}$$

subject to the constraint that

$$\sum_l c_l \leq C \tag{7.2}$$

$$f_l \leq c_l \tag{7.3}$$

where γ is the sum of the requirements and f_l is the flow in link l. Minimizing T_a is equivalent to minimizing the total delay, T, which is γT_a. This simplifies the notation slightly and this is done in the discussion below.

A simple approach is to assign capacities to the links in proportion to their flows. Thus, link l with flow f_l, is assigned capacity c_l, where c_l is given by

$$c_l = \frac{f_l}{\sum_l f_l} C \tag{7.4}$$

This results in all links having the same utilization and minimizes the maximum utilization of a link. This leads to a "fair" assignment where delays are similar in all parts of the network. This proportional assignment rule, however, does not minimize T.

It is possible to minimize T by using Lagrangean relaxation [Fis81] as described below. Recall that in a relaxation of a problem it is possible to remove some constraints or replace the objective function by another (presumably simpler) function which bounds the optimum value. In Lagrangean relaxation, some of the constraints are replaced by adding terms to the objective function. The added terms form a penalty for violating the removed constraints and the optimal solution to the relaxed problem is a bound on the value of the optimal solution to the original problem. Here, since the objective is to minimize cost or delay, the optimal solution to the relaxed problem is a lower bound. In this case, as will be apparent, the lower bound is exact.

To relax the problem, the constraint on total capacity is replaced by an additional term in the objective function. Specifically, there is the following form for the relaxed problem:

$$\text{Minimize } z = \sum_l \frac{f_l}{c_l - f_l} + L\left(\sum_l c_l - C\right) \tag{7.5}$$

$$f_l \le c_l \tag{7.6}$$

where L is known as the Lagrange multiplier. If L is large enough, it is a penalty for violating the constraint on total capacity. If the sum of the c_l exceeds C, the term in brackets is positive and if multiplied by L increases the value of the objective to be minimized. Therefore, values of c_l are sought which do not violate the constraint. If L is too big, however, it is possible to make this new objective function smaller by letting the sum of the c_l become strictly less than C, thus minimizing the new objective function but not the original one. As L increases from zero, the sum of the c_l decreases as the first term is traded in the objective against the second. There is a unique value of L which makes the sum exactly equal to C. This is the value sought along with the corresponding values of the c_l. Finding all this turns out to be easy to do.

First, find the minimum of the (new) objective assuming that L is given. This is done by finding D_i, the derivative of z with respect to each of the c_l and setting the derivative to 0. Thus,

$$D_i = \frac{-f_l}{(c_l - f_l)^2} + L = 0 \tag{7.7}$$

Solving for c_l the solution is,

$$c_l = f_l + \sqrt{\left(\frac{f_l}{L}\right)} \tag{7.8}$$

The objective now is to find the value of L for which the sum of the c_l minimizing z is exactly equal to C. Thus,

$$\sum_l c_l = \sum_l \left(f_l + \frac{1}{\sqrt{L}}\sqrt{f_l}\right) = C \tag{7.9}$$

Solving for L the solution is,

$$L = \frac{(\sum_l \sqrt{f_l})^2}{(C - \sum_l f_l)^2} \tag{7.10}$$

and so,

$$c_l = f_l + \frac{C - \sum_l f_l}{\sum_l \sqrt{f_l}} \sqrt{f_l} \tag{7.11}$$

Thus, each link is first assigned capacity for its flow and then the remaining capacity is assigned in proportion to the square root of the flow. This is known as the square root assignment law.

Using the same Lagreangean relaxation technique, it is possible to solve the more general problem of minimizing delay subject to a constraint on total cost when cost is a linear function of capacity (i.e., when the cost of the link l, is given by $d_l c_l$).

In this case

$$c_l = f_l + \frac{D - \sum_l f_l d_l}{\sum_l \sqrt{\frac{f_l}{d_l}}} \sqrt{\frac{f_l}{d_l}} \tag{7.12}$$

which reduces to the result above when d_l equals 1.

Adding a fixed cost, d_{l0}, to each link does not complicate the problem significantly. In this case, since it is necessary to spend the fixed cost of each link, simply replace the cost constraint, D, by

$$D' = d - \sum_l d_{l0}$$

Another extension which can be made is to minimize T^p, rather than just T. The advantage of this is that as p increases, there is a fairer allocation of capacity. Specifically, as p grows large, the objective is dominated by the largest link delay. In the limit as $p \to \infty$, the optimal solution is to make all the link delays equal.

Again, the Lagrangean relaxation technique described above can be used to obtain the optimal link capacities. In this case,

$$c_l = f_l + \frac{D - \sum_l f_l d_l}{\sum_l \left(f_l d_l^p\right)^{\frac{1}{p+1}}} \left(\frac{f_l}{d_l}\right)^{\frac{1}{p+1}} \tag{7.13}$$

which reduces to the previous expressions for $p = 1$. Therefore, the general form of the relationship is that spare capacity is assigned (after each link is given capacity equal to its flow) to link l in proportion to

$$\left(\frac{f_l}{d_l}\right)^{\frac{1}{p+1}}$$

In the limit, as $p \to \infty$, this is a constant, that is, the spare capacity is divided equally among all the links. Notice also that for p equal to zero, this reduces to the proportional assignment rule.

An implementation of the capacity assignment algorithms described above is given in the procedure CapAsgL below. It takes as input the arrays Flow and Var-Cost, the link flows and cost per unit capacity, respectively. It also takes Power, the power that `delay` is raised to in the objective, and `MaxCost`, the limit on cost. It returns a vector, `Lcap`, with the selected link capacities. The function `pow(b,p)` is built-in and returns b^p.

If `CapAsgL` is called with `VarCost`, equal to one for all links, and `MaxCost` is set to the capacity limit, it will return capacities based on a limit of the total capacity. If it is called with `Power` equal to zero, it will return a proportional assignment. If it is called with `Power` set to a large value (say, 100), it will return a minimax capacity assignment (i.e., an assignment which minimizes the maximum link delay). It returns `TRUE` if a feasible capacity assignment is found and `FALSE` otherwise.

Example 7.1. Consider the 5 city network shown in Figure 7.1. Table 7.1 gives the link endpoints, flows, and cost per unit traffic. There are four capacity assignment procedures—proportional, square root (minimize delay given capacity) min—cost (minimize delay given cost), and minimax (minimize the maximum delay given capacity). The results are shown in Table 7.2 The total flow is 21. When capacity is constrained, it is limited to 42. Thus, the average utilization is 0.5. Therefore, in the proportional assignment all utilizations are 0.5. The average delay is 0.286, the cost is $46,800, and the individual delays vary from 0.167 to 1.000, a factor of 6.

With the use of the square root assignment rule and the same capacity limit, the average delay has gone down by about 7%. The cost has also gone down slightly (by

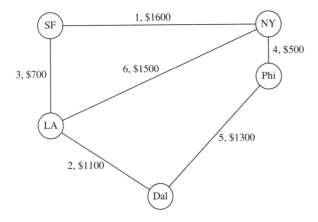

FIGURE 7.1

TABLE 7.1
Capacity assignment
example (input)

Endpoints		Flow	Cost/Cap
SF	LA	3	700
SF	NY	1	1600
LA	NY	6	1500
LA	Dal	2	1100
NY	Phi	4	500
Phi	Dal	5	1300

TABLE 7.2

Capacity assignment example (output)

	Link	Flow	Cost	Cap	Util	Delay	TotCost
		Cap = 42	Power = 0 (proportional)				
SF	LA	3.000	700.0	6.000	0.500	0.333	4200.00
SF	NY	1.000	1600.0	2.000	0.500	1.000	3200.00
LA	NY	6.000	1500.0	12.000	0.500	0.167	18000.00
LA	Dal	2.000	1100.0	4.000	0.500	0.500	4400.00
NY	Phi	4.000	500.0	8.000	0.500	0.250	4000.00
Phi	Dal	5.000	1300.0	10.000	0.500	0.200	13000.00
Total				42.000		0.286	46800.00

	Link	Flow	Cost	Cap	Util	Delay	TotCost
		Cap = 42	Power = 1 (square root)				
SF	LA	3.000	700.0	6.358	0.472	0.298	4450.59
SF	NY	1.000	1600.0	2.939	0.340	0.516	4701.97
LA	NY	6.000	1500.0	10.749	0.558	0.211	16123.36
LA	Dal	2.000	1100.0	4.742	0.422	0.365	5215.96
NY	Phi	4.000	500.0	7.877	0.508	0.258	3938.73
Phi	Dal	5.000	1300.0	9.335	0.536	0.231	12135.68
Total				42.000		0.266	46566.28

	Link	Flow	Cost	Cap	Util	Delay	TotCost
		Cost = 46800	Power = 1 (min-cost)				
SF	LA	3.000	700.0	7.340	0.409	0.230	5138.24
SF	NY	1.000	1600.0	2.657	0.376	0.603	4251.99
LA	NY	6.000	1500.0	10.193	0.589	0.238	15289.75
LA	Dal	2.000	1100.0	4.827	0.414	0.354	5309.74
NY	Phi	4.000	500.0	9.930	0.403	0.169	4965.02
Phi	Dal	5.000	1300.0	9.112	0.549	0.243	11845.26
Total				44.060		0.253	46800.00

	Link	Flow	Cost	Cap	Util	Delay	TotCost
		Cap = 42	Power = 100 (minimax)				
SF	LA	3.000	700.0	6.500	0.462	0.286	4550.01
SF	NY	1.000	1600.0	4.462	0.224	0.289	7139.43
LA	NY	6.000	1500.0	9.524	0.630	0.284	14286.17
LA	Dal	2.000	1100.0	5.486	0.365	0.287	6034.58
NY	Phi	4.000	500.0	7.510	0.533	0.285	3755.00
Phi	Dal	5.000	1300.0	8.518	0.587	0.284	11073.08
Total				42.000		0.285	46838.27

	Link	Flow	Cost	Cap	Util	Delay	TotCost
		Cap = 42	(equal)				
SF	LA	3.000	700.0	7.000	0.429	0.250	4900.00
SF	NY	1.000	1600.0	7.000	0.143	0.167	11200.00
LA	NY	6.000	1500.0	7.000	0.857	1.000	10500.00
LA	Dal	2.000	1100.0	7.000	0.286	0.200	7700.00
NY	Phi	4.000	500.0	7.000	0.571	0.333	3500.00
Phi	Dal	5.000	1300.0	7.000	0.714	0.500	9100.00
Total				42.000		0.531	46900.00

roughly 0.5%), but this need not have been the case. The utilizations of the links with larger flows have gone up, and the utilizations of the links with smaller flows have gone down. At first, it is easy to think this would increase the average delay, but it does not. When it is examined closely, it is apparent that the utilizations of the links with less flow go down by more than the amount by which the utilizations of the links with more flow go up. For example, the SF-NY link has less than one unit of capacity added, relative to the proportional assignment, but this is a relatively large fraction of the total capacity of this link and so the link utilization goes down by 16%. On the other hand, the LA-NY link loses over one unit of capacity but its utilization goes up by only 5%.

Another factor contributing to the reduction in delay, in this case, is that the links with larger flows have larger capacities, and therefore smaller service times. Thus, increasing their utilization does not increase delay by as much as it does on the links with smaller flows. This effect is an artifice of the M/M/1 delay model being used, which assumes that all the capacity is in a single server. If an M/M/m model was used, this effect would be less pronounced. With the M/M/1 model, the variation in delay is from 0.211 (for the link with the highest capacity and highest utilization) to 0.516 (for the link with the lowest capacity and lowest utilization), a factor of roughly 2.5.

The next step is to optimize based on cost, holding to a budget of $46,800 so that it is possible to compare, again, with the proportional assignment. The delay has gone down to 0.253, a reduction of 11.5% relative to the proportional assignment. This is also less than the capacity based assignment, but the cost is higher. Also, delays vary by over a factor of 3.5.

Next, the power (of delay in the objective) is set to 100 in order to minimize the maximum delay. It is apparent that, relative to the square root rule, cost has gone up slightly (by 0.5%), and the average delay has too (by 7%), but the delays are all within 2% of one another.

One of the most striking observations is that the differences among all these approaches are rather small. They all work reasonably well. At a higher utilization, expect to see larger differences, but the utilization of 0.5 actually used is quite realistic.

All of these approaches share a common method. They allocate more capacity to links with more flow. They differ only in how they distribute the spare capacity. In contrast to this, suppose the capacity is simply divided evenly among all the links. This may seem silly at first, but it is actually a widely used practice and is, in fact, the only choice if the flows are not known in advance. In this case, it is apparent that it does make a difference. The links with the largest flow also have the largest delay and the average delay is now almost twice the minimum obtainable at the same cost and capacity. Indeed, had the flows varied over a wider range, it is even possible that this approach would yield an infeasible result. Therefore, it is seen that it is important to be sensitive to the flow in some way.

Here is one last extension of this approach. In reality, cost does not vary linearly with capacity. A more realistic model is that cost is a concave function of capacity. Consider the problem of minimizing delay subject to a constraint on cost, where cost is a concave function of capacity. One approach is to model cost as a constant, plus a linear function. This is easily done as was discussed before, but it does not capture the full flavor of the problem and may lead to mediocre capacity assignments. A

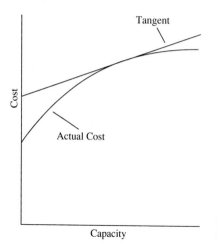

Cost

Tangent

Actual Cost

Capacity

FIGURE 7.2
Concave cost versus capacity function.

better approach [Ger73] is to solve a sequence of linear problems where the cost is approximated by the linear function, which is tangent to the concave function at the current capacity (see Figure 7.2).

In general, this approach only leads to a local minimum, not a global optimum. For the special case where $d_l(f_l)$, the cost of link l at capacity f_l, is of the form $d_{l0} + d_l f_l^p$, it is still possible to find a global minimum [Ger73].

Begin with an initial (feasible) assignment of link capacities, c_l, and replace the link cost function by a constant, plus a linear function tangent to the real cost function at c_l. Then solve the problem again with these new cost functions and obtain new capacity assignments. Then iterate this process until no significant progress (in terms of the average delay) is made.

The `CapAsgC` procedure that follows is an implementation of the previous algorithm, starting with capacities equal to the flows. This is not a feasible solution but is a reasonable point to begin computing a linearized estimate of the link costs. The procedure calls upon the `CapasgL` procedure described to solve the linearized problems as they arise. At each iteration, proceed to compute the slope and intercept of the tangent to the cost-capacity curve at the most recent estimate of capacity for each link. Thus, a link, l, with capacity c_l, fixed cost d_{0l} and variable (capacity dependent) cost coefficient d_{1l} has a cost of

$$a = d_{0l} + d_{1l}(c_l^p)$$

The slope of the linearized estimate of the cost function is the derivative of this function with respect to f_l; i.e.,

$$s = d_{1l} p c_l^{p-1}$$

and the y-intercept (fixed component) of the estimate is

$$y = a - s c_l$$

Example 7.2. Again consider the example shown in Figure 7.1 and Table 7.1, but with the following changes to the specifications of the problem. First, cost is now of the form

$$d_l + d_l(f_l^{.5})$$

where d_l is the Cost/Cap in Table 7.1. Thus, the fixed cost is equal to the variable cost coefficient, and cost now goes up with the square root of capacity rather than linearly. The cost limit has also been reduced to $30,000 in order to keep the example meaningful; it is now possible to obtain significantly more capacity for the same amount of money.

Table 7.3 summarizes the results of an optimization minimizing delay subject to a cost limit. Only 4 iterations were required for convergence. Note that the capacities tend to increase as the iterations proceed. This is not surprising since our initial estimate of the slope was made at c_l equal to f_l, that is, at a lower value of capacity. This overestimated the cost of the network. As we take advantage of the economy of scale available from the concave cost function, a lower estimate of cost is obtained for the same capacity, freeing up money to buy more capacity within the same budget. Note also that it is possible to purchase significantly more capacity for significantly less money using this cost function.

The power used, 0.5, is in fact representative of economies of scale actually available in realistic line tariffs. It is apparent, therefore, how important it is to try to make use of this economy of scale. A network with significantly better cost-performance is obtainable.

7.1.2 Discrete Capacities

Now turn to the more realistic case where each link is comprised of one or more channels. Each channel has a speed (capacity) selected from among a given set. Assume that one or more channels of each speed may be used within a link comprised of multiple channels, but that different channel speeds may not be mixed within the same link. This is not a critical assumption; the procedures described below can be easily expanded to allow for mixed speed links. The restriction to homogeneous links,

TABLE 7.3
Concave cost vs capacity (output)

			Power = 0.5	Cost Limit = 30000			
Delay 0.142	Caps: 10.511	3.179	14.629	6.420	15.027	13.084	
Delay 0.098	Caps: 14.422	4.235	17.986	8.577	21.064	16.430	
Delay 0.095	Caps: 14.851	4.331	18.099	8.778	21.800	16.598	
Delay 0.095	Caps: 14.886	4.336	18.065	8.788	21.874	16.577	

Link		Flow	Cost	Cap	Util	Delay	TotCost
0	1	3.000	700.0	14.886	0.202	0.084	3400.73
0	2	1.000	1600.0	4.336	0.231	0.300	4931.52
1	2	6.000	1500.0	18.065	0.332	0.083	7875.37
1	4	2.000	1100.0	8.788	0.228	0.147	4360.91
2	3	4.000	500.0	21.874	0.183	0.056	2838.50
3	4	5.000	1300.0	16.577	0.302	0.086	6592.96
				84.525		0.095	30000.00

however, is realistic and it simplifies both the implementation and the discussion below.

```
/*  Continuous Capacity Assignment Algorithm (Linear Costs)  */
void <- CapAsgL( nl , Flow , VarCost , Power , MaxCost , *Lcap )
     dcl Flow[nl] , VarCost[nl] , Lcap[nl] , Lfract[nl]
  PowerI <- 1 / (Power+1)
  LcostSum , LfractSum <- 0
  for_each ( link )
     LcostSum += VarCost[link] * Flow[link]
     Lfract[link] <- pow( Flow[link]/VarCost[link] , PowerI )
     LfractSum +=
        pow( Flow[link] * pow( VarCost[link] , Power ) , PowerI )
  LfractSum /= ( MaxCost - LcostSum )

  for_each ( link )
     Lcap[link] <- Flow[link] + Lfract[link] / LfractSum

/*  Continuous Capacity Assignment Algorithm (Concave Costs)  */
void <- CapAsgC( nl , Flow , FixedCost , VarCost ,
                  Power , Cpower , MaxCost , *Lcap )
     dcl Flow[nl] , FixedCost[nl] , VarCost[nl] , Lcap[nl] ,
         Slope[nl]

  CurrentCostLimit <- MaxCost
  for_each ( link )
     CurrentCostLimit -= FixedCost[link]
  CapAsgL( nl , Flow , VarCost , Power ,
        CurrentCostLimit , Lcap )
  CurrentDelay <- FindCurrentDelay( nl , Flow , Lcap )
  PreviousDelay <- INFINITY
  while (((PreviousDelay - CurrentDelay)/CurrentDelay) > EPSILON)
     PreviousDelay = CurrentDelay;
     CurrentCostLimit = MaxCost;
     for ( link = 0 ; link < nl ; link++ )
        cost <- FixedCost[link] +
              VarCost[link] * pow( Lcap[link] , Cpower )
        Slope[link] <- VarCost[link] * Cpower *
                     pow( Lcap[link] , Cpower-1 )
        cost -= Slope[link] * Lcap[link]
        CurrentCostLimit -= cost
     CapAsgL( nl, Flow, Slope, Power, CurrentCostLimit, Lcap )
     CurrentDelay <- FindCurrentDelay( nl , Flow , Lcap )

float <- FindCurrentDelay( nl , Flow , Lcap )
     dcl Flow[nl] , Lcap[nl]

  delay , TotFlow <- 0
  for_each ( link )
     TotFlow += Flow[link]
     delay += Flow[link] / ( Lcap[link] - Flow[link] )
  return( delay / TotFlow )
```

There are a given set of channel speeds, s_k, and associated costs, d_{lk}, for each link. The objective is to minimize cost subject to a constraint on the total delay on all the links. Alternatively, it is possible to minimize delay subject to a cost constraint. So, formally, the problem is to

$$\text{Minimize } z = \sum_{l,k} n_{lk}d_{lk} \qquad (7.14)$$

$$\text{subject to: } \sum_{l} f_l t(c_l, f_l) \le T \qquad (7.15)$$

Where T is a constraint the total delay suffered by all flow. T is, of course, closely related to $\bar{T}$ the average end-to-end delay suffered by a requirement. Specifically,

$$T = \gamma \bar{T}$$

where γ is the sum of all requirements. The discussion which follows in this section is simplified by dealing with T rather than $\bar{T}$.

Thus, the objective is the n_{lk}, the number of channels of speed, k, for each link, l. c_l is the capacity assigned to link l, which is simply $n_{lk}s_k$. $t(c_l, f_l)$ is the delay in link l, given its flow and capacity. T is the delay limit. Our constraint to homogeneous links also limits n_{lk} to be non-zero for exactly one k for each l.

Now look at two algorithms. The first guarantees an optimal solution but, in the worst case, requires an exponential amount of time. The second obtains an approximate solution in polynomial time and gives a bound on the distance between the solution found and the optimum. It is now convenient to solve the form of the problem where cost is minimized, subject to a constraint on the average delay. The techniques below can be suitably modified to solve the problem of minimizing delay subject to a cost limit. In fact, they produce a set of **undominated** cost-delay pairs. A cost-delay pair, (c_1, d_1) is said to be **dominated** by another pair, (c_2, d_2), if $c_1 > c_2$ and $d_1 > d_2$. In this case, the first pair has nothing to recommend it over the second, and the first pair can be eliminated from any further consideration.

7.1.2.1 SERIAL MERGE. The first method is due to [Zad73], and is applicable to other types of optimization as well. It is a form of dynamic programming [V.82]. The method makes use of the idea of dominance among partial solutions. A partial solution is defined as an assignment of capacities to a subset of the links. Two partial solutions are comparable if they make assignments to the same links. Solution A dominates solution B if A is both cheaper than B and has less total delay (sum of delay times flow in the assigned links).

The algorithm begins by considering all possible assignments for the first link. This includes all reasonable multiples of each of the channel speeds. The following rules are used to limit the candidate capacities for each link:

1. Since there are no heterogeneous links (i.e., links with different speed channels), it is only necessary to consider combinations with a single channel speed.

2. It is not necessary to consider any combinations which provide less capacity than the link flow.

3. It is not necessary to consider any combinations which provide so little capacity that the link utilization gets too high and the link delay is unacceptable. While the primary constraint is on the average delay, it may also be desirable to limit the delay in any given link. Doing so limits the variation in delay among node pairs and provides a fairer allocation of capacity, where all users experience similar delays. This rule actually subsumes the preceding one, since the acceptable utilization is never above 100%. We assume that there is a parameter, `MaxUtil`, which specifies the maximum acceptable link utilization.

4. It is not necessary to consider any combinations which provide so much capacity that the link utilization falls below an acceptable minimum (i.e., which is not cost effective). We assume that there is a parameter, `MinUtil`, which specifies the minimum acceptable link utilization.

5. It is not necessary to consider any dominated combinations.

> **Example 7.3.** Suppose 9600 bps links cost $1000/mo and 56000 bps links cost $2500/mo. The link flow is 15000, the minimum acceptable utilization is 25% and the maximum acceptable utilization (based on delay considerations) is 75%. A single 9600 bps channel has less capacity than the flow. Two 9600 bps channels just miss the upper utilization limit (at 78%). Three 9600 bps channels are within the utilization bounds, but at a cost of $3000, are dominated by one 56000 bps channel which is cheaper and has lower delay. One 56000 bps channel is a feasible candidate, meeting all the constraints. Two 56000 bps channels fail based on low utilization. So, the only viable candidate in this case is one 56000 bps channel.

Both algorithms, as will be shown, are begun by computing the set of undominated capacities for each link. This is done by considering multiples of each speed channel. If the constraint is relaxed on heterogeneity of channel speeds within a link, then combinations containing different speed channels would also be considered.

Start from the smallest channel speed and work up. For convenience, only consider multiples of each speed channel until the speed of the next largest channel is reached. For example, if we are considering 9600 bps channels and 56000 bps channels, we would only consider multiples of up to six 9600 bps channels. This is done because of the assumption that there is an economy of scale present in cost versus capacity, that is, that the 56000 bps channel costs significantly less than six 9600 bps channels. Thus, it is preferable to use the higher speed channel once that amount of capacity is required. Strictly speaking, however, if it is desirable to absolutely guarantee an optimal solution, it is necessary to solve a sequence of knapsack problems to determine the best combination of speeds for each capacity. This would significantly slow down the algorithm.

The list of combinations for each link is maintained as a queue of candidates. A candidate is a structure containing

(cost, delay, type, multiplicity)

which gives the cost, delay contribution (i.e., link delay times link flow), channel type (giving the speed), and channel multiplicity, respectively. Refer to the elements of a candidate, c, as c → cost, c → delay, etc. The routine SetCands returns a queue of undominated candidates satisfying the utilization constraints. A queue is chosen because it offers more flexibility in adding elements to the candidate set.

As candidates are generated, they are checked for feasibility in terms of their utilization and inserted into the queue in ascending order of cost, using the Insert-Cand routine. InsertCand returns TRUE if the candidate is not dominated by any candidate currently in the queue.

It is also possible that the candidate being inserted may dominate one already in the queue. When a candidate is dominated it is removed from the queue and the memory associated with it is freed. The routines MakeCand and FreeCand allocate and free memory associated with a candidate. Similarly, the routines MakeQueue and FreeQueue allocate and free all the memory associated with a queue. MakeQueue also initializes the queue to an empty state. While the details of memory allocation are not ordinarily a concern, it is critical here because this algorithm will run out of memory quickly if unused storage is not recovered.

The arrays, costs, and speeds give the nspeed choices of channel speeds and costs for this link. flow gives the link flow. MinUtil and MaxUtil constrain the link utilization. The function DelayF gives the link delay contribution (flow times delay) for a link with flow f and capacity c. In our implementation, DelayF is simply

$$d = \frac{f}{c - f} \tag{7.16}$$

but any function at all will do. We do not rely on properties of the delay function although we do implicitly assume that delay decreases as c increases.

```
/////////////////////////////////////////
// Select candidates for a given link  //
/////////////////////////////////////////

queue <- set_cands( nspeed , costs , speeds , flow ,
                    MinUtil , MaxUtil )
     dcl costs[nspeed] , speeds[nspeed] ,
         CandQueue[queue,cand] , Comb[cand]

  q <- MakeCQueue()
  for_each ( i , nspeed )
    if ( i < nspeed )
       MaxMult <- ( caps[i+1] / caps[i] )
    else
       MaxMult = 1 + flow / ( MinUtil * caps[i] ) )
    for_each( j , MaxMult )
       cap = j * caps[i];
       if ( (cap >= flow/MaxUtil) & (cap <= flow/MinUtil) )
           Comb = MakeCand( j * costs[i] ,
                            DelayF( flow , cap )
```

```
                                        i ,
                                        j )
                if ( InsertCand( CandQueue , Comb ) == FALSE )
                    FreeCand( Comb )
        return( q );

/////////////////////////
// Insert a candidate //
/////////////////////////

boolean <- InsertCand( CandQueue , Comb )
        dcl CandQueue[queue,cand] , Comb[cand]
    cur <- CandQueue->front
    if ( cur == Φ )      // First candidate
        Push( q , cand )
        return( TRUE )
    if ( cand->cost < cur->cost ) {
        while ( (cur != Φ) & (cand->delay<=cur->delay) )
            prev <- cur           // New cand dominates existing one
            cur <- cur->next
            FreeCand( prev )
        cand->next <- cur
        q->front <- cand
        if ( cur == Φ )
            q->rear <- cand
        return( TRUE )
    else
        while ( (cur->next!=Φ) & (cur->next)->cost<=cand->cost) )
            cur <- cur->next
        if ( cur->delay <= cand->delay )
            return( FALSE )      // New cand is dom'd by existing one
        next = cur->next
        cur->next <- cand
        cur <- next
        while ( (cur!=Φ) & (cand->delay<=cur->delay) )
            prev = cur;          // New cand dominates existing one
            cur = cur->next
            free( prev )
        cand->next <- cur
        return( TRUE )
```

Example 7.4. Assume there is a network with three links. Table 7.4 summarizes the speeds, costs and flows to be considered. Note it is unimportant what the endpoints of the links are since the link flows are already chosen (presumably by having already selected a routing).

Table 7.4 also shows the output from the first capacity assignment algorithm. It begins by showing the surviving capacity choices for each link. Assume the allowable range of utilization is from 0.333 to 0.667. For Link 1 with a flow of 3, at least 4 channels of Type 1 (capacity $= 1$) would be required to bring the utilization below the maximum. By this point, however, it is cheaper to use Type 2 links, which only cost

TABLE 7.4
Example for discrete link capacity assignment

T = 1.5	MinUtil = 0.333	MaxUtil = 0.667

Link	Flow	COSTS Type = 1 cap = 1.0	Type = 2 cap = 3.5
1	3	700	1700
2	1	1600	4600
3	6	1500	2600

Candidates for Link 1:

Cost	3400	Delay	0.750	Type	2	Mult	2
Cost	5100	Delay	0.400	Type	2	Mult	3

Candidates for Link 2:

Cost	3200	Delay	1.000	Type	1	Mult	2
Cost	4600	Delay	0.400	Type	2	Mult	1

Candidates for Link 3:

Cost	7800	Delay	1.333	Type	2	Mult	3
Cost	10400	Delay	0.750	Type	2	Mult	4
Cost	13000	Delay	0.522	Type	2	Mult	5
Cost	15600	Delay	0.400	Type	2	Mult	6
Cost	18200	Delay	0.324	Type	2	Mult	7

Merge of Links 1 and 2:

Cost	6600	Delay	1.750	Type	1	Mult	2
Cost	8000	Delay	1.150	Type	2	Mult	1
Cost	9700	Delay	0.800	Type	2	Mult	1

Merge with Line 3:

Cost	14400	Delay	3.083	Type	2	Mult	3
Cost	15800	Delay	2.483	Type	2	Mult	3
Cost	17500	Delay	2.133	Type	2	Mult	3
Cost	18400	Delay	1.900	Type	2	Mult	4
Cost	20100	Delay	1.550	Type	2	Mult	4
Cost	22700	Delay	1.322	Type	2	Mult	5
Cost	25300	Delay	1.200	Type	2	Mult	6
Cost	27900	Delay	1.124	Type	2	Mult	7

Link	Flow	Cost	Cap	Util	Delay	TotCost
1	3	1700	10.5	0.286	0.133	5100
2	1	4600	3.5	0.286	0.400	4600
3	6	2600	17.5	0.343	0.087	13000
Avg-Total:			31.5		0.132	22700

17/7 as much as Type 1 links and provide 3.5 times as much capacity. Thus, the only surviving combinations are Type 2 channels. At least 2 such channels in order to bring the utilization below the maximum. Two such channels provide 7 units of capacity, a utilization of 3/7 ($= 0.429$) and a delay contribution of

$$d = \frac{3}{7 - 3} = 0.75$$

The cost is $(2)(1700) = 3400$. Similarly, 3 Type 2 links cost 5100, with a capacity of 10.5 and a delay of $3/(10.5 - 3)$. Table 7.4 also shows the two undominated choices for Link 2 and the 5 undominated choices for Link 3.

We return to this example after describing the rest of the algorithm.

Given an acceptable set of candidate capacities for each link, the objective is to select a specific candidate for each to minimize delay at a given cost, or to minimize cost at a given delay. In fact, all undominated (cost,delay) pairs will be found. This is done via serial merges.

A serial merge of two lists of candidates is a procedure whereby elements for the merged list are built by selecting pairs of elements, one from each of the lists being merged. The elements in the new list have costs and delays which are the sums of the costs and delays of the elements forming it. Therefore, retain all newly formed elements not dominated by other elements in the new list.

In the previous example in Table 7.4, Lists 1 and 2 both have 2 elements. Thus, there are 4 (i.e., 2 times 2) possible pairs which can be formed. Of these pairs, 3 are undominated and are shown in the result of merging Lists 1 and 2. The fourth pair, formed from the second element of List 1 and the first element of List 2, has cost 8300 and delay 1.400, and is dominated by the retained element with cost 8000 and delay 1.15.

An element in a merged list is represented by a structure

(cost, delay, parent, last)

where cost and delay are the cost and delay associated with the newly formed element. Parent points back into the first list and last points into the second list. Together, they identify the pair forming the new element.

The routine, Smerge, below carries out the serial merge of two lists. Smerge works on two lists. The first is the result of a previous serial merge. The second is a list of candidates associated with a single link. Thus, each serial merge adds a selection for one more link to a partial solution.

Conceptually, given a list of elements which are all undominated partial solutions with capacity assignments for links 1 through $m - 1$ and a list of assignments for link m, a new list is formed whose elements are assignments for links 1 through m. Smerge takes two inputs—BaseQ, which is the list of partial solutions thusfar, and CandQ, which is the list of candidates for the current link.

Smerge first forms a list of elements by adding the first element in CandQ to each of the elements in BaseQ. The elements of BaseQ are sorted in increasing order of cost and decreasing order of delay. This creates the elements of the newly formed list, q. The routine `AppendElement` carries out this function. This list, q, becomes the current result of the serial merge.

Next, `AppendElement` is called again to make a list, q2, adding the second element of CandQ to each element of BaseQ. The routine `Merge` is then called to produce a single list, q3, from q and q2. `Merge` simply passes through q and q2, which are both sorted in ascending order of cost, and produces a single list of undominated elements, also sorted. It is simple to check for dominance because the lists are sorted.

q3 then becomes the new q, representing the result of this serial merge. `AppendElement` is then called to produce a list using the third element in CandQ, and `Merge` is called again. The process continues until all elements of CandQ have been considered.

Example 7.5. The result of this process is illustrated in Table 7.4 in the list headed "Merge with Link 3." The first three elements of this list (identified by "Type 2 Mult 3") were formed by appending the first candidate of the Link 3 list to each of the three elements in the list resulting from the merge of Link 1 and Link 2. As it turns out, none of these elements is dominated by any elements generated later and thus all appear in the list shown in Table 7.4.

Next, a list is formed appending the second element of CandQ to each element in the list, resulting from the merge of Link 1 and Link 2. This list is then merged with the current list (the three elements mentioned above). Only two of these elements survive. The third, with cost 17000 and delay 2.500, is dominated by the element with cost 15800 and delay 2.483.

The result of the final serial merge is shown in Table 7.4. There are 8 surviving candidates, trading cost against delay. Thus, of the $(2)(2)(5) = 20$ possible combinations, which can be formed from the three lists, only 8 survived undominated.

Suppose the objective is to minimize cost subject to a constraint that the average delay not exceed 0.150. The solution given by the sixth element in this list is then optimal and is shown at the bottom of the table.

The complexity of this procedure, both in terms of the time and space required, is clearly a function of the size of the list produced. In the worst case, to merge L lists, l, of size N_l, the size of the final list can be as large as

$$\prod_l N_l$$

Dominance, however, keeps the list much smaller. The effectiveness of dominance is a function of the particular data dealt with, specifically, the strength of the economy of scale and the tightness of the constraints. In practice, this approach can only be expected to work for problems of modest size; eventually, the exponential rate of growth overwhelms the dominance.

It is possible, however, to bring the problem back under control at the expense of some loss of accuracy by extending the notion of dominance to allow one solution to dominate another, if it is better based on one criterion (e.g., delay) and almost as good based on the other (e.g., cost). For example, if the objective is to minimize cost subject to a delay constraint, it is possible to allow a solution with cost 1000 and delay 1.1 to dominate a solution with cost 997 and delay 1.3. If there is an allowable cost tolerance of δ in checking dominance in each merge, and there are L merges, the final solution will be within $L\delta$ of the optimum. Thus, it is possible to trade off complexity against accuracy.

```
/////////////////
// Serial Merge //
/////////////////
queue <- Smerge( BaseQ , CandQ )
     dcl BaseQ[queue,elem] , CandQ[queue,cand] ,
         q[queue,elem] , q2[queue,elem] , q3[queue,elem]
         c[cand] , e[elem]

  c <- CandQ->front
  q = AppendElement( BaseQ , c )
  for_each ( c , Candq->front )
    q2 <- AppendElement( BaseQ , c )
    q3 <- Merge( q , q2 )
    q <- q3
    FreeQueue( q2 )
  return( q )

/////////////////////////
// Append to an element //
/////////////////////////
queue <- AppendElement( qIn , Cand )
     dcl qIn[queue,elem] , qOut[queue,elem]
         Cand[cand]

  qOut <- MakeQueue()
  for_each ( e , qIn )
    e2 <- MakeCand( e->cost + cand->cost ,
                    e->delay + cand->delay ,
                    e ,
                    Cand )
    Push( qOut , e2 )
  return( qOut )

/////////////////////
// Merge two lists //
/////////////////////
queue <- Merge( q1 , q2 )
{ q1[queue,elem] , q2[queue,elem] , q[queue,elem]

  q <- MakeQueue();
  e1 <- q1->front
  e2 <- q2->front
  while ( (e1 != Φ) && (e2 != Φ) )
    if ( e1->cost < e2->cost )
```

```
        if ( e1->delay <= e2->delay )   // e2 dominated
            e <- e2 ;  e2 <- e2->next ;  free( e )
        else                             // e1 to merged list
            e <- e1 ;  e1 <- e1->next ;  Push( q , e )
    else if ( e2->cost < e1->cost )
        if ( e2->delay <= e1->delay )   // e1 dominated
            e <- e1 ;  e1 <- e1->next ;  free( e )
        else                             // e2 to merged list
            e <- e2 ;  e2 <- e2->next ; Push( q , e )
    else
        if ( e1->delay <= e2->delay )   // e2 dominated
            e <- e2 ;  e2 <- e2->next ;  free( e )
        else                             // e1 dominated
            e <- e1 ;  e1 <- e1->next ;  free( e )

while ( e1 != Φ )   // rest of list1 to merged list
    e <- e1 ;  e1 <- e1->next ;  Push( q , e )
while ( e2 != Φ )   // rest of list 2 to merged list
    e <- e2 ;  e2 <- e2->next ;  Push( q , e )

return ( q )
```

The top level of the capacity assignment algorithm is given next. set_cands is called for each link and a serial merge is called to merge the resulting list of candidates with the partial solution thusfar obtained. Next, the first (least expensive) solution satisfying the constraint on total delay is found. Finally, the actual solution is recovered by backtracking using parent and cand.

```
/////////////////////////////////////////////////
// Discrete Capacity Assignment via Serial Merge //
/////////////////////////////////////////////////
int <- CapAsg3 ( nl ,  Flow , Costs , Caps ,    TotDelayLimit ,
                nspeed , Cap , Lcost , MinUtil , MaxUtil )

    dcl Flow[nl] , Costs[nspeed,nl] , Caps[nspeed]
        Lcap[nl] , Lcost[nl]

    Lcand[1] <- set_cands( nspeed , 1 , Costs , Caps ,
                            Flow[1] , MinUtil , MaxUtil )
    if ( Lcand[1]->front = Φ) return ( FALSE )
    q <- MakeQueue()
    for_each ( c , Lcand[1] )
        e <- MakeElem( c->cost , c->delay , Φ , c )
        Push( q , e );
    for_each ( link = 2 , nl )
        Lcand[link] <- set_cands( nspeed , link , Costs ,
                        Caps , Flow[link] , MinUtil , MaxUtil )
        if ( Lcand[link]->front == Φ ) return ( FALSE )
        q = Smerge( q , Lcand[link] )
```

```
// Find the cheapest element satisfying the delay constraint

e = q->front
while ( (e->next != Φ) & (e->delay > TotDelayLimit )
    e = e->next

// Backtrack the solution

for_each ( link = nl , 1 )
    Cap[link] <- Caps[e->cand->type] * e->cand->mult
    Lcost[link] <- e->cand->cost
    e <- e->parent
return ( TRUE )
```

Thus, if the objective is to minimize cost subject to a constraint that the total delay not exceed 1.50, the sixth member of the final Merge list in Table 7.4 is the solution. It is the least expensive solution with a delay below the constraint.

> **Example 7.6.** Examining the table, it is found that the last link (Link 3) uses 5 channels of Type 2 (i.e., has capacity 17.5). This solution points back (pointer not shown in the table, but infer it based on the difference in cost) at the third member of the Merge of Links 1 and 2 list. This explains that the next to last link (Link 2) uses one channel of Type 2. Finally, this latter member points back at the second member of the Link 1 list. Thus, Link 1 uses one Type 2 channel. Therefore, the entire solution is recovered via backtracking.

7.1.2.2 GREEDY ALGORITHM. The approach above guarantees an optimal solution but can take an inordinate amount of time. Alternatively, by strengthening the dominance criterion, it can yield results which are too approximate. Another approach, described below, guarantees a solution in a very reasonable amount of time and also gives a bound on the quality of the solution it obtains (which is not, in general, optimal.) The approach is originally due to [Whi72] and is also described in [V.82].

The method relies on Lagrangean relaxation, previously described, in conjunction with Kleinrock's approach to the continuous capacity assignment problem. Suppose the objective is to minimize cost subject to a constraint on total capacity. It is possible, alternatively, to ignore the constraint and just minimize cost. This is easy to do. Simply form the list of candidates for each link, as done before, and then pick the first member (with the lowest cost and highest delay) in each list. If this satisfies the delay constraint, the best solution has been reached. If, as is unfortunately more likely, the constraint has been violated, the algorithm can be encouraged to find a solution with a smaller delay by modifying the objective function to take delay into account as follows. It is possible to minimize

$$z(L) = \sum_{l,k} d_{lk} n_{lk} + L \sum_{l} t(f_l, c_l) \tag{7.17}$$

where L is the Lagrange multiplier. The larger L is, the more the algorithm will be willing to trade higher link cost for lower delay.

Given L, the problem is still easy to solve. It is only necessary to examine each link, l, separately and find the value of k which minimizes

$$z_l = d_{lk}n_{lk} + Lt(f_l, c_l) \tag{7.18}$$

This involves evaluating z_l for each value of k and keeping the best one.

All that is left is to find the best value of L. This can be done via a search over a sufficiently large interval, say from L (equal 0) to M. If M is very large, choose the highest cost and lowest delay possible.

So, it is possible to find a good capacity assignment by minimizing $z(L)$ for various values of L and keeping the best feasible assignment (i.e., the least cost assignment which also satisfies the delay constraint). In fact, it is possible to find the optimal assignment and prove that it is optimal by doing this, as is illustrated here.

Let $P(L)$ be the problem, given a value for L. As we have just seen, it is easy to find the optimal solution to $P(L)$ via a simple search. Let $T(L)$ be the delay associated with this solution and $D(L)$ be the link cost. So,

$$z(L) = D(L) + LT(L) \tag{7.19}$$

$P(L)$ is solved by considering each link, l, separately. Suppose that the optimal solution for link l and Lagrange multiplier value L is

$$z_l(L) = D_l(L) + LT_l(L) \tag{7.20}$$

Now consider what happens if L is increased to a larger value, L'. It is understood that since none of the capacity choices is dominated, as we go through the candidate capacity choices in decreasing order of cost, delay increases. Thus, it is not possible to decrease both delay and cost at the same time. As L increases, it is possible to leave the capacity choice alone or try to improve the solution by trading more cost for less delay. Therefore,

$$D_l(L') \geq D_l(L) \tag{7.21}$$

and

$$T_l(L') \leq T_l(L) \tag{7.22}$$

Hence, adding the results for all l, results in

$$D(L') >= D(L) \tag{7.23}$$

and

$$T(L') <= T(L) \tag{7.24}$$

Thus, D is an increasing function of L. The procedure then could be to start with L at a small value (maybe even 0) and see if it is possible to satisfy the delay constraint. If not, it is possible to increase L, trading some cost for an improvement in delay. Eventually, the smallest value of L is found which forces us to satisfy the delay constraint.

One way of carrying out this search is to do a binary search for L over the interval 0 to M, where M is given by

$$M = \max_l \frac{d_{lK}n_{lK} - d_{l1}n_{l1}}{t(f_l, c_{l1}) - t(f_l, c_{lK})} \tag{7.25}$$

where $t(f_l, c_{lk})$ is the delay assuming capacity choice k is selected, and K is the total number of capacity choices.

If the search terminates with the delay exactly equal to the constraint, then the optimal solution has been found, since $D(L)$ is an increasing function of L. If, as is more likely, the search terminates with $D(L)$ less than the constraint, there is a "gap" and no certainty as to whether or not the optimal solution has been found.

In general, when the search terminates, there are two values of L, L_1 and L_2, such that $T(L_1) < T$ and $T(L_2) > T$. The solution for L_1 is accepted as L_1 is the largest value of L found for which $T(L)$ is less than or equal to T. Hence, it is the value which gives the solution with the lowest cost. It is apparent that $D(L_2)$ is a lower bound on the value of the optimal solution since $D(L_2)$ is the optimal solution for T equal to T_2 and D is a non-decreasing function of T.

A closer look at what is happening gives more insight into the process. For each link, l, we are minimizing

$$d_l + Lt_l$$

If l is currently at capacity choice 1 (with minimum cost, d_{l1}, and maximum delay, t_{l1}) there is the option of increasing the capacity to choice k, with cost d_{lk} and delay t_{lk}. The change in cost is

$$\delta_d = d_{lk} - d_{l1} \tag{7.26}$$

and the change in delay is

$$\delta_t = t_{l1} - t_{lk}. \tag{7.27}$$

Thus, if

$$\delta_d/\delta_t < L$$

the objective function decreases and capacity k is preferable to capacity 1. In fact, the largest capacity for which this is true is preferred. Thus, given a value of L, it is possible to trade cost for delay until no link gives less delay at a rate better than L units of cost per unit of delay.

This leads us to another algorithm, described below, which finds exactly the same solutions as the Lagrangean relaxation procedure above, but is somewhat simpler to implement and has the added advantage of producing an entire curve of cost versus delay.

To begin, find all undominated capacity choices for each link as before. Then set all links to the minimum capacity choice. If the delay constraint is met, the optimal solution has been found. If not, the objective is to "buy" better delay at the lowest cost per unit of delay. Therefore find for each link, l, the value of k which offers less delay at the lowest incremental cost per unit. Refer to the current value of the capacity choice for l as `Ccurrent[l]`, and to the value of k just found as `Cnext[l]`. The ratio of change of cost to change of delay is `Ratio[l]`. Then find the link, l, with the smallest `Ratio[l]` and increase its capacity to `Cnext[l]`. Next, recompute `Cnext[l]` and `Ratio[l]`. The algorithm proceeds, greedily trading increased cost for improved delay until the delay constraint is met.

In the implementation that follows, a linear search is used to find the best link. A slightly more complex but more efficient implementation is to use a heap to hold the ratios. The heap is be initialized after computing the initial ratio for each link, updated whenever a link's capacity was upgraded, and popped to find the next link to upgrade.

Note that there need not exist a feasible solution. It is possible, even with all the links at the highest capacity, that the delay constraint is not met. In this case, the algorithm returns FALSE.

Example 7.7. Look again at the example in Table 7.4. Table 7.5 shows the eight undominated solutions obtained by the serial merge algorithm in the preceding subsection. The leftmost columns in the table show the cost, delay, and component channels for each combination. As expected, the costs increase and the delays decrease.

In the rightmost columns, the results of the greedy algorithm are shown, in particular, δ_d, δ_t and the ratio. The eight solutions are also plotted in Figure 7.3. The serial merge procedure finds all eight of these solutions. The greedy algorithm finds only seven. Solution 3 with cost 17500 and delay 2.133 is not found by the greedy algorithm. The reason it is not found is that there is no value of L for which this solution is optimal, that is, the cost-delay function is not convex.

TABLE 7.5
Greedy capacity assignment

Dsgn	Cost	Delay	Link 1	Link 2	Link 3	Del-D	Del-T	Ratio
1	14400	3.083	2×3.5	2×1.0	3×3.5	—	—	—
2	15800	2.483	2×3.5	1×3.5	3×3.5	1400	0.600	2333
3	17500	2.133	3×3.5	1×3.5	3×3.5	1700	0.350	4857*
4	18400	1.900	2×3.5	1×3.5	4×3.5	2600	0.583	4460
5	20100	1.550	3×3.5	1×3.5	4×3.5	1700	0.350	4857
6	22700	1.322	3×3.5	1×3.5	5×3.5	2600	0.228	11404
7	25300	1.200	3×3.5	1×3.5	6×3.5	2600	0.122	21311
8	27900	1.124	3×3.5	1×3.5	7×3.5	2600	0.076	34211

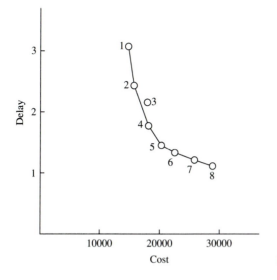

FIGURE 7.3

```
///////////////////////////////////////////////////////
// Greedy Algorithm for discrete capacity assignment //
///////////////////////////////////////////////////////
int <- greedyca( nl , Flow , Costs , Caps , TotDelayLimit ,
          nspeed , Cap , Lcost , MinUtil , MaxUtil )
   dcl Flow[nl] , Costs[nspeed,nl] , Caps[nspeed] ,
       Cap[nl] , Lcost[nl] , Ccurrent[nl] , Cnext[nl] ,
       Ratio[nl]

   // Initialization

   TotDelay <- 0
   for_each ( link , nl )
      Lcand[link] <- set_cands( nspeed , link , Costs , Caps ,
                           Flow[link] , MinUtil , MaxUtil )
      if ( Lcand[link]->front = Φ ) return( FALSE );
      FindNext( Lcand[link]->front , *Cnext[link] ,
              *Ratio[link] );
      Ccurrent[link] <- Lcand[link]->front
      TotDelay += Ccurrent[link]->delay

   // Upgrade capacity until delay constraint is satisfied

   while ( TotDelay > TotDelayLimit )
      link <- FindBestLink( nl , Ratio )
      if ( link == -1 ) return( FALSE )
      TotDelay += ( Cnext[link]->delay - Ccurrent[link]->delay )
      Ccurrent[link] <- Cnext[link]
      FindNext( Ccurrent[link] , *Cnext[link] , *Ratio[link] )

   // Record the solution

   for_each ( link , nl )
      c <- Ccurrent[link];
      Cap[link] <- Caps[c->type] * c->mult
      Lcost[link] <- c->cost
   return( TRUE )

/////////////////////////////////////////////////
// Find the link with the best cost/delay ratio //
/////////////////////////////////////////////////
int <- FindBestLink( Ratio )
   dcl Ratio[nl]

   BestLink <- -1
   BestRatio <- 0
   for_each ( link , nl )
      if ( Ratio[link] > BestRatio )
         BestRatio <- Ratio[link]
         BestLink <- link
   return( BestLink )
```

```
/////////////////////////////////
// Find the next capacity choice //
/////////////////////////////////

void <- FindNext( Clist , *Cnext , *ratio )

  *ratio <- 0
  for_each ( cand , Clist - Clist->front )
     r <- ( Clist->delay - cand->delay ) /
         ( cand->cost - Clist->cost )
     if ( r > *ratio )
        *ratio <- r
        *Cnext <- cand
```

The ratios are computed based on the difference in cost and delay relative to the last solution accepted by the greedy algorithm. Thus, solution 4 shows a delta of 2600 in cost and 0.583 in delay. Except for solution 3, these ratios are monotonically increasing. In Figure 7.3, the curve of D versus T is convex, except for Solution 3. The greedy algorithm, by its nature, finds only the solutions on the convex hull of this curve. These correspond to solutions which are optimal for some value of L.

All undominated values, however, are optimal for some value of T. Thus, the greedy algorithm fails to find an optimal solution when the optimal value is not on the convex hull. In the example presented here, T is 1.5 and the greedy algorithm does indeed find it. The only times when the greedy algorithm would fail to find the optimum here are when T is between 2.483 and 2.133. In those cases, it would choose solution 4 instead of solution 3 (which it cannot find).

Another useful feature exhibited by Figure 7.3 is that the convex hull itself (the line segments connecting the solutions on the hull) is a lower bound on the optimal solution for any value of T. Thus, while the greedy algorithm cannot guarantee an optimal solution, it does give a good bound on the value of the optimum.

7.2 MESH NETWORK TOPOLOGY OPTIMIZATION

Now turn to the problem of determining the topology of a mesh network. Thus the objective is to determine which locations to directly connect. This decision cannot, however, be made in a vacuum. It is closely related to decisions on what speed links to use and how to route traffic through the network. In particular, the requirement is sufficient capacity to carry the load routed onto each link.

The mesh network, topology optimization problem is clearly difficult. Even component subproblems like routing and capacity assignment have been shown to be difficult. Integer programming techniques have been used to produce optimal solutions or tight bounds on optimal solutions only for problems of modest size (a few dozen nodes) [Gav90]. For problems of practical size, virtually all approaches have been heuristics.

7.2.1   Branch Exchange

Mesh topologies can be produced by starting with a topology, usually a feasible topology, and then locally modifying it by adding or dropping links. The add or drop procedure is usually in some sense greedy, maximizing or minimizing some figure of merit. It is possible, for example, to start with a minimal spanning tree with adequate capacity to carry the entire load at a reasonable level of delay. Note that this is not in general a minimum cost network since there might be a large number of parallel channels in each link. Then, links are added to the network, attempting to reduce the cost while still satisfying the requirements on throughput and delay. It is possible to greedily add the link that saves the most money (by allowing us to reduce the multiplicity of other links).

Alternatively, there is the choice to add links which minimize the ratio of change in cost to change in delay [Mar78a], that is, compute for each link, l, the reduction in delay produced by adding link l. This involves rerouting some traffic. Then divide this delta by the cost of the link. This gives a figure of merit for the link. The lower the ratio, the better the link. This involves doing a routing. If the routing is not done carefully, we may get only a poor estimate of the link's value. It is even possible, if the routing procedure is stopped too soon, that it may appear as though adding a link actually increases delay. However, the more careful the routing procedure the longer the evaluation will take and the smaller the number of changes it is possible to analyze in a given amount of time.

Another possibility is to start with a complete graph and then identify a link to drop. Again, it is possible to use a greedy algorithm. In this case, we may choose the cost per bit as a figure of merit, that is, we do a routing, find the flow in each link and then compute the ratio of cost to flow for each link. Then remove the link with the largest ratio. This process is repeated until no further progress can be made.

It is possible to use both of these algorithms simultaneously to exchange one link for another, replacing links with poor figures of merit by links with better ones. This approach can be extended still further, exchanging several links for one another, simultaneously, along the lines suggested by Lin.

In each case, select candidate links using the figures of merit. Then tentatively add and delete the chosen links, redo the routing and reevaluate the network. If an improvement is obtained in terms of cost and/or delay, the exchange is accepted. If not, it is rejected and another is tried. Clearly, an exchange which lowers both cost and delay should be accepted. One which lowers one criterion at the expense of the other may or may not be accepted, depending on the circumstances. If the current network does not meet the delay constraint, it is possible to accept an exchange which lowers delay even if it increases cost. Conversely, working well within the delay constraint, it is possible to accept an exchange which increases delay if it also reduces cost. It is also possible to work based on the ratio of cost to delay. An approach [GP89] which has worked well in practice is to specify lower and upper bounds on delay. If the current delay is below the lower bound, try dropping a link in order to reduce cost. If the delay is above the upper bound, try adding a link to reduce delay. If delay is between the bounds, exchanges are considered. It is also possible to accept a limited number of random moves or manual inputs in an attempt to explore more of the solution space.

Branch exchange is a very general approach and and is widely used. It can accommodate additional constraints. That is, we can check that the maximum number of hops remains within a specified limit by starting with a network satisfying this constraint and refusing any exchange which destroys this property. It also has the advantage of proceeding incrementally. Thus, it can be used to refine a network which already exists, adding or deleting a few links to adjust to changes in traffic over time without completely redesigning the network.

There are, however, two major drawbacks to branch exchange. First, it requires that routing be done in the inner loop of the exchange procedure. Since routing is itself typically $O(N^3)$, this tends to make simple branch exchange procedures $O(N^5)$, which is prohibitive for networks of moderate or large size. Second, it is inherently a local process and tends to find a local minimum somewhere in the region of the starting solution. Thus, it is sensitive to the starting solution. If a poor starting solution is used, a poor final solution may result. One way of dealing with this latter problem is to attempt many solutions starting form different points. This tends to overcome the problem of locality but makes the overall procedure even slower.

One means of accelerating the process is to guide the selection of candiate links to add and delete by an algorithm which is faster than a complete routing. One such approach, known as the Cut Saturation Algorithm was described in [GP89]. This algorithm is based on the observation that in a network with effective routing, congested links form cuts (i.e., if a link is congested it is because there is no better alternative for the traffic using it). A cut whose links are all heavily loaded is referred to as a **saturated cut**. These saturated cuts can be used to guide a branch exchange procedure. Specifically, links are added across saturated cuts and links are deleted on one side of such cuts. This should lead to either an increase in throughput, a decrease in delay, or both. There can be additional reductions in cost with an attempt to delete costly links and add inexpensive ones.

The key, then, is to identify saturated cuts. So, if a saturated cut is defined as a cut whose links are at least as heavily used as the links in any other cut, this turns out to be quite easy. Kruskal's minimum spanning tree algorithm can be used, with link utilizations used lengths in the algorithm. Kruskal's algorithm is run, adding $N - 2$ links to the tree and forming two components. The links in the cut separating these two components is the desired point to cut. A detailed description of this procedure is given in [MGE74].

7.2.2 The MENTOR Algorithm

Thus, it would appear that any reasonable algorithm used to select a mesh topology would be, at best, of $O(N^5)$ since there are $O(N^2)$ candidate links to consider, and consideration of a link would involve doing a routing, which requires finding paths between all pairs of nodes, itself an $O(N^3)$ procedure. Even if N is moderate, say between 50 and 100, this leads to a substantial amount of computation. If, in addition, the objective was to iterate such a procedure within a larger one, which considered different switch locations or device types for example, the computation can easily become prohibitive.

There is, however, an $O(N^2)$ algorithm due to Kershenbaum, Kermani and Grover [KK91], which is called MENTOR (MEsh Network Topology Optimization and Routing). This low complexity is achieved by doing implicit routing within the inner loop of the procedure which evaluates links, as described below. There are other procedures [FC72] [MGE74] [Han73] [Mar78b] [MS86] for finding mesh network topologies, and some of these procedures may find solutions which are several percent better than MENTOR does. The MENTOR algorithm is unique, however, in its ability to find topologies quickly enough to be imbeddable in other more general purpose algorithms, and to function in the inner loop of interactive network design tools, where the user may wish to examine many alternate network architectures in a small amount of time.

The MENTOR algorithm is appropriate for the design of many types of communications networks because it does not rely on characteristics of any particular networking technology or architecture but instead relies on basic network design principles. Specifically, it tries to find a network with all the following characteristics:

1. Requirements are routed on relatively direct paths.
2. Links have a reasonable utilization. It is not desirable for link utilization to get so high that performance (loss, delay) suffers, and it is not desirable for utilization to become so low that the link is not cost effective.
3. Relatively high capacity links are used, thereby allowing us to benefit from the economy of scale generally present in the relationship between capacity and cost.

These three objectives are, to some extent, contradictory. For example, a desire to create well utilized, high capacity links generally involves detouring traffic to indirect paths in order to aggregate it. Nevertheless, the MENTOR algorithm trades these objectives off against one another to create low cost, effective networks.

The algorithm begins by finding a center of mass, C, for the network. This center of mass is defined as the node which minimizes the quantity

$$M_i = \sum_j c_{ij} w_j \tag{7.28}$$

where c_{ij} is the cost of connecting nodes i and j, and w_j is the weight of node j, defined to be the total requirements to and from node j, that is,

$$w_j = \sum_k r_{jk} + r_{kj} \tag{7.29}$$

Thus, the median is the node best suited to bring traffic to, in the sense that traffic should move as little as possible to get there. This step is $O(N^2)$ because it involves only forming N sums of N elements each and then finding the smallest of N values.

The next step in the procedure is to identify the "backbone" network nodes. These nodes are the locations of the tandem switches in the network. Node selection algorithms were discussed extensively in Chapter 5 when the concentrator location problem was examined. Some of the algorithms given there (e.g., the add and drop algorithms) can be adapted to select backbone node locations. There can also be an additional term to account for the variation in the cost of the backbone as we add or drop a given backbone site.

Often, however, it is more appropriate, for the sake of speed, to use a simpler procedure. The MENTOR algorithm uses a thresholding algorithm based on the nodes' weights. In particular, there is W, a threshold on node weight, w_j, as defined above. Any node whose weight exceeds W is selected as a backbone node. In many situations, such as when there are "key" sites which are major sources and destinations of traffic, deciding in this way is quite appropriate.

Next, given R, a parameter which specifies a radius of cost around a backbone node, all remaining nodes within (cost) radius R of a designated backbone node are declared to be "local" nodes (i.e, not backbone nodes). In many cases, if the designated backbone nodes cover the network, all nodes then are designated either backbone or local and this phase of the algorithm is complete.

If not, continue to select the most appropriate as yet undesignated node as a backbone node, as follows. A figure of merit, F_j, is defined for each undesignated node, j, as

$$F_j = P_c \frac{c_{iC}}{D} + (1 - P_c)\frac{w_j}{W} \tag{7.30}$$

where D is the diameter of the network (i.e., the "distance" (in terms of cost) between the two most distant nodes), and P_c is a parameter controlling the relative importance of weight and distance in controlling the selection of backbone nodes.

This node selection procedure works best when there are natural "leaders" in the network. On the other hand, when weight is more uniformly distributed among nodes, this procedure degenerates to a placement of nodes at the centers of somewhat randomly placed circles of radius R. This leads to a reasonable covering of the network by backbone nodes which is neither very good nor very bad. In this case, it may be appropriate to use more traditional k-median finding algorithms [Har75] [HM79].

It is also possible to allow the user to specify some, or all, of the backbone node locations manually. Finally, it is possible to set the threshold fairly high, allowing the "clear" backbone nodes to be selected and then use a k-median algorithm to finish the job. One way or another, there is now a set of backbone nodes with which to proceed. Indeed, since this whole algorithm is so fast, it is also possible to iterate over several choices of backbone node sets.

The median finding algorithm just described is then run again to select a new median, only considering the backbone nodes. The original median, C, is discarded (as a median).

Next, a tree is formed to interconnect the nodes. The tree has the following characteristics (see Figure 7.4):

1. It is a spanning tree on the backbone nodes. Each local node is connected directly to the closest (in terms of cost) backbone node.
2. The tree has short links (i.e., it is like a minimal spanning tree).

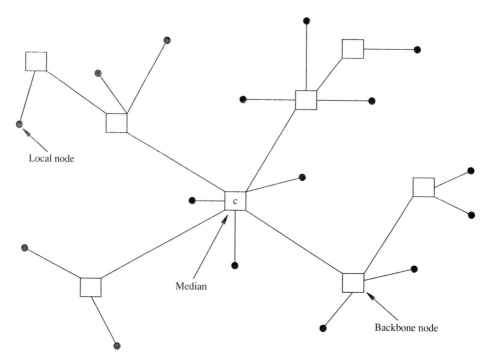

FIGURE 7.4
Spanning tree for MENTOR.

3. The paths in the tree are short. In this sense, it is like a shortest path tree.

If it is possible to find a tree with all these characteristics, then there is a possibility of creating a network which satisfies the three objectives stated above. The tree aggregates traffic, allowing there to be a benefit from economy of scale. It has relatively direct paths. It is possible to attempt to size the links so that they are reasonably utilized, neither too high nor too low. Finally, if the links are short, even poorly utilized links will not waste too much money.

The general problem of finding a tree which minimizes the sum of the shortest paths between all pairs of nodes is itself a difficult problem. If at the same time the objective is to minimize the total length of the links in the tree, the problem also becomes ill defined as it is now not clear how much of one objective to trade for another. Finally, this tree is not itself the network desired, but rather, simply a starting point for the network. We therefore content ourselves with solving a closely related problem which is much easier to deal with. The objective is to minimize a combination of tree length (sum of the lengths of the links in the tree) and path length to the median.

This problem can be approached with a heuristic, which can be thought of as a modification to both Prim's and Dijkstra's algorithms. Observe that Prim's and Dijkstra's algorithms, for minimal spanning trees and shortest path trees, respectively, are nearly identical. In Prim's algorithm, start at a designated node i (in this case the median) and attempt to label other nodes, j, with

$$L_j = d_{ij}$$

where d_{ij} is the distance (or cost) between nodes i and j. This selects trees with short links. In Dijkstra's algorithm, nodes are labeled with

$$L_j = L_i + d_{ij}$$

This selects trees with short paths. A hybrid algorithm can be created by labeling nodes with

$$L_j = \alpha L_i + d_{ij}$$

where α is between 0 and 1. For α equal to 0 this is Prim's algorithm and for α equal to 1 this is Dijkstra's. The larger α is, the more attention is given to path length, as opposed to tree length. When the triangle inequality holds (i.e., when the cost of a link (A,B) is no greater than the sum of the costs of links (A,C) and (C,B), for all A, B, and C), when α is 1 we get a star centered at the median.

Note that these two objectives are not entirely inconsistent. Trees with short paths also tend to have short links, but the two goals are not identical either. In practice, computational experience with this algorithm has shown that values of α between 0.2 and 0.5 tend to yield trees with both short links and short paths. The following example, illustrates both the hybrid algorithm and the relationships among the characteristics of different trees.

Example 7.8. Figure 7.5 shows part of a network. The costs of some of the links are shown. Assume that links now shown have cost greater than 17. Let us consider what happens when the hybrid tree finding algorithm is run for various values of α. Suppose that the median finding algorithm is run and that node D is the median found.

With α equal to 0, the algorithm finds a minimal spanning tree: (A,B), (B,C), (C,D), (D,E), (E,F), (F,G). This tree has length 60. The total length of the paths between all pairs of nodes is 560.

With α equal to 1, the algorithm finds the shortest-path tree rooted at the median, which is a star rooted at D. Note that if the link costs obey the triangle inequality (i.e., if the length of any link is no greater than the sum of the lengths of the other two sides of any triangle it is part of) this shortest-path tree will always be a star, rooted at the

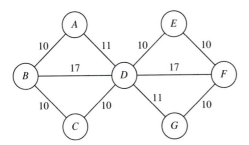

FIGURE 7.5
Hybrid tree.

median. This tree has a length of 76. The total length of the paths between all pairs of nodes is 380. Thus, the tree length has increased by 12.7% and the path length has decreased by 32.4%.

With α equal to 0.5, the median D is given a label of 0 and all the other nodes have labels of ∞. D is brought into the tree and is scanned. It offers labels of

$$(0.5)L_D + c_{Dj} = c_{Dj}$$

to all other nodes and they all accept the new labels which are better than the ∞ they currently have. Thus, nodes A, B, C, E, F and G receive labels of 11, 17, 10, 10, 17, and 11, respectively. Next, the out-of-tree node with the smallest label is brought into the tree and scanned. C and E are tied; let's say we choose C. C is scanned and offers labels of

$$(0.5)L_C + c_{Cj} = 5 + c_{Cj}$$

to all out-of-tree nodes, j. Only B receives a better label (15 in place of 17) and is relabeled. Next, E (which also has a label of 10) is brought into the tree and scanned. E gives F a label of 15. Node A (with a label of 11, tied with G) is brought in next and scanned. No other nodes are relabeled. Then, nodes B and F are brought in and, again, no other nodes are relabeled. Thus, the hybrid tree has been formed:

$$(A, D), (B, C), (C, D), (D, E), (D, G), (E, F)$$

This tree has length 62 and total path length 452. This is an increase of only 3.3% in the tree length relative to the MST, and an increase of 18.9% in the total path length, relative to the shortest path tree (star at the median). Thus, this tree exhibits many of the good properties of both other trees.

This tree finding algorithm has complexity $O(N^2)$, the same as that of Prim's and Dijkstra's algorithms. The effort to bring a node into the tree is $O(N)$ and this is done $O(N)$ times. As with the median finding procedure, it is possible to iterate this procedure to explore different trees.

Given a tree, centered at median node M, the objective is to consider adding a direct link between each pair of nodes. A link is added between any pair of nodes which has accumulated a sufficient amount of traffic to "properly" utilize the link (i.e., which will fill the link up sufficiently to make it cost effective).

There are many possible ways of making this determination. The simplest, and the one described in [KK91], is to use a slack parameter, s, to make the determination. If the capacity of a link is C and a maximum utilization of ρ is permitted, then the usable capacity is ρC. If the slack is s, then add a link (A,B) if $r_{AB} > \rho C(1 - s)$.

A more sophisticated approach would be to let s be a function of the ratio of the direct (A,B) cost and the distance (in terms of cost) between A and B in the tree selected; the smaller this ratio the more slack that would be tolerated. It is also possible to iterate this part of the algorithm using different values of s.

One important consideration in implementing this part of the algorithm is the freedom in actually bifurcating (splitting) traffic among multiple routes. This is a function of the specific routing procedure used in implementing the network. The three most common cases are:

1. The routing can, effectively, create any desired bifurcation. This is the case if any reasonably effective dynamic routing procedure is used, or if routing is done via an explicit table stored in each node telling what percentage of traffic to each destination should go out over each link.

2. No bifurcation is possible. This is an extreme case which is approximated by the simplest static routing procedures. It may also be a reasonable model for routing procedures which can bifurcate traffic but are not load sensitive.

3. Bifurcation is done on a requirement by requirement basis. All the traffic in a given requirement must follow a single path, but different requirements between the same pair of nodes may follow different paths.

The algorithm can handle all of these cases with only slight modifications. The first case is the simplest. Add the (A,B) link if r_{AB} is greater than $\rho C(1-s)$. If $r_{AB} > \rho C$, it is possible to add a second channel to the (A,B) link if there is sufficient traffic to justify it. It is possible to do this either on the basis of average utilization or on the basis of the utilization of the last channel in the link. In the first case, a second channel would be added if $r_{AB} > 2\rho C(1-s)$. In the second case, the second channel would be added if $r_{AB} > \rho C(2-s)$. This reasoning extends to the general case of m links. In [KK91] the first approach is used.

If no bifurcation is possible, then only add a link if it can handle all the offered traffic. Thus, if r_{AB} is between $\rho C(1-s)$ and ρC, a link with a single channel is added. If it is between $2\rho C(1-s)$ and $2\rho C$, a link with two channels is added, etc. In this case, only evaluation based on the average utilization makes sense.

In the final case, it is necessary to solve a bin packing problem to determine if any channels can be loaded to an acceptable level. If so, these channels are added. This increases the complexity of the overall procedure somewhat, but ususally, either the number of requirements per link is small (giving rise to a fast bin packing), or large (allowing us to essentially ignore the bin packing problem). Thus, in practice, this does not usually increase running time significantly.

In general, some channels are added to carry some of the offered traffic, and the remaining offered traffic overflows to a less direct path. In the two extreme cases, no traffic overflows or all of the traffic overflows. When traffic overflows, the algorithms tries to use the most direct path possible, backtracking into the tree as little as necessary.

In Figure 7.6, the situation is illustrated for (A,E) traffic, showing part of the tree. Any (A,E) traffic not carried by a direct (A,E) link would overflow to either A-B-E or A-D-E. The simplest rule, which is again used in [KK91], is to overflow to the cheaper path. Thus if the cost of an (A,B) channel plus a (B,E) channel is smaller than the cost of an (A,D) channel plus a (D,E) channel, then the first path would be used.

The intermediate node added to the path is called a detour point. Thus, in the previous example, node B would be the detour point for (A,B) traffic. The determination of the appropriate detour point for each pair of nodes is done simply by comparing the per channel costs of the two candidate 2-hop paths (e.g., A-B-E and

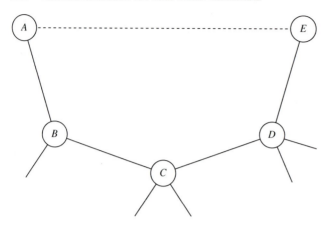

FIGURE 7.6
Considering (*A*,*E*).

A-D-E above). Since this decision is only a function of the tree selected and the costs of the links, it can be made for all node pairs before any decisions are made about which direct links to include. This helps by simplifying the decision on the order in which to consider the node pairs.

The overflow traffic joins the original offered traffic and becomes indistinguishable from it. It is possible that traffic overflowing onto a less direct path might overflow again onto a still less direct path. In the most extreme case, the traffic might detour all the way back to the tree path. Thus, in Figure 7.6, the (*A*,*E*) traffic would be carried on the *A-B-C-D-E* path if none of the appropriate intermediate links were added.

Note that not all links are appropriate. In Figure 7.6, if it were decided to detour to *A-B-E* then the (*A*,*E*) requirement becomes an (*A*,*B*) requirement, in tandem with a (*B*,*E*) requirement. Thus, the (*A*,*D*) link is no longer appropriate for this requirement, and even if the (*A*,*D*) link is added, it will not be used to carry any (*A*,*E*) traffic. This is done not because it is optimal, but rather because it dramatically simplifies the algorithm and speeds it up. In particular, it solves two enormous problems—how to route the traffic and how to order the nodes pairs for consideration.

Routing is done implicitly. The (*A*,*E*) traffic "becomes" (*A*,*B*) and (*B*,*E*) traffic. Thus, all traffic is routed directly; no explicit routing need be done. If the decisions made were recorded, it is possible to recover the actual routes by remembering how much of each requirement overflowed and through which intermediate node.

The algorithm proceeds by sequentially considering each node pair and deciding whether or not to add a direct link between the pair of nodes. If a link is added, a decision is also made as to how many channels to put in the link, and how much traffic to allow the link to carry. Whatever traffic (if any) is not carried by the direct link, overflows via the shorter detour as described above.

Given the simple decision rules being used, there are many equivalent possible sequences for considering the node pairs. The only restriction is that if node pair (*w*,*z*) contributes to the traffic offered to node pair (*x*,*y*), then it is important to consider

(w,z), before considering (x,y), so that when (x,y) is considered it will be evident exactly how much traffic is being offered to it. Thus, in the previous example, if B is the detour point for (A,E) traffic, it is important to consider (A,E) before considering (A,B), or (B,E). It is not, however, necessary to consider (A,E) before (A,D), or (D,E), since the (A,E) traffic can not overflow onto either of these pairs.

In order to produce a valid sequence of the node pairs, it is necessary to do a topological sort [c.f. Knuth] of the directed graph representing the dependencies among the node pairs. A topological sort takes G, an acyclic graph, (a directed graph with no directed cycles) as input and returns an ordering of the nodes such that if (X,Y) is an arc in G, then X precedes (not necessarily immediately) Y in the ordering.

Consider the dependency graph shown in Figure 7.7. Suppose there is a chain A-B-C-D-E as in Figure 7.6, and it has been decided (e.g., based on trying to deviate traffic as little as possible) that (A,E) traffic overflows to A-B-E. Similarly, suppose that (A,D) and (B,E) would overflow to A-C-D and B-D-E, respectively. All other pairs overflow to the obvious places (i.e., there is no choice for them). Figure 7.7 represents all the dependencies which arise. Note that the "nodes" in Figure 7.7 are node pairs in the original graph.

A topological sort proceeds by outputting pairs with no dependencies (i.e., pairs with no arcs leading into them). At least one such pair must exist or else the dependency graph would contain a cycle. It is clear that the dependency graph cannot possibly contain any cycles, since all arcs in it lead from pairs which are more hops apart to pairs which are fewer hops apart in the original tree (in this case a chain).

Thus, we begin by identifying pairs with no dependencies, in this case (A,E), (A,D), and (C,E). It is possible to pick any or all of these as the first pairs in the sequence to consider. Once (A,E) has been considered, (B,E) has no further dependencies and can be considered. Similarly, once (A,D) has been considered, (A,C) has no further dependencies. Proceeding along these lines, it is possible to find many node pair sequences. One valid sequence is:

$(A, E), (A, D), (B, E), (A, C), (B, D), (C, E), (A, B), (B, C), (C, D), (D, E)$

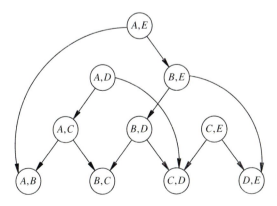

FIGURE 7.7
Dependencies along the A,E path.

There are many others. All will give rise to the same solution if the simple slack rule given above is used for determining when to put in direct links. In fact, if this simple rule is used, it is not necessary to even track dependencies directly. It would be sufficient to consider node pairs in order of the number of hops in the tree, largest number first. As mentioned above, the only dependencies are pairs with a larger number of hops on pairs with a smaller number of hops. With more complex rules for selecting direct links, it becomes more important to keep track of explicit dependencies.

Given a valid node pair sequence, now consider which links to actually include (i.e., which pairs to connect directly by a link). As this is done, we also decide which traffic to offer to the link. As mentioned before, this is dependent upon details of the routing used in the network. Assume, for simplicity, that dynamic routing is used and therefore, a link is offered to all the traffic that homes to it, that is, all traffic which either was originally between the pair, or which "became" traffic between the pair as it was diverted through a detour point.

For example, if the above node pair sequence was used and the link (A,E) was not included, but the link (B,E) was included, then link (B,E) would be offered both the (A,E) traffic (which homed to the link because it was already decided to use node B at a detour point for (A,E) traffic), and (B,E) traffic, which is naturally offered to (B,E). As mentioned above, the criterion for including a link is whether it has been offered enough traffic to make it cost effective, based on an acceptable amount of slack.

An implementation of the MENTOR algorithm is given below along the lines described above. The inputs to the algorithm are:

nn	The number of nodes in the network
Cost	An $nn \times nn$ (symmetric) array of costs for connecting each pair of nodes
Req	An $nn \times nn$ array or requirements between each pair of nodes
Wparm, Rparm, Dparm	The parameters used to select backbone nodes
Alpha	The parameter used to select the tree
Slack	The parameter used to select direct links
ChanCap	The usable capacity of a channel
FDXflag	H for half duplex; F for full duplex

The outputs from the algorithm are:

Bkbn	1 for backbone nodes; 0 for local nodes
nlinks	The number of links (tree and direct) selected
Ends	Endpoints of the selected links
Lmult	Number of channels (multiplicity) of selected links

SetWeights sets the weight of each node to the sum of the requirements into and out of the node. SetMedian finds the node whose (requirement weighted) distance to the other nodes is minimum. The Flag parameter, passed to SetMedian, tells it whether to consider all nodes or just backbone nodes as candidates as a median.

`Thresh` chooses the backbone nodes based on parameters `Wparm` (weight parameter), `Rparm` (radius parameter), and `Dparm` (distance versus weight parameter). All these parameters are normalized—`Wparm` by the usable capacity, `Rparm` by the diameter (i.e., largest link cost) of the network, and `Dparm` between 0 and 1. `Thresh` returns an array of length `nn` with 1 in positions where backbone nodes are selected and 0 elsewhere. Note that it is possible to modify `Thresh` to also take `Bkbn` as input, and to thereby allow the user to specify mandatory and prohibited backbone node locations.

`PrimDijk` is the hybrid Prim-Dijkstra algorithm which finds the tree, forming the skeleton of the backbone. It returns an array, `Pred`, of length `nn`. `Pred` gives the predecessor of each node in the tree, rooted at the median. By convention, the median is its own `Pred`. The parameter `Alpha` controls how starlike the tree is; an `Alpha` of zero yields an MST.

`SetDist` finds the shortest path between each pair of nodes in the tree found. The predecessors in these paths are used to orient the search for dependencies in the sequencing algorithm. The distances themselves are not used in this implementation, but could be used in a more sophisticated version of it, where tree distances were used in the direct link selection process. `SetDist` is an $O(nn^2)$ algorithm which fills in the distance and predecessor arrays by "growing" them one row and column at a time. Specifically, the row and column corresponding to each node added in preorder (based on the tree) is filled in. The distances and predecessors are recorded between this node and all other nodes already considered. This is easily done since the paths to and from the node are the same as those for its predecessor (which by the definition of preorder, has already been considered), plus the one additional link joining the node to its predecessor in the tree.

`Sequence` produces a valid sequence of node pairs suitable for the direct link selection procedure. It also produces an `nn` by `nn` array, `Home`, with the detour point for each node pair. `Sequence` does a topological sort, first building up the number and identity of the dependencies of each pair (specifically the number of pairs dependent upon each pair, and the set of pairs dependent upon it). Then, it outputs pairs with no remaining dependencies. `Home` is set to the detour point for each pair, in this case, simply by choosing the smaller detour from among the predecessors (in the tree) of each pair of nodes. A less sophisticated approach would be to simply order the pairs by hops in the tree. A more sophisticated approach would be to order, dynamically, during the link selection process.

`Compress` selects the direct links, considering the pairs in the order, specified by `sequence`, and accepting links with sufficient offered load. Load is overflowed via the detour points. This implementation uses a simple criterion based on the allowable slack. A more sophisticated approach is to base the decision on the direct distance (cost) between the nodes, as compared to the tree distance; the smaller this ratio, the more slack accepted.

Example 7.9. Consider a 10 node mesh network design problem given in Table 7.6 and Figure 7.8. The nodes are 10 of the largest U.S. cities. The costs, while not exactly based on a tariff, are based on distance with a fixed charge added. Thus, the costs are realistic.

TABLE 7.6
MENTOR algorithm example (input)

NN = 10
WPARM = 2.0   RPARM = 0.5   DPARM = 0.5
ALPHA = 0.0   SLACK = 0.2
CAP = 33600   FDXflag = F

Costs:

		NYK	LSA	CHI	HOU	PHL	DET	SDG	DAL	SAN	PHX
0	NYK	1000	8308	3130	5242	1231	2440	8275	5101	5731	7396
1	LSA	8308	1000	6199	5104	8137	6910	1309	4699	4597	2062
2	CHI	3130	6199	1000	3802	2989	1708	6187	3394	4129	5332
3	HOU	5242	5104	3802	1000	5017	4294	4927	1672	1567	4042
4	PHL	1231	8137	2989	5017	1000	2320	8098	4888	5509	7213
5	DET	2440	6910	1708	4294	2320	1000	6895	3976	4690	6037
6	SDG	8275	1309	6187	4927	8098	6895	1000	4558	4402	1915
7	DAL	5101	4699	3394	1672	4888	3976	4558	1000	1747	3646
8	SAN	5731	4597	4129	1567	5509	4690	4402	1747	1000	3538
9	PHX	7396	2062	5332	4042	7213	6037	1915	3646	3538	1000

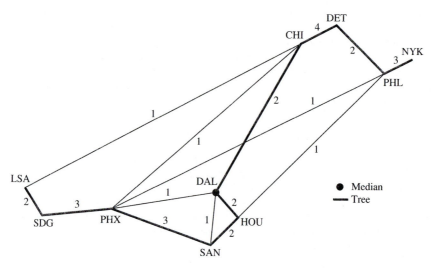

FIGURE 7.8
MENTOR example.

The values of the parameters Wparm, Rparm, Dparm, Alpha, and Slack are all given in Table 7.6. Links are full duplex. Assume 56 Kbps channels utilized at most 60% of the time; thus, the usable capacity is 33,600 bps. There is a uniform requirement of 8000 bps between each pair of nodes (in each direction). While the uniform traffic is not particularly realistic, it makes the example much easier to follow.

Begin by finding a median. In this case, since the traffic is uniform, the median will be the node whose average distance to the other nodes is smallest. Examining Table

7.6, it is apparent that the smallest average (or sum) of costs is in the column (or row) for Dallas. Hence Dallas is chosen as the median.

Next, the threshold algorithm is run to pick backbone nodes. The weight on each node is the total traffic to and from each node. This is 8000 × 18, or 144,000. `Wparm` is 2. When multiplied by the usable capacity, 33,600, this gives 67,200. Thus, all the nodes have weights above the threshold and all are chosen as backbone nodes.

Next, the median finding algorithm is run again, but only considering backbone nodes. Since all the nodes are backbone nodes, again Dallas is chosen as the median.

Next, a spanning tree is selected. Since Alpha is 0, this is a minimum spanning tree. The MST is shown by the heavy lines in Figure 7.8, and happens to be a chain in this case.

Next, a detour point for each node pair is found. Table 7.7, which shows the output of the MENTOR algorithm, gives the detour points for each pair in the column labeled `home`. The tree links have no detour points and thus have a '-' in this column. Note that in each case the detour point is an adjacent point in the tree (chain) and corresponds to the smaller detour. For example, node 9 is the detour point for the (0,6) pair because $cost_{09}+cost_{96}$ (NYK-PHX-SDG) is 9311, which is less than $cost_{04}+cost_{46}$ (NYK-PHL-SDG), which is 9329.

Given the detour points, next, the node pair sequence is found by doing a topological sort. Of course, the endpoints of the chain, nodes 0 and 1, are the first pair. The second pair chosen is (0,2), which has no dependencies. Actually, (0,2) had no dependencies, even at the beginning, since 4 is the detour point for the (0,1) requirement. So, (0,2) could even have been chosen as the first pair. The implementation given below chooses pairs in lexicographic order (ascending numerical order of the nodes in the pairs) and, therefore, chooses (0,1) first. The sequencing algorithm continues, selecting pairs without remaining dependencies, and the ordering shown in Table 7.7 results.

Finally, the direct link selection algorithm, `Compress`, is run. Since the usable capacity is 33,600 and the slack is 0.2, we will accept a direct link whenever a pair has accumulated 26880 (0.8 × 33600) bps of load. Since r_{ij} is 8000 bps for individual pairs, the initial pairs are not expected to qualify and, indeed, they do not. Table 7.7 shows how much load is offered to each pair, which pairs take the overflow, and how much load has accumulated on each overflow pair. For example, the (0,1) requirement does not justify a direct link and overflows to (0,4) and (1,4). The (0,4) requirement, and the (1,4) requirement, are now each 16000. Since the requirements are symmetric, do not distinguish between the (1,4) and (4,1) requirements. Had the requirements not been symmetric, the example would have been somewhat more cumbersome because they would have to be kept track of separately. When the (0,3) requirement, which is still 8000 bps, nothing having been added to it, overflows to (0,4) and (3,4), the (0,4) requirement becomes 24000 (up from 16000).

When the 26th pair is reached, (0,4), a requirement of more than 26880 is finally encountered and put in a direct link. In fact, this is a tree link and there would have been a direct link for it even if the load were below the threshold (as long as there was some load) because there is no overflow traffic from tree links. Since there is 72000 bps of accumulated load, and the usable capacity is 33600, 3 channels are needed. Had this not been a tree link only 2 channels would have been put in, carrying 67200 bps. The remaining 5800 bps is below the threshold and would not have justified the third channel. It would have overflowed via the detour point.

The (7,8) pair is not a tree link and is offered 48000 bps. This justifies a single channel carrying 33600 bps. The remaining 14400 overflows to (3,7) and (7,8).

TABLE 7.7
MENTOR algorithm example (output)

i	j	home	r[i,j]	r[i,h]	r[h,j]	lmult	cost
0	1	4	8000	16000	16000		
0	2	5	8000	16000	16000		
0	3	4	8000	24000	16000		
0	6	9	8000	16000	16000		
0	7	4	8000	32000	16000		
0	8	4	8000	40000	16000		
1	3	8	8000	16000	16000		
1	7	6	8000	16000	16000		
2	4	5	8000	24000	16000		
3	5	2	8000	16000	32000		
3	6	9	8000	16000	24000		
4	6	9	8000	16000	32000		
5	6	2	8000	40000	16000		
5	8	2	8000	48000	16000		
6	8	9	8000	40000	16000		
1	4	5	16000	24000	32000		
0	5	4	16000	56000	48000		
0	9	4	16000	72000	32000		
4	8	3	16000	32000	32000		
1	8	9	16000	24000	32000		
6	7	9	16000	56000	24000		
2	3	7	16000	24000	24000		
3	9	8	16000	48000	48000		
2	8	7	16000	40000	24000		
1	5	2	24000	32000	72000		
0	4	-	72000			3	3693
4	9	5	32000			1	7213
3	4	7	32000			1	5017
1	9	6	24000	40000	80000		
1	2	6	32000			1	6199
5	9	2	8000	80000	16000		
4	7	5	16000	64000	24000		
1	6	-	40000			2	2618
2	6	9	16000	32000	96000		
4	5	-	64000			2	4640
5	7	2	24000	104000	64000		
2	9	7	32000			1	5332
6	9	-	96000			3	5745
2	5	-	04000			4	6832
2	7	-	64000			2	6788
7	9	8	24000	48000	72000		
7	8	3	48000	38400	62400	1 1747	
8	9	-	72000			3	10614
3	7	-	38400			2	3344
3	8	-	62400			2	3134
						Total Cost	72916

The cost of each link (all channels in the link) is given in the cost column in Table 7.7 as is the total network cost, 72916. This is not necessarily the best possible cost, but it is reasonable.

Note that the network is relatively sparse. This is characteristic of the MENTOR algorithm. In some cases, a denser, more reliable network is preferred. This could be achieved by increasing the `Slack` and thereby encouraging more direct links. If `Slack` is increased only slightly, it is possible to get a more reliable network which is only slightly more costly. Also, by increasing the number of direct links there tends to be a reduction in the number of hops (and hence the delay), and an increase in the total amount of usable slack capacity in the network (by putting the slack on a larger number of links).

```
/////////////////////////////////////////////////////////////////
//                    MENTOR ALGORITHM                          //
/////////////////////////////////////////////////////////////////
void <- Mentor( nn , Cost , Req , Wparm , Rparm , Dparm ,
               Alpha , Slack , ChanCap , FDXflag ,
               *Bkbn , *nlinks , *Ends , *Lmult )
     dcl Cost[nn,nn], Req[nn,nn], Weight[nn], Pred[nn],
         Ends[nn,2], Lmult[nn], spDist[nn,nn], spPred[nn,nn],
         Seq[nn*(nn-1)/2]

 // Set the node weights

 SetWeights( nn , Req , *Weight )

 // Find the network median of all the nodes

 Median <- SetMedian( nn , Cost , 0 , Bkbn , Weight )

 // Select backbone nodes

 Thresh( nn , Median , Wparm , Rparm , Dparm , Weight ,
         Cost , Cap , *Bkbn )

 // Find the network median, only considering backbone nodes

 Median <- SetMedian( nn , Cost , 1 , Bkbn , Weight )

 // Find the spanning tree

 PrimDijk( nn , Median , Alpha , Bkbn , Cost , *Pred )

 // Find the distances and predecessors within the tree

 SetDist( nn , Median , Pred , Cost , *spDist, *spPred )
```

```
    // Find the sequence for considering node pairs

    Sequence( nn , spPred , Cost , *Seq , *Home )

    // Select backbone links

    Compress( nn , Seq , Req , Slack , Cap , Home , FDXflag ,
             *nlinks , *Ends , *Lmult )
//////////////////////////
//  Set node weights  //
//////////////////////////
void <- SetWeights( nn , Req , *Weight )
    dcl Req[nn,nn], Weight[nn]

  Weight <- 0
  for_each( i , nn )
     for_each( j , nn )
        Weight[i] += Req[i][j]
        Weight[j] += Req[i][j]

////////////////////////
//  Find a median  //
////////////////////////
int <- SetMedian( nn , Cost , Flag , Bkbn , Weight )
    dcl Cost[nn,nn], Bkbn[nn], Weight[nn]

  BestMoment <- INFINITY
  for_each( i , nn )
     if ( !Flag || (Bkbn[i]==1) )   // Only consider local
        moment <- 0                 // nodes if Flag is 0.
        for_each( j , nn )
           moment += Cost[i][j] * Weight[j]
        if ( moment < BestMoment )
           BestMoment <- moment
           median <- i
  return( median )

/////////////////////////////////////////////////////
//  Threshold backbone node assignment algorithm  //
/////////////////////////////////////////////////////
void <- Thresh( nn , median , Wparm , Rparm , Dparm ,
                Weight , Cost , Cap , *Bkbn )
    dcl Weight[nn], Cost[nn,nn], Bkbn[nn], Merit[nn]

  // Find the cost of connecting the worst pair of nodes
  radius <- 0
  for_each( i , nn )
     for_each( j , nn ; j > i )
        radius <- max( radius , Cost[i][j] )
```

```
// Identify nodes whose weight is above the threshold
Wparm *= Cap
nassg <- 0   // # assigned
for_each( i , nn )
   if ( Weight[i] >= Wparm )
      nassg++
      Bkbn[i] <- B    // Backbone
      merit[i] <- -INFINITY
   else
      Bkbn[i] <- U    // Unassigned
      merit[i] <- Dparm * Cost[i][median] / radius +
                  (1-Dparm) * Weight[i] / Wparm


// Identify nodes within the radius of a backbone node
radius *= Rparm
for_each ( i , nn )
   if ( Bkbn[i] = B )
      for_each ( j , nn )
         if ( (Bkbn[j]= U ) & ( Cost[i][j] <= radius) )
            Bkbn[j] <- L   // Local
            nassg++
            break


// Assign remaining unassigned nodes
while ( nassg < nn ) {
   BestMerit <- F0
   for_each ( i = 0 , nn )
      if ( (Bkbn[i] = U) & (merit[i] > BestMerit) )
         bestI <- i
         BestMerit <- merit[i]
   Bkbn[bestI] <- B
   nassg++
   for ( i = 0 ; i < nn ; i++ ) {
      if ( (Bkbn[i] = U) & (Cost[i][bestI] < radius) )
         Bkbn[i] <- L
         nassg++


/////////////////////////////
//  Tree finding heuristic  //
/////////////////////////////
void <- PrimDijk( nn , median , alpha , Bkbn , Cost , *Pred )
     dcl Bkbn[nn], Cost[nn,nn], Pred[nn], label[nn], InTree[nn]


  Pred <- median
  InTree <- FALSE
  for_each ( i , nn )
     label[i] <- Cost[i][median]
  InTree[median] <- TRUE
```

```
while ( TRUE )
   BestLabel <- INFINITY
   for_each ( i , nn )
      if ( Bkbn[i] & !InTree[i] & label[i]<BestLabel )
         besti <- i
         BestLabel <- label[i]
   if ( BestLabel = INFINITY ) break
   InTree[besti] <- TRUE
   for_each ( i , nn )
      if ( !InTree[i] &
           label[i] > alpha*label[besti] + Cost[i][besti]) )
         label[i] <- alpha*label[besti] + Cost[i][besti]
         Pred[i] <- besti

/////////////////////////////////////////////////
//  Find the order in which to consider node pairs  //
/////////////////////////////////////////////////
void <- Sequence( nn , spPred , Cost , *Seq , *Home )
   dcl spPred[nn,nn], Cost[nn,nn], Seq[np] ,
       Home[nn,nn], nDep[np], Dep1[np], Dep2[np], Pair[np]

// Set up the pair identifiers
npair <- 0
for_each ( i , nn  ;  j , nn )
   Pair[npair] = MakePair( nn , i , j )
   nDep[npair] <- 0
   npair++

// Choose the detour points
// and initialize the dependencies
ndeps <- 0
for_each ( p , npair )
   SplitPair( nn , Pair[p] , *i , *j )
   p1 <- spPred[i,j]   // Pred in shortest path
   p2 = spPred[j,i]    //    through the tree
   if ( p1 = i )   // Tree link
      h <- -1       // No home allowed
   else if ( p1 = p2 )   // 2-hop path; only one possible home
      h = p1;
   else  // Choose cheaper detour
      if ( Cost[i,p1]+Cost[p1,j] <= Cost[i,p2]+Cost[p2,j] )
         h <- p1
      else
         h <- p2
   Home[i,j] <- h
   if ( h == -1 )
      Dep1[ Pair[p] ] <- -1  // Identifiers of pairs dependent
      Dep2[ Pair[p] ] <- -1  //    on this pair (None here)
      continue

   pairih <- MakePair( nn , i , h )
```

```
      Dep1[ Pair[p] ] <- pairih
      nDep[pairih]++
      pairjh <- MakePair( nn , j , h )
      Dep2[ Pair[p] ] <- pairjh
      nDep[pairjh]++


  //  Find the node pair sequence
  //  First include all pairs with no dependencies
  nseq <- 0
  for_each ( p , npair )
     pp <- Pair[p]
     if ( nDep[pp] = 0 )
           Seq[nseq++] <- pp


  //  Now remove dependencies and include newly independent pairs
  iseq <- 0
  while ( iseq < nseq )
     p <- Seq[iseq]
     iseq++
     d <- Dep1[p]
     if ( (d>=0) & (nDep[d]=1) )
        Seq[nseq] <- d
        nseq++
     else
        nDep[d]--
     d <- Dep2[p]
     if ( (d>=0) & (nDep[d]==1) )
        Seq[nseq] <-= d
        nseq++
     else
        nDep[d]--


/////////////////////////////////
//  Form a pair from two nodes  //
/////////////////////////////////
int MakePair( int nn , int i , int j )
  if ( i < j )
    return( nn * i + j )
  else
    return( nn * j + i )


/////////////////////////////////////////
//  Find the component nodes in a pair  //
/////////////////////////////////////////
void SplitPair( int nn , int pair , int *i , int *j )
   i <- pair / nn
   j <- pair % nn
```

```
/////////////////////////////////////////
//  Find shortest paths through the tree  //
/////////////////////////////////////////
void <- SetDist( nn , Median , Pred , Cost , *spDist, *spPred )
      dcl Pred[nn], Cost[nn,nn], spDist[nn,nn], spPred[nn,nn],
          PreOrder[nn], InTree[nn]
  // Find a preordering of the nodes
  n <- 0
  InTree <- FALSE
  InTree[Median] = TRUE;
  PreOrder[n++] = Median
  while ( n < nn )
     for_each ( i , nn )
        if ( (InTree[i]=FALSE) & (InTree[Pred[i]]=TRUE) )
           InTree[i] <- TRUE
           PreOrder[n++] <- i

  // Traverse the nodes in preorder, setting the distances
  for_each ( i , nn )
     j <- PreOrder[i]
     p <- Pred[j]
     spDist[j,j] <- 0
     for_each ( k , i )
        l <- PreOrder[k]
        spDist[j,l] <- spDist[p,l] + Cost[j,p]
        spDist[l,j] <- spDist[j,l]

  // Set the predecessors
  for_each ( i , nn )
     for_each ( j , nn )
        spPred[i,j] <- Pred[j]
     spPred[i,i] <- i
  for_each ( i , nn )
     if ( i = Median ) continue
     p <- Pred[i]
     spPred[i,p] <- i
     while ( p != Median )
        pp <- Pred[p]
        spPred[i,pp] <- p
        p <- pp

///////////////////
//  Select Links  //
///////////////////
void <- Compress( nn , Seq , Req , Slack , Cap , Home , FDXflag ,
      *nl , *Ends , *Lmult )
     dcl Seq[npairs], Req[nn,nn], Home[nn,nn], Lreq[nn,nn],
         Ends[2,nl], Lmult[nl]

  Lreq <- Req
```

```
npairs <- nn * (nn-1) / 2
nl <- 0

for_each ( p , npairs )

   SplitPair( nn , Seq[p] , *n1 , *n2 )
   h <- Home[n1,n2]
   if ( FDXflag = 'F' )
      load = max( Lreq[n1,n2] , Lreq[n2,n1] )
   else
      load = Lreq[n1,n2] + Lreq[n2,n1]
   if( load >= Cap )
      mult <- floor( load / Cap )
      load -= mult * Cap
   else
      mult <- 0

   // Don't overflow from tree links
   // or links with sufficient load
   if ( ((h=-1) & (load>0)) || (load >= (1-Slack)*Cap) )
      mult++
      overflow12 , overflow21 <- 0
   else
      overflow12 <- max( 0 , Lreq[n1,n2] - mult*Cap )
      overflow21 <- max( 0 , Lreq[n2,n1] - mult*Cap )

   // Add a direct link
   if ( mult > 0 )
      Ends[nl] <- n1 , n2
      Lmult[nl++] <- mult

   // Overflow via detour point
   if ( overflow12 > 0 )
      Lreq[n1,h] += overflow12
      Lreq[h,n2] += overflow12
   if ( overflow21 > 0 )
      Lreq[n2,h] += overflow21
      Lreq[h,n1] += overflow21
```

EXERCISES

7.1. Show that for $L_2 > L_1$, if $T_2 < T_1$ then $D_2 > D_1$.

7.2. Using the result from the exercise above, show that $D_2 > D_1$.

7.3. Derive Equation 7.12.

7.4. Derive Equation 7.13.

7.5. Find the average time delay given by the expression in Equation 7.12.

7.6. Find the average time delay given by the expression in Equation 7.13.

7.7. What is the computational comlplexity of the convex hull algorithm?

7.8. Find the spanning tree for the data in Table 7.7 when $\alpha = 0.5$.

TABLE P7.1

Requirements: (bps):

	A	B	C	D	E	F	G
A	0	6083	4054	4851	6207	8685	4030
B	7213	0	4358	5968	3748	3503	5337
C	4663	5208	0	8900	6212	7594	6878
D	7602	7681	7937	0	3911	6752	4888
E	5081	8503	6118	5406	0	6640	7712
F	8589	8219	8199	7047	7550	0	6491
G	5335	5133	4201	7961	5495	5781	0

Costs:

	A	B	C	D	E	F	G
A	400	2836	1110	1767	566	2965	1513
B	2836	400	2133	1633	2706	758	2786
C	1110	2133	400	1198	1003	2255	1627
D	1767	1633	1198	400	1609	1884	1563
E	566	2706	1003	1609	400	2850	1385
F	2965	758	2255	1884	2850	400	3047
G	1513	2786	1627	1563	1385	3047	400

7.9. Draw the dependency graph for the pairs and homes given in Figure 7.8. Give the sequence resulting from choosing reverse lexicographic ordering of the node pairs.

7.10. Try to find a better solution to the example given in Table 7.6.

7.11. Consider the problem described in Table P7.1. Use the MENTOR algorithm to find a mesh network, using alpha = 0.2 and slack = 0.2. Assume all nodes are forced to be backbone nodes. Assume half duplex 56 kbps links with a maximum allowable utiliation of 0.6.

7.12. Consider the problem described in Table P7.1 again. Use the MENTOR algorithm to find where to locate the backbone nodes. Assume 56 kbps links with a maximum utilization of 0.6 and parameters: wparm = 2 , rparm = 0.5 , and dparm = 0.5. (Do not find the links.)

BIBLIOGRAPHY

[FC72] Frank, H., and W. Chou: "Topological optimization of computer networks," *Proceedings of the IEEE*, 60:1385–1397, 1972.

[Fis81] Fisher, M. L.: "The lagrangean relaxation method for solving integer programming problems," *Management Sci.*, 27:1–18, 1981.

[Gav90] Gavish, B.: "Backbone network design tools with economic tradeoffs," *ORSA Journal on Computing*, 2:236–252, 1990.

[Ger73] Gerla, M.: "The Design of Store-and-Forward Networks for Computer Communications," PhD thesis, UCLA, Los Angeles, January 1973.

[GP89] Suruagy Monteiro, J., Gerla, M. and Pazos, R.: Topology design and band-width allocation in ATM networks. *IEEE JSAC*, 7:1253–1262, 1989.

[Han73] Hansler, E.: "An experimental heuristic procedure to optimize a telecommunications network under nonlinear cost functions." In *Proc. Seventh Annual Princeton Conf. Inform. Sci Syst.*, volume 7, pages 130–137, 1973.

[Har75]   Hartigan, J.: *Clustering Algorithms*, John Wiley & Sons, New York, 1975.

[HM79]   Handler, G. Y., and P. Mirchandani: *Location on Networks*, MIT Press, Cambridge, Mass., 1979.

[KK91]   Grover, G., Kershenbaum, A. and P. Kermani: "Mentor: An algorithm for mesh network topological optimization and routing," *IEEE Trans. Commun.*, 39:503–513, 1991.

[Kle75]   Kleinrock, L.: *Queueing Systems, Volume 1: Theory*, Wiley-Interscience, New York, 1975.

[Mar78a]   Maruyama, K.: "Designing reliable packet switched communication networks," *Proceedings of the IEEE ICCC*, pages 493–498, 1978.

[Mar78b]   Maruyama, K.: "Designing reliable packet-switched communication networks," *Proceedings of the IEEE ICCC*, pages 493–498, 1978.

[MGE74]   Chou, W., M. Gerla, H. Frank and J. Eckl: "A cut saturation algorithm for topological design of packet switched communication networks," In *Proceedings of IEEE National Telecommunication Conference*, pages 1074–1085, December 1974.

[MS86]   Monma, C. L., and D. L. Sheng: "Backbone network design and performance analysis: A methodology for pkt switching networks," *IEEE J. Select. Areas Commun.*, 4:946–965, 1986.

[Sch77]   Schwartz, M.: *Computer Communication Network Design and Analysis*, Prentice-Hall, Englewood Cliffs, NJ, 1977.

[V.82]   DeNardo, E. V.: *Dynamic Programming*, Prentice-Hall, Englewood Cliffs, N.J., 1982.

[Whi72]   Whitney, V. K. M.: "Lagrangean Optimization of Stochastic Communication System Models," *Proceedings of the Symposium on Computer Networks and Teletraffic*, pp. 385–395, Polytechnic Institute of Brooklyn, Brooklyn, N.Y., 1972.

[Zad73]   Zadeh, N.: "Construction of efficient tree networks: The pipeline problem," *Networks*, 3:1–31, 1973.

CHAPTER

8

NETWORK
RELIABILITY

8.1 INTRODUCTION

Most of this text has been devoted to describing ways to design networks that meet a given set of requirements at minimum cost. This, naturally, led to techniques for reducing the slack capacity in the networks. In practice, however, it is important that a network contain some slack to allow it to function even if some of its component nodes or links have failed. (Note: Throughout this chapter, unless otherwise explicitly stated, the word **component** is used to mean a specific node or link in the network, not a connected piece of a graph.) Slack, however, should be judiciously placed to maximize the improvement in reliability for a given investment in additional facilities. Thus, the techniques discussed carry forward into this problem as well. The idea is not just to add slack to the network, but to add it where it will do the most good, and only in places where it is needed to meet a given reliability objective.

The first step in approaching this problem is to define quantitative measures of reliability so that it is possible to assess when a network is sufficiently reliable, and to compare alternative topologies with one another.

The following discussion describes a network model. As before, the model of a network is as a set of facilities (i.e, nodes and links). A probability, p_i, is associated with each facility, i. This is the probability that the facility i is working. This probability is based on the mean time between failures (MTBF) and the mean time to repair (MTTR) of the facility. Specifically, if f_i is the MTBF and r_i is the MTTR (see Figure 8.1), then

$$p_i = 1 - \frac{r_i}{f_i} \qquad (8.1)$$

FIGURE 8.1
Definition of MTBF and MTTR.

q_i is the probability that facility i is not working. Clearly, q_i is just $1 - p_i$.

Within this discussion assume that all facilities fail independently of one another. Thus, the probability that both facilities i and j are working is

$$p(ij) = p_i p_j \tag{8.2}$$

and the probability that at least one of facilities i and j is working is

$$p(i|j) = p_i + p_j - p_i p_j \tag{8.3}$$

This is a very simple model. There is no concern about correlations among component failures, which might occur as a result of a software failure or power outage. There is no consideration paid to the distribution of time between failures or time to repair. The mean of these quantities is not even taken into account, only the value of p_i which is a function of their ratio. Nevertheless, this model gives rise to meaningful results which can be used to evaluate networks and compare the reliability of one network with another.

The simplest measure of the reliability of a network is the probability that it is connected. This is the probability that all nodes are working and that there is a spanning tree of working links. Thus, $P_C(G)$, the probability that a network represented by graph G is connected, is given by

$$p_C(G) = \prod_{i=1}^{N} p_i \bigcup_{T \text{ a tree}} \prod_{j \in T} p_j \tag{8.4}$$

where components 1 through N are assumed to be the nodes.

It is evident, even for this simple measure and even with all the simplifying assumptions made, that there is a computational problem here. The last term in Equation 8.4 is the probability that at least one tree is operational. This involves enumerating all the trees in G, which is an exponential amount of effort. It involves taking the union of all these events, which makes things even worse. Fortunately, it is often possible to avoid explicit enumeration. Nevertheless, the problem that remains is still a difficult one and so relatively simple measures of reliability are accepted. Energy is spent simplifying assumptions, especially when the reliability analysis is carried out inside of a design procedure which is evaluating alternate topologies.

The nodes enter into the above expression in a simple way. For the network to work according to the above definition, all the nodes have to work. Thus, the node probabilities enter the expression as a single product. Also, this factor is not a function of the network's topology and so cannot be changed by the design procedures previously discussed. Therefore, many discussions of network reliability focus only on link failures and assume perfectly reliable nodes. Finally, since node failures cannot be compensated for by augmenting the topology, in practice the nodes are often made very reliable by including redundancy within the nodes themselves. Thus, the approximation of perfectly reliable nodes may be a reasonable one.

There are other measures of network reliability, however, which take node failures into account. One is the probability that all working nodes are connected. In this case, node failures enter into the computation of network reliability in a more complex way. When a node fails, all the links incident on it, by definition, fail too. This, in turn, may lead to working nodes becoming disconnected from one another. This measure is discussed in the next section.

Sometimes, a subset of the nodes is designated as critial (e.g., the host computers in the network or its backbone) and the network's reliability is defined as the probability that these critical nodes can all communicate. This problem is known as the **k-terminal reliability problem**. An important special case of this is the **2-terminal reliability problem** (i.e., the probability that a given pair of nodes can communicate). In these cases, the effect of node failures is again non-trivial. Although it is not demanded that non-critical nodes work, their failure affects the probability that the critical nodes can communicate. In this case, $P_C(G, S)$, the probability that all nodes in set S (the critical nodes) can communicate is given by

$$p_C(G, S) = \prod_{i \in S} p_i \bigcup_{T \text{ a tree on } S} \prod_{j \in T} p_j \tag{8.5}$$

which is a generalization of Equation 8.3.

The most realistic measures of reliability are traffic related (e.g., the probability that the surviving portion of the network can still carry a given fraction of the traffic). Such measures involve the component failure probabilities, as well as the routing. It is apparent that the problem of determining if a network satisfies a given traffic matrix is itself a non-trivial problem that involves the computation of optimal routes in the most general case.

On the other hand, if the routes are given in advance (e.g., when fixed routing is used), the problem is greatly simplified because it only involves computing the probability that the given routes are working. Even when several routes are available for each reqirement, the 2-terminal reliability problem is still quite reasonable. This will be discussed in a later section.

8.2   RELIABILITY OF TREE NETWORKS

Large networks are usually comprised of a collection of local access networks tied together by a backbone. The local access networks are often trees. Thus, the analysis of the reliability of a large network can often be accomplished by breaking down the problem into several problems on trees, and a single problem on a relatively small mesh backbone. Most of the nodes in the network are in the local access portion. Also, centralized networks, where all nodes communicate to or through a single central point, are often entirely trees. It is therefore important to be able to efficiently analyze the reliability of tree networks.

Since there is only a single path between every pair of nodes in a tree, the failure of any component node or link in a tree disconnects the network. Therefore, trees are relatively unreliable. This makes the analysis of just how unreliable a tree is all the more important because if a requirement for reliability is present, it is most likely to be violated in the tree portion of the network. Fortunately, this same feature that makes trees unreliable also makes them easier to analyze. Therefore, it is possible to find out a great deal about the reliability of trees.

The thing that makes the analysis of the reliability of general networks complex is that there are many paths between each pair of nodes. Any path would allow the pair of nodes to communicate and so it is necessary, in some sense, to enumerate the probabilities of all possible paths working.

However, if the network is a tree this problem is eliminated. There is only one path between every pair of nodes and so it is relatively easy to compute the reliability of tree networks. For example, the probability of a tree network, T, being connected is simply the probability that all components are working; i.e.,

$$p_C(T) = \prod_{i=1}^{N} p_i \prod_{j=1}^{M} p_j \tag{8.6}$$

where there are N nodes, i, and M links, j. Here, the union of many events in Equation 8.3 is replaced by a single product since the network contains only one tree.

Equation 8.6 can be evaluated iteratively. The reliability of a tree comprised of a single node is simply the reliability of that node. As a link and another node is added to the tree, the probability of the network remaining connected is multiplied by the probabilities that the new node and link are also working. Continuing in this manner, the probability that the entire tree is connected is the product that all the nodes and links in the tree are working.

Equivalently, it is possible to evaluate Equation 8.2 recursively. If i is a leaf (pendant node) of a tree, T, and j is the link connecting i into T, then

$$p_C(T) = p_i p_j p_C(T - i) \tag{8.7}$$

where $T - i$ is the tree with node i and link j removed. Thus, it is possible to evaluate the reliability of T in terms of the smaller tree, $T - i$. Continuing in this manner, the problem is broken down to one of finding the reliability of a single node.

This recursive view is useful in evaluating other measures of reliability for trees [KS73]. The tree is modeled as a **rooted tree** [Knu73]. A rooted tree (Fig. 8.2) has one distinguished node, r, which is called the **root**. The root may be the central site in a centralized network or the backbone node tying the tree into the rest of a multilevel distributed network. Although the links are bidirectional from the point of view of communications, computationally they are viewed as directed towards the root. If the tree contains a link, (i, j) and j is on the path from i to r, then it is said that j is a **predecessor** of i and i is a **successor** of j. Note that each node (except r) has

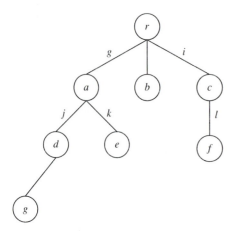

FIGURE 8.2
A rooted tree.

a unique predecessor but that nodes can, in general, have more than one successor. Nodes with no successors are leaves.

Suppose the objective is to find $p_c(i)$, the probability that a node, i, can communicate with r. If the predecessor of node i is node k and the link between them is link j, then

$$p_c(i) = p_i p_j p_c(k) \qquad (8.8)$$

Finally

$$p_c(r) = p_r \qquad (8.9)$$

$E(r)$, the expected number of nodes communicating with r can now be found by summing $p_c(i)$ over all i,

$$E(r) = \sum_i p_c(i) \qquad (8.10)$$

This calculation is $O(Nd)$, where N is the number of nodes in the tree and d is its average depth (number of levels). We can do much better, however, by viewing this recursion in a slightly different way. The expected number of nodes communicating with any node, i, in general, and with the root, r, in particular, is simply the sum of the expected number of nodes communicating with each of the successors of i, multiplied by the probabilities of these successors communicating with i; i.e.,

$$E(i) = p_i + p_i \sum_k p_{j_k} E(k) \qquad (8.11)$$

where $E(k)$ is the expected number of nodes communicating with k (including k itself), j_k is the link connecting k to its predecessor and the sum is taken over nodes which are successors of i. Starting from the leaves and working toward the r, it is possible to evaluate $E(r)$ in $O(N)$ steps, since Equation 8.11 must be used N times, (once to evaluate $E(k)$ for each k) and Equation 8.11 is itself of $O(1)$ (constant order). Note that if i is a leaf, it has no successors and the sum disappears. Equation 8.11 then becomes

$$E(i) = p_i$$

for i a leaf.

Example 8.1. Consider the tree shown in Figure 8.2. Suppose that $p_i = 0.9$ for all nodes, i, and $p_j = 0.8$ for all links, j. Then

$$E(b) = E(e) = E(f) = E(g) = 0.9$$

$$E(c) = 0.9 + (0.9)(0.8)E(f) = 1.548$$

$$E(d) = 0.9 + (0.9)(0.8)E(g) = 1.548$$

$$E(a) = 0.9 + (0.9)(0.8)E(d) + (0.9)(0.8)E(e) = 2.663$$

$$E(r) = 0.9 + (0.9)(0.8)E(a) + (0.9)(0.8)E(b) + (0.9)(0.8)E(c) = 4.580$$

This same result can be obtained using Equations 8.8–8.10.

$$p_c(r) = 0.9$$

$$p_c(a) = p_c(b) = p_c(c) = (0.9)(0.8)p_c(r) = 0.648$$

$$p_c(d) = p_c(e) = (0.9)(0.8)p_c(a) = 0.467$$

$$p_c(f) = (0.9)(0.8)p_c(c) = 0.467$$

$$p_c(g) = (0.9)(0.8)p_c(d) = 0.336$$

$$E(r) = (0.9) + (0.648) + (0.648) + (0.648) + (0.467) + (0.467) + (0.467) + (0.336) = 4.580$$

The same recursive methodology can be used to compute more complex measures of reliability. Thus, if the objective is to find $EPR(r)$, the expected number of node pairs communicating through the root, r, there is:

$$EPR(r) = \sum_{i,k} p_r p_{j_i} E(i) p_{j_k} E(k) + \sum_i p_r p_{j_i} E_i \qquad (8.12)$$

where the sum runs over all pairs of successors, i and k, of r, but r is not counted as a member of a pair.

If the objective is to find $EP(r)$, the expected pairs communicating in a tree network where nodes can communicate directly through the shortest path in the tree (i.e., without going through r if they are in the same subtree), two quantities must be maintained. As before, it is necessary to keep track of $E(i)$, the expcted number of nodes communicating with node i. Now, also keep track of $EP(i)$, the expected number of pairs communicating in the tree rooted at i.

$E(i)$ is computed as before using Equation 8.11. $EP(i)$ is computed in a manner similar to Equation 8.11 except a term is added for the pairs commmunicating via the successors of each node. Thus,

$$EP(i) = \sum_{m,k} p_i p_{j_m} E(m) p_{j_k} E(k) + \sum_m p_i p_{j_m} E_m + EP(m) \qquad (8.13)$$

Example 8.2. Consider the tree shown in Figure 8.2 again, and again suppose that $p_i = 0.9$ for all nodes, i, and $p_j = 0.8$ for all links, j. Then

$$EPR(r) = 0.9.8.8(E(a)E(b) + E(a)E(c) + E(b)E(c))$$

$$= 0.576(2.196 * 0.9 + 2.196 * 1.548 + 0.9 * 1.548) = 3.899$$

$$E(b) = E(d) = E(e) = E(f) = 0.9$$

$$E(a) = 0.9 + (0.9)(0.8)E(d) + (0.9)(0.8)E(e) = 2.196$$

$$E(c) = 0.9 + (0.9)(0.8)E(f) = 1.548$$

$$E(r) = 0.9 + (0.9)(0.8)E(a) + (0.9)(0.8)E(b) + (0.9)(0.8)E(c) = 4.24368$$

This same result can be obtained using Equations 8.10–8.12:

$$p_c(r) = 0.9$$

$$p_c(a) = p_c(b) = p_c(c) = (0.9)(0.8)p_c(r) = 0.648$$

$$p_c(d) = p_c(e) = (0.9)(0.8)p_c(a) = 0.46656$$

$$p_c(f) = (0.9)(0.8)p_c(c) = 0.46656$$

$$E(r) = (0.9) + (0.648) + (0.648) + (0.648) + (0.46656) + (0.46656) + (0.46656) = 4.24368$$

An implementation of the recursive algorithm for tree network reliability is given below. It begins by calling `PostOrder` to produce an ordering of the nodes starting from the leaves and working back toward the center. There are many ways of implementing `PostOrder`. The simplest is to do a breadth first search of the tree, as described in Chapter 3, recording the nodes in the order they are visited. This produces a preordering of the nodes, where the root of each subtree is visited before the rest of the nodes in the subtree. Then, the ordering is reversed to produce a postorder. A preliminary pass over the links can form an adjacency list for each node if such a list does not already exist. `PostOrder` also returns `PredNode` and `PredLink`, the node and link preceding each node in the path back to the center. The entire procedure is $O(N)$.

Given the postorder, the next step is to apply the appropriate equations (e.g. Equations 8.11–8.13) at each of the nodes, visiting the nodes in postorder. A "state" for each node is kept giving the relevant reliability information associated with this node. In this case, values are kept of the expected number of node pairs communicating through each node (directly) and through the center. These values are initialized to zero and updated for the predecessor of each node as the algorithm proceeds. Similarly, the expected number of nodes communicating with each node is maintained.

Note that it is important to update the elements of the state vector in the proper order. In particular, update the pair information before updating the expected number information. If this is not done, the nodes in the subtree being considered would be counted as pairs communicating with themselves.

This same approach can also be used to compute other reliability measures [KS73] by keeping additional information in the state vector. Since each node is visited once, the entire procedure is $O(N)$.

```
/*-----------------------------
   Find the reliability of a tree
   --------------------------*/
float <- TreeRel( center , Ends , Pnode , Plink , *Econn ,
                 *EPcomm , *EProot )
     dcl Ends[N][2]    /* Endpoints of the links in the tree    */
```

```
   Pnode[N],       /* Prob. that each node works.          */
   Plink[N],       /* Prob. that each link works.          */
   PostOrder[N],   /* Nodes in post-order                  */
   PredNode[N],    /* Predecessor node of each node        */
                   /*    in the tree                       */
   PredLink[N],    /* Predecessor link of each node        */
                   /*    in the tree                       */
   Ecomm[N],       /* Expected number of nodes             */
                   /*     communicating with i             */
   EPcomm[N],      /* Expected number of pairs             */
                   /*     communicating in the subtree     */
                   /*     rooted at i                      */
   EProot[N]       /* Expected number of pairs in the      */
                   /*     subtree rooted at i communicating */
                   /*     via i                            */

post_order( center , Ends , PostOrder , Pred )

Pconn <- Pnode[center]
for each ( i , N )      /* Initialize */
   Ecomm <- Pnode
   EPcomm <- 0
   EProot <- 0
for each ( i , N-1 ){/* Visit nodes (not root) in postorder */
   j <- PostOrder[i]
   p <- PredNode[j]
   l <- PredLink[j]
   Pconn <- Pconn * Pnode[j] * Plink[l]
   EPcomm[p] += Ecomm[j] * Ecomm[p] * Plink[l] + EPcomm[j]
   EProot[p] += Ecomm[j] * Ecomm[p] * Plink[l]
               + EProot[j] * Pnode[p] * Plink[l]
   Ecomm[p]  += Pnode[p] * Plink[l] * Ecomm[j]
   return( Pconn )
```

This same methodology can be extended further in a number of ways. First, other reliability measures can be considered. It may be desirable to know the probability that all working nodes can communicate. This is a measure that focusses more on the reliability of the network topology (which can be altered by the procedures discussed in this book) rather than the reliability of the network nodes (which cannot be altered by changing the network topology). Similarly, the objective may be to find the expected number (or fraction) of working node pairs which can communicate. These measures are discussed in [KS73].

Another generalization is to consider traffic weighted measures. It is a serious problem if a node which is the source or destination of a lot of traffic is separated from the rest of the network. Likewise, it is more serious if a major portion of the network, as opposed to a single node, is split off by a failure. The latter case is, to some extent, dealt with by measures that compute the expected number of nodes or node pairs communicating. The former measure is not. It is straight forward, however, to extend the measures and algorithms above to such cases by associating a weight with each node.

Thus, if w_i is the weight (e.g., total traffic) associated with node i, it is possible to extend Equation 8.11 to compute the expected weight of the nodes communicating with i:

$$E(i) = p_i w_i + p_i \sum_k p_{j_k} E(k) \tag{8.14}$$

The measures for node pairs can be similarly extended.

Finally, these methods can be extended to "tree-like" networks. Often, the reliability of a tree is unacceptable but so is the cost of a general mesh network. It is possible then to augment a tree by adding a small number of links to increase reliability without greatly increasing cost. The simplest example of this is the use of rings (simple cycles) both as backbones and as local access topologies.

A cycle is just a chain (which is a tree) with a single link added to it. Its reliability can be analyzed by breaking the problem into two cases, both of which can be analyzed. Let C be a network which is a single simple cycle (see Figure 8.3) and

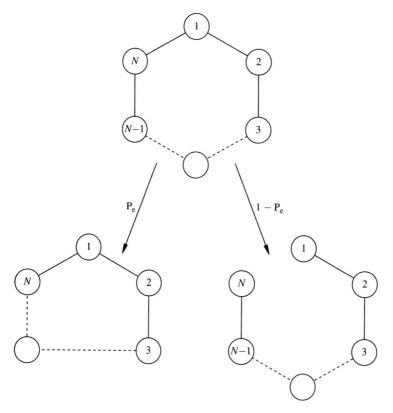

FIGURE 8.3

Analysis of a ring.

let e be an edge in C. $p_c(C)$, the probability that C is connected, can be computed as

$$p_c(C) = p_e p_c(C/e) + (1 - p_e)p_c(C - e) \tag{8.15}$$

where (C/e) is C with edge e contracted (i.e., the endpoints of e are collapsed to a single node) and $C - e$ is C with e removed. Note that in (C/e) a compound node is formed. The probability of this node working is the product of the probabilities of its component nodes working.

Therefore, the problem of evaluating a ring with N nodes is replaced by the evaluation of a chain (a tree), and the evaluation of another ring (with one fewer node). If $T(N)$ is the order of complexity to evaluate a ring of N nodes, then

$$T(N) = T(N - 1) + O(N) \tag{8.16}$$

and

$$T(1) = O(N) \tag{8.17}$$

So, $T(N)$ is $O(N^2)$. It is possible to extend this approach still further to handle networks with several rings, but the complexity increases by an order of N for every ring added. Thus, this method is practical only for small networks or networks with very few rings. This approach is an example of analyzing a network by transforming it into one or more simpler networks. This method is explored further in the following section.

8.3 MESH NETWORK RELIABILITY ANALYSIS

Now turn to the problem of analyzing the reliability of general mesh networks, given the probabilty of failure of the component nodes and links. For the sake of clarity, begin by considering the simplest problem, assuming that failures occur independently, that nodes are perfectly reliable, and that the primary interest is in the probability that the network is connected.

8.3.1 The Reliability Polynomial

The analysis of the reliability of a network can be thought of as an enumeration of the states in which the network is working and the states in which it has failed. Since each component may be working or failed, there are 2^C possible states for a network with C components (links, in the case where nodes are considered to be perfectly reliable.) Thus, explicit enumeration is not practical except for very small networks.

If all links fail with the same probability, however, we can organize this enumeration and obtain a great deal of information about the network's reliability with comparatively little effort. It is also possible to make important observations about the reliability of networks in general.

Let P_W be the probability that the network is working (i.e., the probability that it is connected in the case being considered here.) P_W is given by

$$P_W = \sum_{k=0}^{N} N_W(k) p^k (1 - p)^{(C-k)} \tag{8.18}$$

where p is the probability that a link is working and $N_W(k)$ is the number of working states (i.e., states where the network is deemed functional) with k links working and $C - k$ links failed. The problem is then to determine $N_W(k)$ for each. Determining $N_W(k)$ exactly is difficult but it is possible to observe a great deal about its value fairly easily. Equation 8.18 is known as the **reliability polynomial** and can be used to determine the reliability of the network as a function of p.

First,

$$N_W(k) \leq A_{ck} = \frac{C!}{k!(C - k)!} \tag{8.19}$$

If $N_W(k)$ is replaced by A_{ck} in Equation 8.18, there is a binomial distribution; i.e:

$$1 = \sum_{k=0}^{N} A_{ck} p^k (1 - p)^{(C-k)} \tag{8.20}$$

This represents the case where the network is considered to be working in all states. In this case, P_W is one. While this case is not realistic, the distribution of probability mass among the terms on the right hand side of Equation 8.20 explains a great deal about the range of values that P_W can take. For example, if the network is 2–connected (i.e., cannot be disconnected by the failure of fewer than two links), then:

$$P_W \geq 1 - \sum_{k=2}^{N} N_W(k) p^k (1 - p)^{(C-k)} \tag{8.21}$$

For p close to 1, the terms on the right hand side fall off very quickly as a function of k. Specifically, if $Cp(1 - p) < 1$, then the terms fall off faster than $1/k!$. Thus, most of the probability mass is in the first few terms of the sum in Equation 8.20. As p gets smaller or C gets larger, this probability mass spreads over a larger number of terms.

This leads to a widely used rule of thumb for the design of reliable networks. Bounds like the one in Equation 8.21 are computed for small values of k until one that is acceptable is found and then make the network k-connected.

This is, of course, very conservative. It would also be sufficient to make the values of $N_W(k)$ large enough for each value of k. It is, however, hard to determine $N_W(k)$ in general and so, in practice, designers often rely on connectivity, which is easier to determine.

Similarly, it may be possible to obtain an upper bound on reliability from the reliability polynomial. If a network has N nodes and M links, then

$$N_W(k) = 0 \quad \text{for } k > M - (N - 1) \tag{8.22}$$

since at least $N - 1$ links are needed to connect the network. When M is close to N, this bound is significant. The most extreme case is when the network is a tree. In this case only one term (the term with all links working) contributes to the probability that the network is connected. Also, in this case $N_W(C)$ is known; it is one.

Thus, based on these bounds and the relationship between p and the number of nodes, it is possible to form conclusions about how highly connected a network should be. This also focuses efforts to increase the reliability of a given network. The objective is to increase $N_W(k)$ for high k, since these terms contribute the most to P_W. The most obvious way of doing this is to 2-connect the network, thus maximizing $N_W(C)$ and $N_W(C-1)$.

Knowledge of the structure of the reliability polynomial, if not all of its coefficients, also helps to focus analysis efforts. Several of the techinques described in the following sections are guided by properties of the reliability polynomial. Stratified sampling [VF72] based on knowledge of the structure of the reliability polynomial is an effective way of obtaining information about network reliability via simulation without an excessive amount of computation.

8.3.2 State Space Decomposition

One approach to finding the reliability of a general mesh network is to decompose the probability space, considering the case that each component is, alternatively, working or failed. Thus, if N is a network and C is a component (a link here) of N, we have

$$P_W(N) = P(C)P_W(N|C) + (1 - P)(C))P_W(N - C) \qquad (8.23)$$

where $(N|C)$ is N with C working and $(N - C)$ is N with C failed.

This is a statement that the events "C working" and "C failed" are mutually exclusive, and exactly one of them is always true. Thus the probability that N is working can be found by decomposing the possible states of the network into those where C is working and those where C has failed. In the case examined here, C is a link; and the nodes are perfectly reliable. The technique extends naturally to the case of imperfect nodes.

It is possible to use Equation 8.23 recursively to evaluate $P_W(N)$. The recursion can be terminated whenever it is determined that $(N|C)$ is working or $(N - C)$ has failed. This determination is made based on the criterion for the network working. Suppose the objective is to find the probability that two specific nodes, `src` and `dst` can communicate; this is a 2-terminal reliability problem. A network is working, then, if a path comprised entirely of working links exists from `src` to `dst` and has failed if a cut comprised entirely of failed links exists between those nodes. Thus, whenever it is determined that a path or cut exists, it is possible to halt the recursion.

> **Example 8.3.** Consider the network shown in Figure 8.4. The objective is to find the probability that S can communicate with T. Suppose that the nodes are perfectly reliable and that links work with probability 0.9.
>
> A decompositon of the state space along the lines described above is shown in Figure 8.5. The root node (at the top of the figure) represents the entire event space, where the status of all links (working or failed) is open. The probability associated with this node is 1.0. The node itself is open (i.e., the network may be working or it may have failed).
>
> Since this node is still open, it is broken down into two events, the event where link 1 is working and all other nodes are open, and the event where link 1 has failed and all other links are open. These events are the nodes marked 1 and 1', respectively, in Figure 8.5. The probabilities of these two events are 0.9 and 0.1, respectively. Both nodes are open, since the events still do not include either an S,T-path or an S,T-cut and so it is still not known whether or not the network is working.

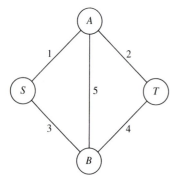

FIGURE 8.4
Two terminal reliabilities.

The next step is to decompose the node marked 1, based on the status of link 2. This creates two more nodes in the state decomposition. These nodes emanate from node 1 and are referred to as nodes 12 and 12', respectively. The first corresponds to the event that links 1 and 2 are both working and the status of all other nodes is still open. This event, at last, contains an S,T-path and is marked with success. The probability associated with it is 0.81. This contributes to the network's probability of success. We mark it with a '+' and close it.

The other node, referred to as 12', corresponds to the event that link 1 is working and link 2 has failed. This node is open and it is necessary to expand it further. It has a probability of 0.09 (0.9 for link 1 working times 0.1 for link 2 failing).

Proceeding in this manner, 25 nodes are generated in the state space tree. The leaves of this tree all correspond to closed nodes, marked with '+' for success and '−' for failure. The probabilities of the nodes marked '+' sum to 0.97848. The probabilities of the nodes marked '−' sum to 0.02152. It is apparent that all of the nodes at levels 1 and 2 (i.e., nodes with 1's and 2's in them) were generated. Only 75% of the eight possible nodes at level 3 were generated. Only 50% of the nodes at level 4 were generated. Finally, only 12.5% of the nodes at level 5 were generated. It is typical that the deeper into the tree traveled, the lower the percentage of nodes actually generated.

The question is whether the pruning done by closing nodes containing paths or cuts is sufficient to overcome the exponential rate of growth of the number of nodes by level. The answer is, unfortunately, usually no. There are several things that can be done, however.

First, it is possible to be more selective about the order of expansion. In the example in Figure 8.5, the links were simply expanded in numerical order. It is not necessary to do this. In fact, it is not necessary to expand the nodes in the same order in different parts of the state space tree. Figure 8.6 shows an alternative state space expansion for the network of Figure 8.4. This expansion choses links to expand based on the heuristic that whenever it was possible to choose a link which immediately formed a path or cut, do so. This results in an expansion with 19 nodes instead of 25. While this heuristic does not guarantee an expansion with a minimum number of nodes, it tends to decrease the number of nodes significantly. It also, of course, requires additional effort to select the link at each step. In the end, however, it is usually not enough to overcome the exponential rate of growth of the number of nodes by level.

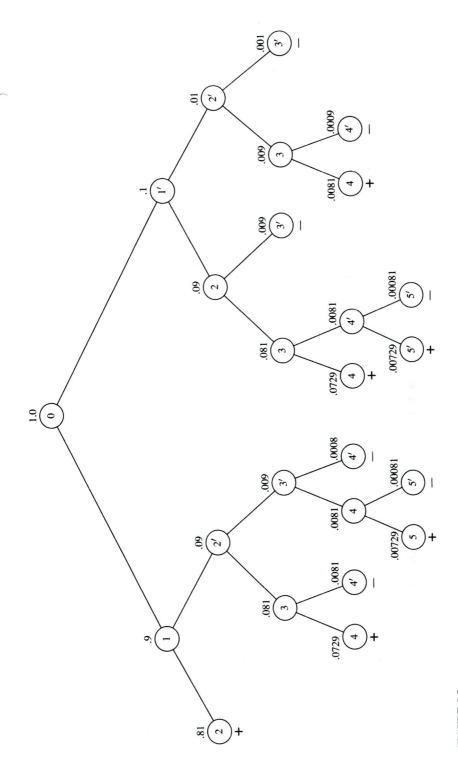

FIGURE 8.5
State space decomposition.

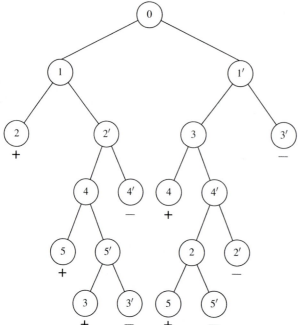

FIGURE 8.6
Tighter decomposition.

It is possible to control the complexity of this procedure, however, if it is sufficient to accept an approximate answer. It is possible to halt the expansion at any point, with some nodes still open. In this case, there are still P_s and P_f, the probabilities of success and failure, respectively. Now there is also P_o, the probability which is still open (i.e., which is not known to be part of either the success or failure probability). Thus, there are bounds on the network's reliability. Specifically, it is understood that the true probability of success is no less than P_s, and is no more than $P_s + P_o$. Similarly, the true probability of failure is between P_f and $P_f + P_o$. As more nodes are expanded, P_o becomes smaller and the bound gets tighter.

A closer examination of Figures 8.5 and 8.6 shows that when the probability of a link working is close to 1.0 that most of the probability is concentrated in a small number of states. In particular, a state's probability falls off quickly as the number of failed states in it increase. Indeed, it is known that PS_{wf}, the probability of a state with w links working, and f states failed is given by

$$PS_{wf} = p^w(1 - p)^f \tag{8.24}$$

where p is the probability of a link failing. This gets small quickly as f increases since $(1 - p)$ is close to zero. Note that this is the same kind of result obtained from examining the reliability polynomial.

This leads to an algorithm where states are expanded based on their probabilities, largest first, in an effort to account for as much probability as quickly as possible. It is important to keep track of Sval, the value of the probability associated with each state and maintain a heap, Vheap, on these values. Vheap is maintained on the values of Sval with the largest value at the top of the heap. Init initializes the heap to empty, Push adds another element to the heap, Pop removes the value from the top of the heap and returns an index of a state, and Empty returns TRUE if the heap is empty.

The states themselves are maintained as bit vectors of length C, the number of components (i.e., the number of links, in this case.) Slinksin[s] has a 1 in position k if link k is working in state s. Similarly, Slinksout[s] has a 1 in position k if link k is failed in state s. Snodes[s] has a 1 in position k if there is a path of working nodes connecting node k to the source, src. Turning on individual bits is facilitated by creation of the bit vector bits which has a 1 in position k of bits[k].

There is, as input, a vector, Plink of the probabilities of each of the links working, and adj, a vector of lists of links adjacent to each node. There are also ends, the endpoints of the links, Pthresh the smallest state probability the algorithm will consider, and (src,dst), the two terminals for which the network's reliability is tested.

Begin with a single state with all links open and with only src connected to itself. The probability value associated with this state is one. Also initialized (to zero) are pWork and pFail, the probabilities of the network working or failing.

Proceed by popping the state with the largest value off the heap. If a path exists from src to dst, which can be tested by checking if the bit corresponding to dst is set to one in Snodes[state], pWork is updated. If not, begin a search for a link (node,node2) such that a path of working links exists from src to node, but no such path exists to node2. If no such link can be found, there is a cut of failed links and pFail is updated. If there is such a link, however, then create two new states, one with the link forced in (i.e., working) and one with the link forced out (i.e., failed).

In addition to returning pWork and pFail, the algorithm also returns the status of each state as '+', '-', or ' ', for working, failed, or open, respectively.

```
/*-------------------------------------------------------
  Find the probability that src and dst can communicate
  -----------------------------------------------------*/

float,float <- mesh_net_rel1( nn , nl , ends , adj , Plink ,
                              Pthresh , src , dst , *status )
  dcl ends[nl,2], Plink[nl], adj[nn,list] , bits[nl,set] ,
      status[NS,char] , Snodes[NS,set], Slinksin[NS,set],
      Slinksout[NS,set], Sval[NS], Vheap[NS,heap]

  for each( i , nl )
    bits[i] <- 2^i

  ns <- 1
  Snodes[ns] <- bits[src]   /* Only src has been reached */
  Slinksin[ns] <- 0
  Slinksout[ns] <- 0
  Init( NS , Vheap , Sval , "MAX" )
  Sval[ns] <- 1.0
```

```
Push( ns , Vheap , Sval )
pWork <- 0.0
pFail <- 0.0
while( ns < MAXNS && !Empty(Vheap) )
   state <- Pop( ns , Vheap , Sval ) /* find the open state */
                                     /* with the max prob    */
   if( Sval[state] < Pthresh ) break
   if( (Snodes[state]&dst) != 0 )   /* (A) Found a path      */
      pWork += Sval[state]          /* Increment success     */
                                    /* probability           */
         status[state] <- '+'       /* Mark this state as a   */
                                    /* success state         */

      else
         node2 <- -1
         for each( node , nn )
            if( Snodes[state] & bits[node] ) /* node          */
                                    /* communicates with src */
               for each( link , adj[node] )
                  node2 <- OtherEnd( node , link )
                  if( ( (bits[link] & Slinksout[state]) = 0) &&
                      ( (bits[node2] & Snodes[state]) = 0) )
                     bestlink <- link
                     bestnode2 <- node2
                     break
         if( node2 = -1 )
            pFail += Sval[state] /* No link found; have a cut. */
            status[state] = '-'  /* Mark this state as a       */
                                 /* failure state              */
      else

         /* Form the state with bestlink forced in */

         ns ++
         Slinksin[ns] <- Slinksin[s] | bits[bestlink]
         Slinksout[ns] <- Slinksout[s]
         Snodes[ns] <- Snodes[s] | bestnode2 /* bestnode2    */
                                    /* has been reached */
         status[ns] <- ' ' /* Status of state is still open */
         Sval[ns] <- Sval[s] * Plink[bestlink]
         Push( ns , Vheap , Sval )

         /* Form the state with bestlink forced out */

         ns++
         Slinksout[ns] <- Slinksout[s] | bits[bestlink]
         Slinksin[ns] <- Slinksin[s]
         Snodes[ns] <- Snodes[s]
         status[ns] <- ' '
         Sval[ns] <- Sval[state] * ( 1 - Plink[bestlink] )
         Push( ns , Vheap , Sval )
```

This procedure can be extended in a straight forward way to the all terminal reliability problems. In this case, start from any node, rather than from a given source. The state space representation and the criterion for failure (cut) remain the same. the only thing that changes is the criterion for success. In this case, there is a demand that all the bits in `Snodes` be set to 1, indicating that all nodes can communicate (i.e., that a tree of working nodes has been found).

Example 8.4. Returning to Figure 8.5, start with state 0 with probability 1.0. This is an open state which generates states 1 and 1′, with probabilities 0.9 and 0.1, respectively. Both of these are also open states. Next, state 1 (with the highest probability is explored, generating states 12 and 12′. State 12 is a success state, so mark it as such and update `pWork`. Continuing in this manner, exploring states in decreasing order of probability, Table 8.1 below is obtained.

Note that by the time state 12′34′ is reached there is a reasonable estimate of the network's reliability. In general, it is possible to trade as much accuracy as necessary for increased speed.

A close examination of Table 8.1 shows that most of the probability mass collected is due to a small number of paths and cuts which are found early in the procedure.

TABLE 8.1
State space decomposition

State	Slinksin	Slinksout	Prob.	Status	pWork	pFail
0	00000	00000	1.0		.00000	.00000
1	00001	00000	.9		.00000	.00000
12	00011	00000	.81	+	.81000	.00000
1′	00000	00001	.1		.81000	.00000
12′	00001	00001	.09		.81000	.00000
1′2	00010	00001	.09		.81000	.00000
12′3	00101	00010	.081		.81000	.00000
1′23	00110	00001	.081		.81000	.00000
12′34	01001	00010	.0729	+	.88290	.00000
1′234	00111	00001	.0729	+	.95580	.00000
1′2′	00000	00011	.01		.95580	.00000
12′3′	00001	00110	.009		.95580	.00000
1′23′	00010	00101	.009	−	.95580	.00900
1′2′3	00100	00011	.009		.95580	.00900
12′34′	01101	00010	.0081	−	.95580	.01710
12′3′4	01001	00110	.0081		.95580	.01710
1′234′	00110	01001	.0081		.95580	.01710
1′2′34	01100	00011	.0081	+	.96390	.01710
12′3′45	11001	00110	.00729	+	.97119	.01710
1′234′5	10110	01010	.00729	+	.97848	.01710
1′2′3′	00000	00111	.001	−	.97848	.01810
12′3′4′	00001	01110	.0009	−	.97848	.01900
1′2′34′	00100	01011	.0009	−	.97848	.01990
12′3′45′	01001	01101	.00081	−	.97848	.02071
1′234′5′	00110	11001	.00081	−	.97848	.02152

Methods based on finding paths and cuts have been used successfully to find estimates of network reliability [VF72] , [Mur92].

8.3.3   Graph Reduction

Another method for finding the reliability of a network, in the sense of whether all or some of the nodes can communicate, is to replace the graph, G, representing the network by one or more simpler graphs with the same reliability. This process is called **graph reduction** and it is useful in the analysis of many others types of problems on networks as well (e.g., circuit analysis).

One of the simplest types of reduction is **parallel reduction**, where two links in parallel are replaced by a single link of equivalent probability. This is illustrated in Figure 8.7a. If there are parallel edges between two nodes and the edges work with probabilities p_1 and p_2, respectively, then they can be replaced by a single edge whose probability of working is equal to the probability that either (or both) of the edges is working. Thus, assuming (as we always do) that the edges fail independently,

$$P_3 = P_1 + P_2 - P_1 P_2 \qquad (8.25)$$

It is possible then to reduce the original graph, G, to a new graph G', with a single edge, with working probability p_3 replacing the two edges in G. G' has the same reliability as G. Thus, it is possible to analyze G' instead of analyzing G. G' is a simpler graph than G; it has one fewer edge. Therefore, the problem of analyzing G has been reduced to that of analyzing G'. This reduction works for the 2-terminal, k-terminal and all-terminal reliability problems.

For the 2-terminal and k-terminal problems, where node a is not a target, the **series reduction** shown in Figure 8.7b can also be used. In this case G is reduced by collapsing two edges in series, and the node between them into a single edge with equivalent probability. Here the new edge works with probability p_3, which is equal to the probability that all of the removed components work; i.e.,

$$P_3 = P_1 P_2 P_a \qquad (8.26)$$

In the all-terminal problem, the previous transformation is not valid. The removed node (node b in Figure 8.7c) must also be able to communicate with the other nodes. Thus, if the original graph, G, is to be transformed into a simpler graph, G', with node b its two adjacent edges replaced by a new edge with probability p_3, it is necessary to also account for the probability of b being able to communicate with the remaining nodes. Thus,

$$P(G) = P(b)P(G'|b) \qquad (8.27)$$

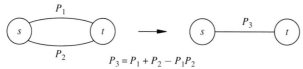

$$P_3 = P_1 + P_2 - P_1 P_2$$

FIGURE 8.7a

Parallel reduction.

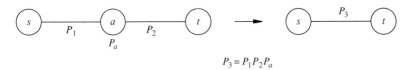

$$P_3 = P_1 P_2 P_a$$

FIGURE 8.7b

Series reduction.

where $P(G)$ is the probability that G is working (i.e., that all nodes can communicate), $P(G')$ is the probability that G' is working (i.e., that all nodes in G' can communicate), and $P(b)$ is the probability that b can communicate with all other nodes in G.

$P(b)$ is just the probability that at least one of the edges adjacent to it is working; i.e.,

$$P(b) = P_b (P_1 + P_2 - P_1 P_2) \tag{8.28}$$

$P(G'|b)$ is the probability that G' is working given that b can communicate with the remaining nodes. The only dependency between $P(G')$ and $P(b)$ is through p_3, the edge replacing the removed chain. Specifically,

$$P_3 = \frac{P_1 P_2}{P_1 + P_2 - P_1 P_2} \tag{8.29}$$

Thus, P_3 is the probability that edges 1 and 2 both work given that at least one of them is working.

Notice that for the case of all-terminal reliability that all the nodes must work in order for the network to work. Thus, the node failure probabilities can be factored out of the problem; i.e.,

$$P(G) = \prod_i P_i P(G| \text{ all nodes working}) \tag{8.30}$$

where the product is taken over all nodes.

The node removal transformation given in Equation 8.27 is valid, in theory, even for the k-terminal reliability problem, but the computation becomes impractical in the general case. $P(b)$ is the probability that node b is working and that it can reach another working node. In the all-terminal case, $P(G')$ is the probability that G'

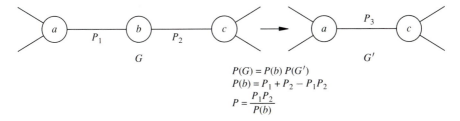

$$P(G) = P(b) P(G')$$
$$P(b) = P_1 + P_2 - P_1 P_2$$
$$P = \frac{P_1 P_2}{P(b)}$$

FIGURE 8.7c

Node removal.

works, which implies that all nodes are working. In the k-terminal case, however, it is possible that one or both of b's neighbors is not a target node. Thus, b might be able to reach one of its neighbors but the neighbor itself might not be working. In this case, it is necessary to trace a path back to target nodes and compute the probability of b being able to reach a target node. This is possible in specific cases, but the number of states which must be examined, and hence the amount of computation, grows exponentially with the number of edges involved. Thus, the node removal transformation is of limited usefulness for the k-terminal problem. This makes the k-terminal problem more difficult to deal with than either the 2-terminal or all-terminal problems.

Another very simple transformation is the **pendant node transformation** where a pendant node (i.e., a node of degree 1) is factored out of the graph along with its incident edge. The node, i, and edge, j, are removed and $P(G)$ is multiplied by $p_n p_l$. This transformation is meaningful for when n is a target node. If n is not a target node, it and its incident edge can be removed without changing $P(G)$.

These series, parallel and pendant transformations, can be applied repeatedly as long as there are opportunities to do so. In some cases, the entire graph can be reduced in this manner. A graph which is reducible using just these transformations is called **series parallel reducible** [Col87]. These transformations are themselves of polynomial order. Each reduces the number of edges in the graph, and so the number of such transformations is of polynomial order. Thus, the entire procedure for analyzing a series parallel reducible graph is of polynomial order.

[Col87] gives the criterion for a graph to be series parallel reducible. Such graphs are called **2-trees**. A graph with two nodes and a single edge is the simplest 2-tree. More complex 2-trees are formed by adding a node to an existing 2-tree. First, two nodes, i and j, already directly connected by an edge in the 2-tree are identified. Then a more complex 2-tree is formed by adding a node, k, and connecting it to both i and j as shown in Figure 8.8. It should be clear that (except for k-terminal reliability problems) any such node k can be removed via a series transformation and that the resulting edge can then be removed via a parallel transformation. Thus, 2-trees can be completely analyzed via series parallel reductions.

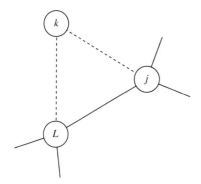

FIGURE 8.8
Augmenting a 2-tree.

Unfortunately, many interesting graphs are not 2-trees. In fact, it is apparent by the nature of the above augmentation procedure that a 2-tree with n nodes will have at most $2n - 3$ edges. Any subgraph of a 2-tree is also a 2-tree, and can also be reduced via series, parallel, and pendant transforms.

The graph reduction technique can be extended to handle a wider class of graphs by allowing additional, more complex reductions. These are discussed in detail in [Sho92], [SW85] and [Col87]. One such transformation, which in fact allows any graph to be analyzed with perfect nodes, is **edge factoring**. Edge factoring is illustrated in Figure 8.9. The idea, similar to state space decomposition, is to replace G by two other graphs, G' and G'', with some edge, e, working and failed, respectively. Therefore,

$$P(G) = p_e P(G') + (1 - p_e) P(G'') \qquad (8.31)$$

In the case of G', where e is working, we **contract** edge e, merging its endpoints into a single node whose probability is the product of the working probabilities of these two endpoints. In the case of G'', where e has failed, simply delete e from the graph. Both G' and G'' have one fewer edge than G. Also, it may now be possible (as it is in Figure 8.9) to apply further series and parallel reductions.

The difficulty with edge factoring is that it creates two subproblems. If it is necessary to apply edge factoring to these subproblems, there will be four. Thus, the number of subproblems grows exponentially in the number of edge factorings. In many practical situations, where the network is relatively sparse, it may be possible to analyze the entire network with only a few edge factorings. This was the case with near-tree networks discussed in the previous section.

At the other extreme, parts of the network (e.g., the backbone) may be very dense. In this case, none of these reduction techniques work very well. It is still possible to obtain reasonable bounds on the reliability of the network, however, by transforming the graph.

Suppose G contains a dense subgraph, G'. It is possible to contract G' to a single point and obtain a much simpler graph, G'', from which it is possible to then obtain both an upper and a lower bound on the probability of G being connected (see Figure 8.10). Let $P(G')$ be the probability that G' is connected with no help from any other edges in G. Thus $P(G')$ is the probability that G' is connected, because G' is a separate graph. If G' is small, say less than eight nodes, it is not difficult to obtain $P(G)$ using any of the methods described before. Thus,

$$P(G') P(G'') \le P(G) \le P(G'') \qquad (8.32)$$

Both bounds are obtained by observing that G' is or is not connected; and if G' is not connected then neither is G. The lower bound is obtained by (pessimistically) insisting that G' be connected without any help from the remainder of G. The upper bound is obtained by (optimistically) assuming that G' is connected.

There are two considerations in choosing G'. If $P(G')$ is close to 1.0, the lower and upper bounds are very close to one another and good bounds are obtained on the network's reliability. Thus, it is good to choose G' as a very dense graph. The other consideration is that it should be relatively easy to evaluate $P(G')$. Therefore, it is

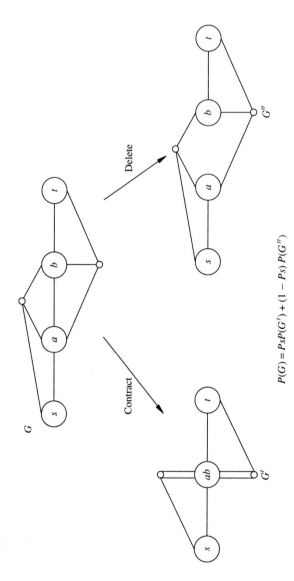

$$P(G) = P_S P(G') + (1 - P_S) P(G'')$$

FIGURE 8.9
Edge factoring.

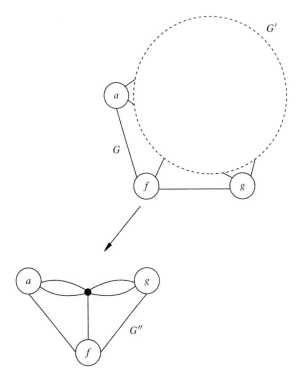

FIGURE 8.10
Subgraph replacement.

preferable not to choose a G' which is too large. In practice, a G' which is a complete graph with four or five nodes usually works well. It is easy to evaluate $P(G')$ for such a graph, and its reliability is high enough to produce good bounds. If the network is very dense, it is possible to carry out this reduction repeatedly, removing a large dense portion of the network.

8.3.4   Connectivity Testing

Examining the reliability polynomial, it is apparent that the low order terms have the greatest effect on the network's reliability. Thus, if there are single edges whose failure disconnects the network, the events where these edges fail dominate the network's probability of failure. This leads to a criterion that the network not contain any single edge whose removal disconnects the network. This measure can also be extended to nodes, that is, that the network not contain any node whose removal disconnects the network. The discussion below focuses on node failures, but the techniques discussed can be applied to edge failures as well.

The notion of connectivity can be generalized. If a network contains k disjoint paths between every pair of nodes the network is **k-connected**. **Disjoint paths** are paths that have no elements in common. Paths are said to be **edge disjoint** (**node disjoint**) if they have no edges (nodes) in common. The network is said to be k-node-connected or k-edge-connected depending on whether the paths are node disjoint or edge disjoint. Node disjoint paths are, of course, also edge disjoint. The following discussion focuses on node disjoint paths.

It is possible to test if a network is 1-connected simply by doing a breadth first search (or a depth first search) as described in Chapter 4. This can be done in $O(E)$, where E is the number of edges in the network. Similarly, it is possible to test if a network is 2-node-connected using the Lowpoint Algorithm described in [AU74]. This too is an $O(E)$ algorithm.

For k greater than two, it is possible to test for k-edge-connectivity by solving a sequence of maximum flow problems. If there is a network, $N = (V, E, C)$, with nodes (vertices) V, edges E and capacities C, it is possible to set up a new network, $N' = (V, E, 1)$, with the same nodes and edges, but with all edge capacities equal to 1. If a single commodity maximum flow problem is now solved for each pair of nodes in N and the maximum flow is found to be greater than or equal to k, then N is k-edge-connected.

It is possible to use a similar method to test for k-node-connectivity. To do so, however, it is necessary to first transform the network. The transformation is shown in Figure 8.11. First, the network is transformed into a directed graph, if necessary, by replacing each undirected edge by two directed edges. Then, each node, n, is split into two nodes, n' and n'' with a directed edge, (n', n'') between them. All edges entering n now enter n' and all edges leaving n now leave n''. This transformation essentially gives each node a "capacity" of one, since any path through the original node n now must go through the new edge, (n', n'') with a capacity of one.

Now if a maximum flow is found between any pair of nodes, i and j, in this newly transformed graph, the maximum number of node disjoint paths will be found from i to j. If a maximum flow problem is solved on all pairs of nodes, then the connectivity of the network will be found. This, however, involves solving $O(N^2)$ maximum flow problems. Even in this graph, where all capacities are one, each maximum flow computation is $O(kE)$, where k is the number of disjoint paths from i to j, and the overall algorithm is $O(kEN^2)$. Even for small k and E of $O(N)$, this is $O(N^3)$ which, while computationally feasible is too complex to use effectively to evaluate candidate topologies in the inner loop of a network design procedure.

These maximum flow problems have a number of important properties which allow us to speed up the computation. First, it is not necessary to actually find the maximum flow. It is possible to stop when the flow is equal to k. Second, since the edges all have capacity one, there is no need to worry about many iterations of the

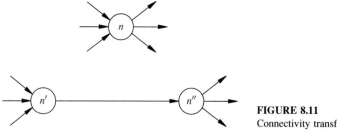

FIGURE 8.11
Connectivity transformation.

inner loop of the Ford-Fulkerson algorithm (see Chapter 4), and therefore it can be simply implemented.

The most important observation, however, is due to Kleitman [D.K69] who showed that it is not necessary to consider all pairs of nodes. Specifically, he showed that

> **Theorem 8.1.** $G = (V, E)$ is k-connected if for any node, $v \in V$, there are k node-disjoint paths from v to each other node, and the graph G', formed by removing v and all its incident edges from G, is $(k\text{-}1)$-connected.

> *Proof.* The premise of the theorem states that there are k node-disjoint paths from v to all other nodes in G. Thus, from the perspective of v, the graph is k-connected. Now consider any other pairs of nodes, x and y. The premise of the theorem states that there are $k - 1$ node disjoint paths between x and y. If any specific $k - 1$ nodes are removed it is possible at worst to destroy all of these paths, $k - 1$ paths between v and x, and $k - 1$ paths between v and y. This still leaves one path between v and x and one path between v and y. These two paths must contain a path between x and y.

Thus, it is only necessary to find k paths from any node, say v_1, to all others, $k - 1$ paths from another node, say v_2 to all others in the graph with v_1 removed, and $k - 2$ paths from v_3 to all others in the graph with v_1 and v_2 removed, etc. This requires $O(kN)$ maximum flow computations and the complexity of the entire procedure is thus reduced to $O(k^2NE)$, which for sparse graphs is $O(k^2N^2)$.

EXERCISES

8.1. Analyze the all-terminal reliability of the graph shown in Figure 8.4 using the state decomposition method.

8.2. Assume that links function with probability 0.997 and nodes are perfectly reliable. Compare the reliability of a chain on 10 nodes with the reliability of a star on 10 nodes on the following bases:
   (*a*) Find the probability that the network is connected.
   (*b*) Find the expected number of node pairs communicating.
   (*c*) Repeat this problem assuming nodes also function with probability 0.997.

8.3. Assume that links function with probability 0.997 and nodes are perfectly reliable. Compare the reliability of a chain on 10 nodes with the reliability of a ring 10 nodes on the following bases:
   (*a*) Find the probability that the network is connected.
   (*b*) Find the expected number of node pairs communicating.

8.4. Analyze the all-terminal reliability of K_4, the complete graph on 4 nodes, using series and parallel reductions. Assume that all nodes have reliability p_n and that all edges have reliability p_e. Give the answer as a function of p_n and p_e.

8.5. Find the reliability of the network shown in Figure P8.1. (The link reliabilities are given in the figure and it is assumed that the nodes are perfectly reliable.) Specifically,
   (*a*) Find the probability that the network is connected.
   (*b*) Find the expected number of node pairs communicating.
   (*c*) Find the expected number of nodes communicating with the root, A.

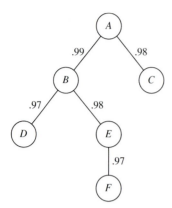

FIGURE P8.1

(*d*) Find the expected number of node pairs communicating through the root.

8.6. A ring network has 8 nodes and 8 links. The links fail independently with probability 0.01.

(*a*) What is the probability that *k* links have failed, for *k* equal to 0, 1, 2, and 3?

(*b*) What is the number of connected networks with 0, 1, 2 and 3 links failed?

(*c*) What is the probability that the network is connected?

(*d*) Repeat parts (*a*), (*b*) and (*c*) for a ring with an added link between two nodes which are 4 hops apart in the original ring.

8.7. An N by N grid is a graph with N^2 nodes, n_{ij} arranged as n rows and n columns with node n_{ij} connected to $n_{(i+1,j)}$ and $n_{(i,j+1)}$. Suppose there is a 3 by 3 grid with edges which fail with probability 0.1.

(*a*) Estimate the probability that the network is connected.

(*b*) Show that the graph is 2-node-connected.

(*c*) What is the smallest number of edges that need be added to make it 3-node-connected? and 4-node-connected?

BIBLIOGRAPHY

[AU74] Hopcroft, J., A. Aho, and J. Ullman: *The Design and Analysis of Computer Algorithms*, Addison-Wesley, Reading, Mass., 1974.

[Col87] Colburn, C. J.: *The Combinatorics of Network Reliability*, Oxford University Press, New York, 1987.

[D.K69] Kleitman, D.: "Methods of investigating connectivity of large graphs," *IEEE Transactions on Circuit Theory (Corresp.)*, CT–16:232–233, 1969.

[Knu73] Knuth, D. E.: *The Art of Computer Programming Volume I: Fundamental Algorithms*, Addison-Wesley, Reading, Mass., 1973.

[KS73] Kershenbaum, A., and R. Van Slyke: "Recursive analysis of network reliability," *Networks*, 3:81–94, 1973.

[Mur92] Murray, K.: "Path and Cutset Based Bounds for Network Reliability Analysis," PhD thesis, Polytechnic University, Brooklyn, New York, May 1992.

[Sho92] Shooman, A. M.: "Exact Graph-Reduction Algorithms for Network Reliability Analysis," PhD thesis, Polytechnic University, Brooklyn, New York, May 1992.

[SW85] Satyanarayana, A., and R. K. Wood: A linear time algorithm for computing *k*-terminal reliability in series parallel networks, *SIAM Journal of Computing*, 14:818–832, 1985.

[VF72] VanSlyke, R., and H. Frank: "Network reliability analysis: Part i," *Networks*, 1:279–290, 1972.

HEAP AND QUEUE MANIPULATION ROUTINES

```
#define LINFINITY 999999999  /*  Infinity (long)        */
#define TRUE      1
#define FALSE     0

/*======================================*/
/* HEADER FOR HEAP MANIPULATION ROUTINES */
/*======================================*/

typedef long  VTYPE;
typedef struct _HEAP {
    int    size;       // Number of elements in the heap
    int    *Hocc;      // Hocc[i]=j means Hval[j] is in position
                       //    i in the heap
    int    *Hpos;      // Hpos[i]=j means Hval[i] is in position
                       //    j in the heap
    VTYPE  *Hval;      // Hval is the vector of values in the
                       //    heap
} HEAP;
typedef HEAP *PHEAP;
```

```
/*================================================================*/
/*                      HEAP UTILITIES                            */
/*  The heap is formed with the smallest element on top           */
/*  Heap values are of type VTYPE (long here)                     */
/*================================================================*/



/*=========================================*/
/*  Bubble down the value in position pos   */
/*=========================================*/
void bubble_down( PHEAP h , int pos )
{ int nh, n, p, p1, p2;



  n = h->size;
  nh = n/2 - 1;
  while( pos <= nh ) {
     p1 = 2*pos+1;
     p2 = 2*pos+2;
     if( (p2 >= n)
        || (h->Hval[h->Hocc[p1]] <= h->Hval[h->Hocc[p2]])) p = p1;
     else
        p = p2;
     if( h->Hval[h->Hocc[pos]] <= h->Hval[h->Hocc[p]] )
        break;
     heap_swap( h , pos , p );
     pos = p;
  }
}



/*=========================================*/
/*  Bubble up the value in position pos   */
/*=========================================*/
void bubble_up( PHEAP h , int pos )
{ int p;



  while( pos > 0 ) {
     p = (pos-1)/2;
     if( h->Hval[h->Hocc[p]] <= h->Hval[h->Hocc[pos]] )
        break;
     heap_swap( h , pos , p );
     pos = p;
  }
}
```

```
/*=========================*/
/* Free memory from a heap */
/*=========================*/
void free_heap( PHEAP h )
{
  free( h->Hval );
  free( h->Hpos );
  free( h->Hocc );
}


/*==============================================================*/
/*  Pop a value out of the heap (leaving the heap with n        */
/*      elements)                                               */
/*  Note: This returns the position of the top not the value    */
/*==============================================================*/
int heap_pop( PHEAP h )
{ int    occ;

  occ = h->Hocc[0];
  heap_replace( h , occ , LINFINITY );
  return( occ );
}


/*============================================*/
/*  Return the value at the top of the heap   */
/*  without disturbing the  heap              */
/*============================================*/
VTYPE heap_top( PHEAP h )
{
  return( h->Hval[h->Hocc[0]] );
}


/*============================================================*/
/*  Replace the i-th value in the vector of values and adjust */
i/*     the heap                                             */
/*============================================================*/
void heap_replace( PHEAP h , int i , VTYPE val )
{
  if( val <= h->Hval[i] ) {
     h->Hval[i] = val;
     bubble_up( h , h->Hpos[i] );
  }
  else {
     h->Hval[i] = val;
     bubble_down( h , h->Hpos[i] );
  }
}
```

```
/*============*/
/*  Heapsort  */
/*============*/
void heapsort( int n , VTYPE values[] , int permu[] )
{ PHEAP h;
  int    i;

  h = make_heap( n );
  set_heap( h , n , values );
  for ( i = 0 ; i < n ; i++ ) {
     permu[i] = heap_pop( h );
  } /* endfor */
  free( h );
}


/*==========================================================*/
/*  Swap the positions of the elements in pos1 and pos2  */
/*==========================================================*/
void heap_swap( PHEAP h , int pos1 , int pos2 )
{  int occ1 , occ2;

   occ1 = h->Hocc[pos1];
   occ2 = h->Hocc[pos2];
   h->Hpos[occ1] = pos2;
   h->Hpos[occ2] = pos1;
   h->Hocc[pos1] = occ2;
   h->Hocc[pos2] = occ1;
}


/*================================*/
/*  Allocate memory for a heap  */
/*================================*/
PHEAP make_heap( int n )
{ PHEAP h;

  h = (PHEAP)malloc( sizeof(HEAP) );
  h->size = n;
  h->Hval = (VTYPE *) malloc( n * sizeof(VTYPE) );
  h->Hpos = (int *) malloc( n * sizeof(int) );
  h->Hocc = (int *) malloc( n * sizeof(int) );
  return( h );
}


/*==================*/
/*  Set up a heap  */
/*==================*/
void set_heap( PHEAP h , int n , VTYPE vals[] )
{ int i, nh;
```

```
  h->size = n;
  for ( i = 0 ; i < n ; i++ ) {
     h->Hval[i] = vals[i];
     h->Hpos[i] = i;
     h->Hocc[i] = i;
  } /* endfor */

  nh = h->size/2 - 1;
  for ( i = nh; i >= 0 ; i-- ) {
     bubble_down( h , i );
  } /* endfor */
}


/*===============================================================*/
/*                  QUEUE   WITH   MEMBERSHIP                     */
/*===============================================================*/

typedef struct _QWM_QUEUE {
   int     size;        //  Number of elements in the queue
   int     empty;       //  TRUE if queue is empty
   int     front;       //  Position of front of queue
   int     rear;        //  Position of rear of queue
   int     *in_queue;   //  in_queue[i] is TRUE if i is in the
                        //     queue
   int     *members;    //  Items in the queue (indices)
} QWM_QUEUE;
typedef QWM_QUEUE *PQWM_QUEUE;


/*=========================================*/
/*  Allocate the memory for a queue of size n  */
/*=========================================*/
PQWM_QUEUE QWM_allocate_queue( int n )
{  PQWM_QUEUE q;

   q = (PQWM_QUEUE) malloc( sizeof( QWM_QUEUE ) );
   q->in_queue = (int *) malloc( n * sizeof( int ) );
   q->members = (int *) malloc( n * sizeof( int ) );
   q->size = n;
   return( q );
}


/*===========================*/
/*  Test if a queue is empty  */
/*===========================*/
int QWM_empty( PQWM_QUEUE q )
{
   return( q->empty );
}
```

```c
/*================================*/
/*  Free the memory from a queue  */
/*================================*/
void QWM_free_queue( PQWM_QUEUE q )
{
    free( q->in_queue );
    free( q->members );
    free( q );
}

/*================================*/
/*  Initialize a queue to empty   */
/*================================*/
void QWM_initialize_queue( PQWM_QUEUE q )
{   int i;

    q->front = 0;
    q->rear = 0;
    q->empty = TRUE;
    for ( i = 0 ; i < q->size ; i++ )
       q->in_queue[i] = FALSE;
}

/*==========================================================*/
/*  Return the value at the front of the queue and remove it */
/*       from the queue                                      */
/*  (unless the queue was already empty, in which case,      */
/*       return -1)                                          */
/*==========================================================*/
int QWM_pop( PQWM_QUEUE q )
{   int i;

    if ( q->empty )
       return( -1 );
    i = q->members[q->front];
    q->in_queue[i] = FALSE;
    if ( q->front == q->rear )
       q->empty = TRUE;
    else
       q->front = ( q->front < (q->size)-1 ) ? (q->front)+1 : 0;
    return( i );
}

/*==========================================================*/
/*  Add node i to the rear of the queue if it is not already */
/*       in the queue                                        */
/*  Note that a queue-with-membership cannot overflow        */
/*==========================================================*/
void QWM_push( PQWM_QUEUE q , int i )
{
    if ( q->in_queue[i] )
```

```
        return;
    if ( q->empty )
        q->empty = FALSE;
    else
        q->rear = ( q->rear < (q->size)-1 ) ? (q->rear)+1 : 0;
    q->members[q->rear] = i;
    q->in_queue[i] = TRUE;
}
```

INDEX